Forum for Interdisciplinary Mathematics

The *Forum for Interdisciplinary Mathematics* is a Scopus-indexed book series. It publishes high-quality textbooks, monographs, contributed volumes and lecture notes in mathematics and interdisciplinary areas where mathematics plays a fundamental role, such as statistics, operations research, computer science, financial mathematics, industrial mathematics, and bio-mathematics. It reflects the increasing demand of researchers working at the interface between mathematics and other scientific disciplines.

Airat M. Bikchentaev · Fuad Kittaneh ·
Mohammad Sal Moslehian · Yuki Seo

Trace Inequalities

For Matrices and Hilbert Space Operators

Airat M. Bikchentaev
Department of Mathematics, Lobachevskii
Institute of Mathematics and Mechanics
Kazan Federal University
Kazan, Russia

Mohammad Sal Moslehian
Department of Pure Mathematics
Ferdowsi University of Mashhad
Mashhad, Iran

Fuad Kittaneh
Department of Mathematics
The University of Jordan
Amman, Jordan

Yuki Seo
Department of Mathematics Education
Osaka Kyoiku University
Kashiwara, Japan

ISSN 2364-6748 ISSN 2364-6756 (electronic)
Forum for Interdisciplinary Mathematics
ISBN 978-981-97-6522-5 ISBN 978-981-97-6520-1 (eBook)
https://doi.org/10.1007/978-981-97-6520-1

Mathematics Subject Classification: 15A15, 15A45, 15A60, 15B48, 15B57, 46L05, 46L10, 46L30, 46L51, 46L53, 47A05, 47A08, 47A30, 47A50, 47A63, 47A64, 47B10, 47B15, 47C15

This Springer imprint is published by the registered company Springer Nature Singapore Pte Ltd.
The registered company address is: 152 Beach Road, #21-01/04 Gateway East, Singapore 189721, Singapore

Dedicated to our Ph.D. supervisors with respect and affection:

Prof. Anatolij Sherstnev
Prof. Joseph Stampfli
Prof. Assadollah Niknam
Prof. Kichi-Suke Saito

Preface

As mentioned by Albrecht Pietsch [1], the term $\sum_i a_{ii}$ for a square matrix $[a_{ij}]$ has been used since the eighteenth century. The notation "Spur" was introduced by R. Dedekind around 1882 in the theory of algebraic numbers. J. von Neumann defined the trace of a positive operator in a Hilbert space in his influential monographs in 1932. Later, in collaboration with F. J. Murray (1936) and R. Schatten (1946), von Neumann extended this definition to trace class operators.

This book provides a comprehensive and advanced presentation of the most important trace inequalities in the framework of complex matrices and operators acting on complex Hilbert spaces. In fact, the book contains our favorite trace inequalities, hence one can also find in the literature trace inequalities other than those covered in this book. We aim to introduce elegant inequalities with ingenious proofs. Instead of providing generalized versions that may be complicated and lack excellent formulations, we present beautiful and original inequalities. The book is organized as follows.

In Chap. 1, we provide fundamental concepts and preliminary results that will be used throughout the book. Although we assume that readers have a basic background in linear algebra, functional analysis, and operator theory; we briefly discuss significant results such as functional calculus, spectral representation, and the Löwner operator order.

Chapter 2 introduces the concept of a trace and its basic properties. We first focus on the trace of a matrix. It is then studied in a broader background, that is, the unitarily invariant norms on Hilbert space operators as symmetric gauge functions of singular values. Each such norm is defined on a two-sided ideal of the algebra of bounded linear operators acting on a Hilbert space and turns the ideal into a Banach space. Several inequalities, particularly those related to Hilbert–Schmidt and trace norms, are explored.

In Chap. 3, we present trace inequalities for positive semidefinite matrices. These inequalities are closely related to two questions raised by Jean-Christophe Bourin during his investigation of matrix subadditivity inequalities. These questions have been formulated as unitarily invariant norm inequalities. It is evident that for the Schatten p-norms, including the Hilbert–Schmidt norm and the trace norm, every norm inequality is equivalent to a trace inequality. We provide partial answers to the questions using the famous Araki–Lieb–Thirring and Ando–Hiai–Okubo trace inequalities.

Chapter 4 deals with further investigation of the aforementioned questions. We give a complete answer to the first question. Related questions and conjectures are also considered. Various unitarily invariant norm inequalities and majorization relations involving positive semidefinite matrices are investigated.

The study of operator means is a significant area of research in its own right. In Chap. 5, we briefly discuss operator convex and operator monotone functions, and introduce the concepts of operator means. We also study positive and completely positive maps and establish several theorems by employing operator matrices.

Chapter 6 studies the Golden–Thompson trace inequality and its generalization. We explore log-majorization and unitarily invariant norm inequalities, which have developed from the consideration of the Golden–Thompson trace inequality. Furthermore, we investigate the reverse inequalities of the Golden–Thompson trace inequality in terms of the generalized Kantorovich constant and the Specht ratio, which are determined by the maximum and minimum eigenvalues of given matrices. Finally, we study the Hadamard product and multivariable versions, which serve as further generalizations of the Golden–Thompson trace inequality.

Chapter 7 studies matrix trace inequalities on quantum relative entropies. We first show the fundamental properties of the Umegaki relative entropy and the Fujii–Kamei relative entropy as a quantum generalization of the classical Kullback–Leibler divergence between probability distributions. We present two quantum Tsallis relative entropies, which are a one-parameter extension of the Umegaki relative entropy and the Fujii–Kamei relative entropy, respectively. We also estimate relations among the two quantum Tsallis relative entropies.

Traces and weights on C^*-algebras are fundamental tools in operator theory and its applications. Chapter 8 gives trace inequalities on von Neumann algebras. It is interesting to note that the fulfillment of similar inequalities for a weight (or a positive functional) on a von Neumann algebra implies that the given weight (or positive functional) is a trace (or a tracial functional). From this, one can obtain numerous characterizations of the commutativity of von Neumann algebras or C^*-algebras. We also prove that if some analogue of the classical inequalities for the determinant and trace of matrices holds for a positive functional φ on $\mathbb{M}_n$ with $\varphi(I) = n$ then $\varphi = \text{tr}$. We also describe noncommutative probability spaces and present some general inequalities such as Burkholder–Gundy, Rosenthal, and Kolmogorov inequalities. We establish the noncommutative versions of Bennet, Bernstein, Hoeffding, Azuma, McDiarmid, and Chernoff-type inequalities.

This book is intended for researchers and graduate students in mathematics, physics, and engineering. It provides detailed explanations for the majority of the results. We provide numerous exercises and problems to help readers better understand the content and inspire them to explore more advanced topics. Moreover, this book includes hints for the exercises and problems, as well as numerous references to academic research, adding further value.

The first author is supported by the development program of Volga Region Mathematical Center (agreement no. 075-02-2023-944).

Finally, the authors would like to sincerely thank Prof. Rajendra Bhatia for his encouragement, Prof. Jean-Christophe Bourin, Prof. Fedor Sukochev, Prof. Jun-Ichi Fujii, Prof. Fumio Hiai, Prof. Omar Hirzallah, Prof. Mitsuru Uchiyama, Prof. Ghadir Sadeghi, Dr. Ali Talebi, Dr. Ali Dadkhah, and Dr. Saja Hayajneh for their valuable comments and suggestions.

Kazan, Russia — Airat M. Bikchentaev
Amman, Jordan — Fuad Kittaneh
Mashhad, Iran — Mohammad Sal Moslehian
Kashiwara, Japan — Yuki Seo

Reference

1. Pietsch, A., *Operator ideals with a trace*, Math. Nachr. **100** (1981), 61–91.

Contents

Chapter 1
Fundamentals of Matrices and Operators

In this chapter, we collect some basic facts about functional analysis that are used throughout the book. By an operator, we mean a bounded linear operator. All linear spaces in this book are assumed to be over the field of complex numbers, denoted as $\mathbb{C}$. It is well known that there are significant differences between vector space structures over real and complex fields; see [1]. To simplify the notation, we frequently use the two-point compactification $\mathbb{R} \cup \{-\infty, +\infty\}$ and denote ∞ as $+\infty$ when there is no ambiguity.

1.1 Operators on Hilbert Spaces

A linear space $\mathcal{X}$ is a *seminormed space* if it is endowed with a *seminorm*, that is, a function $\|\cdot\| : \mathcal{X} \to [0, \infty)$ satisfying the following properties:

(i) $\|x\| = 0$ if $x = 0$;
(ii) $\|\lambda x\| = |\lambda| \, \|x\|$;
(iii) $\|x + y\| \leq \|x\| + \|y\|$ (the triangle inequality)

for all $x, y \in \mathcal{X}$ and all scalars $\lambda \in \mathbb{C}$. If $\|x\| = 0$ implies that $x = 0$, then $(\mathcal{X}, \|.\|)$ is said to be a *normed space*. Evidently, $d(x, y) = \|x - y\|$ defines a metric on $\mathcal{X}$. If this metric is complete meaning that every Cauchy sequence in the space is convergent, then $(\mathcal{X}, \|\cdot\|)$ is called a *Banach space*. It is remarkable that every finite-dimensional normed space $\mathcal{X}$ is complete, and any two norms $\|\cdot\|_1$ and $\|\cdot\|_2$ on it are equivalent in the sense that there exist positive numbers α and β such that $\alpha\|x\|_1 \leq \|x\|_2 \leq \beta\|x\|_1$ for all $x \in \mathcal{X}$. The dual space $\mathcal{X}'$ of a normed space $\mathcal{X}$ is defined as the Banach space consisting of all continuous linear functionals $f : \mathcal{X} \to \mathbb{C}$.

Example 1.1.1

- The linear space $\mathbb{M}_{m\times n}(\mathbb{C})$, denoted also by $\mathbb{M}_{m\times n}$, of all $m \times n$ matrices equipped with any one of the following norms is a Banach space:

A. M. Bikchentaev et al., *Trace Inequalities*, Forum for Interdisciplinary Mathematics,
https://doi.org/10.1007/978-981-97-6520-1_1

(i) $\|[a_{ij}]\|_p = (\sum_{i=1}^{m}\sum_{j=1}^{n}|a_{ij}|^p)^{\frac{1}{p}}$ $(p \geq 1)$;
(ii) $\|[a_{ij}]\|_m = \max_{1\leq i\leq m}\max_{1\leq j\leq n}|a_{ij}|$;
(iii) $\|[a_{ij}]\|_r = \max_{1\leq i\leq m}(\sum_{j=1}^{n}|a_{ij}|^2)^{\frac{1}{2}}$;
(iv) $\|[a_{ij}]\|_c = \max_{1\leq j\leq n}(\sum_{i=1}^{m}|a_{ij}|^2)^{\frac{1}{2}}$.

- The linear space $C_0(\Omega)$ consisting of all complex-valued functions f defined on a locally compact Hausdorff space Ω vanishing at infinity. This means that for all $\varepsilon > 0$ the set $\{x \in \Omega : |f(x)| \geq \varepsilon\}$ is compact, with the sup-norm

$$\|f\|_\infty = \sup\{|f(t)| : t \in \Omega\}.$$

When Ω is compact, $C_0(\Omega)$ is indeed the space $C(\Omega)$, which consists of all continuous functions on Ω.

A *semi-inner product space* is a complex linear space $\mathscr{H}$ equipped with a function $\langle\cdot,\cdot\rangle : \mathscr{H} \times \mathscr{H} \to \mathbb{C}$, which is called a *semi-inner product*, satisfying

(i) $\langle x+y, z\rangle = \langle x, z\rangle + \langle y, z\rangle$;
(ii) $\langle \lambda x, y\rangle = \lambda\langle x, y\rangle$;
(iii) $\langle y, x\rangle = \overline{\langle x, y\rangle}$;
(iv) $\langle x, x\rangle \geq 0$

for all $x, y, z \in \mathscr{H}$ and all scalars λ. Furthermore, the Cauchy–Schwarz inequality

$$|\langle x, y\rangle| \leq \langle x, x\rangle^{1/2}\langle y, y\rangle^{1/2} \qquad (x, y \in \mathscr{H}) \tag{1.1.1}$$

holds.

If $\langle x, x\rangle = 0$ ensures that $x = 0$, then the pair $(\mathscr{H}, \langle\cdot,\cdot\rangle)$ is called an *inner product space*. In addition, by employing Cauchy–Schwarz inequality (1.1.1), one can show that $\|x\| = \langle x, x\rangle^{\frac{1}{2}}$ defines a norm on $\mathscr{H}$. If the norm $\|\cdot\|$ is complete, then $\mathscr{H}$ is said to be a *Hilbert space*. Every Hilbert space $\mathscr{H}$ admits an orthonormal basis $(e_j)_{j\in J}$ such that $x = \sum_j\langle x, e_j\rangle e_j$. Moreover, the *Parseval identity* $\|x\|^2 = \sum_j|\langle x, e_j\rangle|^2$ holds.

In addition, the *parallelogram identity*

$$\|x+y\|^2 + \|x-y\|^2 = 2\|x\|^2 + 2\|y\|^2 \qquad (x, y \in \mathscr{H}) \tag{1.1.2}$$

and the *polarization identity*

$$\langle x, y\rangle = \frac{1}{4}\sum_{k=0}^{3} \mathrm{i}^k\|x + \mathrm{i}^k y\|^2 \qquad (x, y \in \mathscr{H}) \tag{1.1.3}$$

are valid in every semi-inner product space.

Example 1.1.2

- The space $\mathbb{C}^n = \{(x_1, \dots, x_n) : x_i \in \mathbb{C},\ 1 \le i \le n\}$ is a Hilbert space under the inner product
$$\langle (x_1, \dots, x_n), (y_1, \dots, y_n) \rangle := \sum_{j=1}^{n} x_j \overline{y_j}.$$
In addition, the classical Cauchy–Schwarz inequality reads as follows:
$$\left| \sum_{i=1}^{n} x_i \overline{y_i} \right| \le \left(\sum_{i=1}^{n} |x_i|^2 \right)^{\frac{1}{2}} \left(\sum_{i=1}^{n} |y_i|^2 \right)^{\frac{1}{2}}. \tag{1.1.4}$$
- The Banach sequence space ℓ_p equipped with the norm $\|(x_n)\|_p = \left(\sum_{n=1}^{\infty} |x_n|^p\right)^{1/p}$ is a Hilbert space if and only if $p = 2$. To see this, let us take $x = (1, 1, 0, 0, \dots)$ and $y = (1, -1, 0, 0, \dots)$. Then $\|x\|_p = \|y\|_p = 2^{1/p}$ and $\|x + y\|_p = \|x - y\|_p = 2$. Hence, the parallelogram identity (1.1.2) is not satisfied for $p \ne 2$. However, for $p = 2$, we have
$$\langle (x_n), (y_n) \rangle = \sum_{n=1}^{\infty} x_n \overline{y_n},$$
which turns ℓ_2 into a Hilbert space.

A vector $x \in \mathscr{H}$ is called *orthogonal* to $y \in \mathscr{H}$ if $\langle x, y \rangle = 0$ and we write $x \perp y$. The set
$$M^{\perp} = \{x \in \mathscr{H} : x \perp y \text{ for all } y \in M\}$$
is called the *orthogonal complement* of M. For each closed subspace M of $\mathscr{H}$ and each point $x \in \mathscr{H}$, let y be the *closest point* in M to x, in the sense that $\|y - x\| = \inf\{\|z - x\| : z \in M\}$, and put $z = x - y$. Then $z \in M^{\perp}$ and x can be uniquely represented as $x = y + z$. It follows that $\mathscr{H} = M \oplus M^{\perp}$.

If $\mathscr{H}_1$ and $\mathscr{H}_2$ are Hilbert spaces, then their *Hilbert space direct sum* $\mathscr{H}_1 \oplus \mathscr{H}_2$ is defined as the Cartesian product $\mathscr{H}_1 \times \mathscr{H}_2$ equipped with the inner product
$$\langle (x_1, y_1), (x_2, y_2) \rangle = \langle x_1, x_2 \rangle + \langle y_1, y_2 \rangle.$$

A linear operator $A : \mathscr{H} \to \mathscr{K}$ between Hilbert spaces is called a *bounded operator* if
$$\|A\| := \sup\{\|Ax\| : x \in \mathscr{H}, \|x\| = 1\} < \infty.$$
Note that $\|A\| = \sup\{|\langle Ax, y \rangle| : x, y \in \mathscr{H}, \|x\| = \|y\| = 1\}$. The linearity of A entails that A is bounded if and only if it is continuous. The set of all such operators

is denoted by $\mathbb{B}(\mathscr{H}, \mathscr{K})$, which is a Banach space endowed with the operator norm $A \mapsto \|A\|$.

Throughout the book, $\|\cdot\|$ denotes the operator norm.

The identity operator is denoted by I. The kernel and the range of an operator A are denoted by $\ker A$ and $\mathrm{ran}(A)$, respectively. In addition, the rank of A is defined by $\mathrm{rank}(A) := \dim(\mathrm{ran}(A))$. We use the abbreviation $\mathbb{B}(\mathscr{H})$ for $\mathbb{B}(\mathscr{H}, \mathscr{H})$.

Theorem 1.1.3 (Riesz representation theorem) *Let $\mathscr{H}$ be a Hilbert space and f be a bounded linear functional. Then there exists a unique element $z \in \mathscr{H}$ such that $\|z\| = \|f\|$ and $f(x) = \langle x, z\rangle$ for all $x \in \mathscr{H}$.*

Assume that $\mathscr{H}$ is a Hilbert space. Put $\mathscr{H}_* := \mathscr{H}$ as sets. We endow it with the addition $x +_* y = x + y$, the scalar multiplication $\lambda \cdot_* x = \overline{\lambda} \cdot x$, as well as the inner product $\langle x, y\rangle_* = \langle y, x\rangle$. Then one can easily verify that $\mathscr{H}_*$ is a Hilbert space. For each $x \in \mathscr{H}$, we define the functional $\phi(x) \in \mathscr{H}_*'$ by $\phi(x)y = \langle y, x\rangle_* = \langle x, y\rangle$. It follows from the Riesz representation theorem that $\phi : \mathscr{H} \to \mathscr{H}_*', \ x \mapsto \phi(x)$ is an isometric linear isomorphism.

For any operator $A \in \mathbb{B}(\mathscr{H}, \mathscr{K})$, its *adjoint operator* is defined as the unique operator $A^* \in \mathbb{B}(\mathscr{K}, \mathscr{H})$ such that $\langle Ax, y\rangle = \langle x, A^*y\rangle$ $(x \in \mathscr{H}, y \in \mathscr{K})$. For any $y \in \mathscr{K}, x \mapsto \langle Ax, y\rangle$ gives a bounded linear functional on $\mathscr{H}$. Let A^*y be the unique element in the Riesz representation theorem corresponding to this functional. The adjoint operation $A \mapsto A^*$ has the following properties:

- $(A + \lambda B)^* = A^* + \overline{\lambda} B^*$;
- $(AB)^* = B^*A^*$;
- $(A^*)^* = A$;
- $\|A^*A\| = \|A\|^2$ (C^*-condition)

for all $A, B \in \mathbb{B}(\mathscr{H})$ and $\lambda \in \mathbb{C}$. A complex number λ is called an *eigenvalue* of an operator A if there is a nonzero vector x such that $Ax = \lambda x$.

If $\mathscr{H}$ is of finite dimension n, we can identify $\mathbb{B}(\mathscr{H})$ with the matrix algebra $\mathbb{M}_n = \mathbb{M}_{n\times n}(\mathbb{C})$. The identity matrix is denoted by I_n, and if there is no ambiguity, we represent it by I. For each $i, j = 1, \ldots, n$, the matrix whose (i, j)- entry is 1 and all the other entries are 0 is denoted by E_{ij}. Therefore,

$$A = [a_{ij}] = \sum_{i,j=1}^{n} a_{ij} E_{ij}, \qquad (A \in \mathbb{M}_n).$$

These matrices constitute a basis for $\mathbb{M}_n$ and are called the *matrix units* that enjoy the basic property $E_{ij}E_{kl} = \delta_{jk}E_{il}$, where δ denotes the Kronecker delta defined by

$$\delta_{ij} = \begin{cases} 1, & \text{if } i = j \\ 0, & \text{if } i \neq j. \end{cases}$$

In general, any $m \times n$ matrix A generates an operator T_A from $\mathbb{C}^n$ into $\mathbb{C}^m$ defined by $T_A(X) = AX$, where X is an $n \times 1$ column matrix, and vice versa, every linear operator T from $\mathbb{C}^n$ to $\mathbb{C}^m$ gives us an $n \times m$ matrix A_T, whose j-column is Te_j, where the vectors $e_j = (\delta_{1j}, \delta_{2j}, \ldots, \delta_{nj})$ $(1 \le i \le n)$ form the standard (canonical) basis of $\mathbb{C}^n$. In particular, if $A = [a_{ij}]$, then $a_{ij} = \langle T_A e_j, e_i \rangle$. Since $\langle T_{A^*} e_j, e_i \rangle = \overline{\langle T_A e_i, e_j \rangle}$, we have $A^* = [\overline{a_{ji}}]$. In addition, the matrix $A^t = [a_{ji}]$ is called the *transpose* of A.

An (orthogonal) *projection* is an operator $P \in \mathbb{B}(\mathscr{H})$ such that $P = P^* = P^2$. The range of each projection in $\mathbb{B}(\mathscr{H})$ is a closed subspace of $\mathscr{H}$. Conversely, if M is a closed subspace, we observe that each element x of $\mathscr{H}$ can be uniquely represented as $y + z$, where $y \in M$ and $z \in M^\perp$. Then $Px := y$ is a projection in $\mathbb{B}(\mathscr{H})$, whose range is M and whose kernel is $M^\perp$. If P is a projection, then so is

$$P^\perp := I - P.$$

Evidently, $\ker P^\perp = \operatorname{ran}(P)$ and $\operatorname{ran}(P^\perp) = \ker P$. If $\{P_\lambda\}_{\lambda \in \Lambda}$ is a family of projections, we denote by $\vee_{\lambda \in \Lambda} P_\lambda$ (respectively, $\wedge_{\lambda \in \Lambda} P_\lambda$) the projection from $\mathscr{H}$ onto the closed subspace generated by $\cup_{\lambda \in \Lambda} P_\lambda(\mathscr{H})$ (respectively, onto the subspace $\cap_{\lambda \in \Lambda} P(\mathscr{H})$). Consequently, $(\vee_{\lambda \in \Lambda} P_\lambda)^\perp = \wedge_{\lambda \in \Lambda} P_\lambda^\perp$.

For $x, y \in \mathscr{H}$, the rank-one operator $x \otimes y^*$ is defined on $\mathscr{H}$ by

$$(x \otimes y^*)z = \langle z, y \rangle x. \tag{1.1.5}$$

If x is a unit vector, then $x \otimes x^*$ is a projection. Furthermore, $A(x \otimes y^*) = (Ax) \otimes y^*$ for all $A \in \mathbb{B}(\mathscr{H})$. The linear span of rank-one operators is called the space of *finite-rank operators*, and its closure is called the space of *compact operators* and is denoted by $\mathbb{K}(\mathscr{H})$. Indeed, $\mathbb{K}(\mathscr{H})$ is a closed two-sided ideal of $\mathbb{B}(\mathscr{H})$. The quotient space $\mathbb{B}(\mathscr{H})/\mathbb{K}(\mathscr{H})$ is called the *Calkin algebra.*

An operator $A \in \mathbb{B}(\mathscr{H})$ is called a *self-adjoint operator* (*Hermitian matrix* in the setting of matrices) if $A = A^*$. This is equivalent to that $\langle Ax, x \rangle \in \mathbb{R}$ for all $x \in \mathscr{H}$, since $A = A^*$ if and only if

$$\langle Ax, x \rangle = \langle x, A^*x \rangle = \langle x, Ax \rangle = \overline{\langle Ax, x \rangle}.$$

The real space of all self-adjoint operators is denoted by $\mathbb{B}_{sa}(\mathscr{H})$. It is a norm-closed real subspace of $\mathbb{B}(\mathscr{H})$.

We denote by $\operatorname{Re} A$ and $\operatorname{Im} A$ the *real part* $(A + A^*)/2$ and the *imaginary part* $(A - A^*)/(2\mathrm{i})$ of A, respectively; hence $A = \operatorname{Re} A + \mathrm{i} \operatorname{Im} A$, which is said to be the *Cartesian decomposition* of $A \in \mathbb{B}(\mathscr{H})$. An operator $A \in \mathbb{B}(\mathscr{H})$ is called

- a *skew-self-adjoint operator* (*skew-Hermitian matrix*) if $\mathrm{i}A$ is self-adjoint (Hermitian);
- a *normal operator* if $A^*A = AA^*$;
- a *contraction* if $\|A\| \le 1$;
- a *partial isometry* if $AA^*A = A$;
- an *isometry* if $A^*A = I$;

– a *unitary* if $A^*A = AA^* = I$;
– an *idempotent* if $A^2 = A$.

The spectrum of an operator $A \in \mathbb{B}(\mathscr{H})$ is the set $\text{sp}(A)$ of all complex numbers λ such that $A - \lambda I$ is not invertible in $\mathbb{B}(\mathscr{H})$. It is a compact nonempty set. The spectrum of a matrix $A \in \mathbb{M}_n$ is indeed the set of its eigenvalues, since a matrix, considered as an operator, is one to one if and only if it is surjective. This is due to the *dimension formula* $n = \dim(\ker A) + \dim(\text{ran}(A))$. The *spectral radius* of an operator A is defined by

$$r(A) = \max\{|\lambda| : \lambda \in \text{sp}(A)\}.$$

The *Gelfand–Beurling formula* states that

$$r(A) = \lim_{k\to\infty} \|A^k\|^{1/k} = \inf_k \|A^k\|^{1/k} \tag{1.1.6}$$

(see, for example, [2, p. 349]).

Therefore, $r(A) \leq \|A\|$. Furthermore, the equality holds if A is self-adjoint, since $\|A^{2^k}\| = \|A\|^{2^k}$ for all positive integers k. For $A, B \in \mathbb{B}(\mathscr{H})$,

$$\text{sp}(AB) \cup \{0\} = \text{sp}(BA) \cup \{0\}, \tag{1.1.7}$$

and hence

$$r(AB) = r(BA). \tag{1.1.8}$$

If U is a unitary operator, then $\|U\| = \|U^*U\|^{1/2} = \|I\|^{1/2} = 1$. In addition, U^* a unitary operator. For $\lambda \in \text{sp}(U)$, we have $|\lambda| \leq r(U) \leq \|U\| = 1$ and $\lambda^{-1} \in \text{sp}(U^{-1}) = \text{sp}(U^*)$, so $|\lambda^{-1}| \leq r(U) \leq 1$. Thus, $|\lambda| = 1$. Hence, $\text{sp}(U)$ is a subset of the unit circle $\mathbb{T} = \{\lambda \in \mathbb{C} : |\lambda| = 1\}$.

If A is normal, then $\|A\| = r(A)$, since it follows from Exercise 1.8.4 that

$$\|A\| = \|A^*A\|^{1/2} = r(A^*A)^{1/2} \leq (r(A^*)r(A))^{1/2} = r(A) \leq \|A\|.$$

An operator A is normal if and only if $\|Ax\| = \|A^*x\|$ for all $x \in \mathscr{H}$, since

$$\langle A^*Ax, x\rangle = \langle AA^*x, x\rangle \Leftrightarrow \langle Ax, Ax\rangle = \langle A^*x, A^*x\rangle \Leftrightarrow \|Ax\| = \|A^*x\|.$$

An operator A is an isometry if and only if $\langle Ax, Ax\rangle = \langle x, x\rangle$ or equivalently $\|Ax\| = \|x\|$ for all $x \in \mathscr{H}$.

Definition 1.1.4 For $A \in \mathbb{B}(\mathscr{H})$,

$$W(A) = \{\langle Ax, x\rangle : \|x\| = 1\}$$

is called the *numerical range* of A and $w(A) = \sup\{|\lambda| : \lambda \in W(A)\}$ is defined to be the *numerical radius* of A.

It is known that $W(A)$ is convex and $w(\cdot)$ is a norm on $\mathbb{B}(\mathscr{H})$ with

$$\frac{1}{2}\|A\| \leq w(A) \leq \|A\| \tag{1.1.9}$$

for all $A \in \mathbb{B}(\mathscr{H})$; see [3, Theorem 2.2.1]. Moreover, if A is normal, then $w(A) = \|A\|$. If $A^2 = 0$, then $w(A) = \|A\|/2$; see [4, Chap. 1] for other inequalities for one operator.

An operator $A \in \mathbb{B}(\mathscr{H})$ is said to be a *positive operator* (*positive semidefinite matrix* in the setting of matrices) if it is self-adjoint and $\mathrm{sp}(A) \subseteq [0, \infty)$. In this case, we write $A \geq 0$. If A is invertible and positive, then we write $A > 0$ and say A is positive and invertible (*positive definite matrix* for matrices). For self-adjoint operators $A, B \in \mathbb{B}(\mathscr{H})$, we say $A \leq B$ (respectively, $A < B$) if $B - A \geq 0$ (respectively, $B - A > 0$). We denote by $\mathbb{B}_+(\mathscr{H})$ ($\mathbb{B}_{++}(\mathscr{H})$, respectively) the set of all positive operators (positive invertible operators, respectively). We call $\leq$ the *Löwner order*. In particular, we write $m \leq A \leq M$ instead of $mI \leq A \leq MI$. We discuss the properties of this order in Sect. 1.4.

The following theorem presents some elementary properties of self-adjoint operators.

Theorem 1.1.5 *If* $A \in \mathbb{B}_{\mathrm{sa}}(\mathscr{H})$, *then*

(i) $r(A) = \|A\|$;
(ii) $\mathrm{sp}(A)$ *is a subset of the real line* $\mathbb{R}$;
(iii) at least one of $\|A\|$ *or* $-\|A\|$ *is in* $\mathrm{sp}(A)$ *and* $\mathrm{sp}(A) \subseteq [-\|A\|, \|A\|]$;
(iv)

$$\|A\| = \sup_{\|x\|=1} |\langle Ax, x\rangle| \,.$$

A *Banach algebra* is a complex associative algebra $\mathscr{A}$ that possesses the structure of a Banach space and satisfies $\|AB\| \leq \|A\|\,\|B\|$ for all $A, B \in \mathscr{A}$. A Banach algebra is *unital* when it has an identity element (of norm one) with respect to the multiplication. A *Banach $*$-algebra* is a Banach algebra $\mathscr{A}$ equipped with an *involution* $^* : \mathscr{A} \to \mathscr{A}$ satisfying $(\lambda A + B)^* = \overline{\lambda} A^* + B^*$, $(AB)^* = B^*A^*$, and $\|A^*\| = \|A\|$ for all $A, B \in \mathscr{A}$.

A *C^*-algebra* is a complex Banach $*$-algebra $\mathscr{A}$ such that $\|A^*A\| = \|A\|^2$ for all $A \in \mathscr{A}$. Any C^*-algebra can be realized as a C^*-subalgebra in $\mathbb{B}(\mathscr{H})$ for some Hilbert space $\mathscr{H}$ by the celebrated Gelfand–Naimark theorem; see [5, Theorem 3.4.1]. Moreover, if $\mathscr{A}$ is separable, then $\mathscr{H}$ can be chosen to be separable.

A linear map $\rho : \mathscr{A} \to \mathbb{C}$ is called a *positive linear functional* if it sends positive elements of the C^*-algebra $\mathscr{A}$ to nonnegative numbers. It is said to be a *state* if $\|\rho\| = 1$. If $\mathscr{A}$ is unital, a bounded linear functional ρ on $\mathscr{A}$ is positive if and only if $\rho(I) = \|\rho\|$. If ρ is a positive linear functional on $\mathscr{A}$, then $\langle A, B\rangle = \rho(B^*A)$ defines a semi-inner product on $\mathscr{A}$. Then the Cauchy–Schwarz inequality yields that

$$|\rho(B^*A)| = |\langle A, B\rangle| \leq \langle A, A\rangle^{1/2}\langle B, B\rangle^{1/2} = \rho(A^*A)^{1/2}\rho(B^*B)^{1/2}\,.$$

A linear map $\varphi : \mathscr{A} \to \mathscr{B}$ between C^*-algebras is called a $*$-*homomorphism* if it preserves the product and the adjoint, that is, $\varphi(AB) = \varphi(A)\varphi(B)$ and $\varphi(A^*) = \varphi(A)^*$ for all $A, B \in \mathscr{A}$. It is known that a $*$-isomorphism φ between C^*-algebras is an isometry. A commutative C^*-algebra $\mathscr{A}$ is $*$-isomorphic to the function algebra $C_0(\Omega)$ equipped with the pointwise operations and $f \mapsto \bar{f}$ as the involution, where X is a locally compact Hausdorff space. In particular, X is compact if and only if the algebra $C_0(\Omega)$ is unital. The celebrated Stone–Weierstrass theorem states that a closed $*$-subalgebra of $C(\Omega)$ separating points on X (that is, if $x \neq y$ in X, then there is a function f in the subalgebra such that $f(x) \neq f(y)$) and containing the constant function 1 coincides with $C(\Omega)$.

The locally convex topology generated by seminorms $p_{x,y}(A) = |\langle Ax, y\rangle|$ in which x and y run over $\mathscr{H}$ is called the *weak operator topology* (wot) on $\mathbb{B}(\mathscr{H})$. Therefore, $A_i \stackrel{\text{wot}}{\to} A$ if and only if $\langle A_i x, y\rangle \to \langle Ax, y\rangle$ for all $x, y \in \mathscr{H}$. The locally convex topology generated by seminorms $p_x(A) = \|Ax\|$ in which x runs over $\mathscr{H}$ is called the *strong operator topology* (sot) on $\mathbb{B}(\mathscr{H})$. Hence, $A_i \stackrel{\text{sot}}{\to} A$ if and only if $A_i x \to Ax$ for all $x \in \mathscr{H}$.

In the case of $\dim \mathscr{H} = n$, all norms on $\mathbb{M}_n$ are equivalent. Furthermore, the weak operator topology, the strong operator topology, and the norm topology coincide. Thus,

$$\begin{aligned} A_k = [a_{ij,k}] \stackrel{\|\cdot\|}{\longrightarrow} A = [a_{ij}] &\Longleftrightarrow a_{ij,k} \to a_{ij} \text{ as } k \to \infty \text{ for all } 1 \leq i, j \leq n \\ &\Longleftrightarrow A_k \stackrel{\text{wot}}{\longrightarrow} A \text{ as } k \to \infty \\ &\Longleftrightarrow A_k \stackrel{\text{sot}}{\longrightarrow} A \text{ as } k \to \infty. \end{aligned}$$

1.2 Functional Calculus for Self-adjoint Operators

Assume that $A \in \mathbb{B}_{\text{sa}}(\mathscr{H})$. There exists an isometric $*$-isomorphism from $C(\text{sp}(A))$ onto the closed $*$-subalgebra $C^*(A, I)$ of $\mathbb{B}(\mathscr{H})$ generated by A and I. This isomorphism maps the function $f(t) = t$ to A and the constant function $f(t) = 1$ to I. In particular, it maps the polynomial $p(t) = \alpha_n t^n + \cdots + \alpha_1 t + \alpha_0$ to $p(A) = \alpha_n A^n + \cdots + \alpha_1 A + \alpha_0 I$. According to the Stone–Weierstrass theorem if $f \in C(\text{sp}(A))$, then there exists a sequence of polynomials $\{p_n\}_{n=1}^{\infty}$ such that $\lim_{n\to\infty} p_n = f$ in $C(\text{sp}(A))$ equipped with the sup-norm. Consequently, there is a unique operator $f(A) \in \mathbb{B}(\mathscr{H})$ such that $f(A) = \lim_{n\to\infty} p_n(A)$. The $*$-isomorphism $f \mapsto f(A)$ is called the *continuous functional calculus* for A. Moreover, $\text{sp}(f(A)) = f(\text{sp}(A))$, the so-called *spectral mapping theorem*.

In addition, the operator $f(A)$ is self-adjoint if and only if $f(t) \in \mathbb{R}$ for all $t \in \text{sp}(A)$. Due to the compactness of the spectrum, there exist real numbers m, M such that $\text{sp}(A) \subseteq [m, M]$. By the functional calculus, this is equivalent to $m \leq A \leq M$.

If P is a projection, then $\text{sp}(P) \subseteq \mathbb{R}$ and $P^2 - P = 0$. Therefore, $t^2 - t = 0$ for all $t \in \text{sp}(P)$. Hence, $\text{sp}(P) \subseteq \{0, 1\}$.

If A is an isometry, then $A^*A = I$, whence $(AA^*)^2 = AA^*$. Thus, $\mathrm{sp}(AA^*) \subseteq \{0, 1\}$. It follows from the functional calculus for A^*A that $0 \leq AA^* \leq I$.

If $A \in \mathbb{B}_{sa}(\mathscr{H})$ is a contraction, then the operator $U = A + \mathrm{i}(I - A^2)^{1/2}$ is a unitary operator and $A = (U + U^*)/2$. An immediate consequence of this decomposition and the following polarization-type identity

$$AB = \frac{1}{4}\sum_{k=0}^{3}(-\mathrm{i})^k(A^* + \mathrm{i}^k B)^*(A^* + \mathrm{i}^k B) \tag{1.2.1}$$

is that if φ is a linear functional on a C^*-algebra, then the following conditions are equivalent:

(i) $\varphi(AB) = \varphi(BA)$ (the tracial or cyclicity property);
(ii) $\varphi(A^*A) = \varphi(AA^*)$.

We remark that (1.2.1) has the following equivalent form:

$$A^*B = \frac{1}{4}\sum_{k=0}^{3}\mathrm{i}^{-k}|A + \mathrm{i}^k B|^2.$$

1.3 Spectral Representation for Self-adjoint Operators

In linear algebra, the *spectral decomposition* asserts that if $A \in \mathbb{M}_k$ is a Hermitian matrix with eigenvalues λ_j $(1 \leq j \leq k)$, then there exist orthogonal projections P_j $(1 \leq j \leq k)$ with $\sum_{j=1}^{k} P_j = I$ such that

$$A = \sum_{j=1}^{k}\lambda_j P_j.$$

If the eigenvalues of A are contained in an interval J and $f : J \to \mathbb{R}$ is a continuous function, then

$$f(A) = \sum_{j=1}^{k} f(\lambda_j)P_j,$$

where $f(A)$ is the same as $f(A)$ defined by the functional calculus; see [6].

Next, we generalize the above structure to self-adjoint operators. Let $A \in \mathbb{B}_{sa}(\mathscr{H})$ and f be a nonnegative upper semicontinuous function on $\mathbb{R}$. Recall that a real-valued function f on $\mathbb{R}$ is called *upper semicontinuous* if it is a pointwise limit of a decreasing sequence (f_n) of continuous nonnegative functions on $\mathbb{R}$. By applying the functional calculus, we can construct $f_n(A)$ for each n. Then $(f_n(A))$ constitutes a decreasing and bounded below sequence of self-adjoint operators in $\mathbb{B}(\mathscr{H})$. By the

Vigier theorem [5, Theorem 4.1.1], this sequence converges in the strong operator topology to an operator denoted by $f(A)$. If f is continuous, then $f(A)$ is nothing more than the operator defined by the usual functional calculus.

For each $\lambda \in \mathbb{R}$, we set

$$f_\lambda(s) := \begin{cases} 1 & \text{for } -\infty < s \le \lambda, \\ 0 & \text{for } \lambda < s < \infty. \end{cases}$$

Then f_λ is upper semicontinuous, so we can construct

$$E_\lambda := f_\lambda(A), \tag{1.3.1}$$

which is a projection. The family of projections $\{E_\lambda\}_{\lambda \in \mathbb{R}}$ defined by (1.3.1), called the *spectral family* of A.

Now, the spectral representation theorem reads as follows.

Theorem 1.3.1 (Spectral representation theorem) *Suppose that* $A \in \mathbb{B}_{\rm sa}(\mathscr{H})$. *Let* $m = \min\{\lambda : \lambda \in {\rm sp}(A)\}$, *and let* $M = \max\{\lambda : \lambda \in {\rm sp}(A)\}$. *Then the spectral family has the following properties:*

(i) $E_\lambda \le E_{\lambda'}$ *for* $\lambda \le \lambda'$.
(ii) $E_\lambda = 0$ *for every* $\lambda < m$ *and* $E_\lambda = I$ *for every* $\lambda \ge M$, *and* $E_\mu \to E_\lambda$ *in the strong operator topology as* $\mu \to \lambda^+$.
(iii) for every continuous complex-valued function f *defined on* $\mathbb{R}$, *it holds that*

$$f(A) = \int_{m-0}^{M} f(\lambda)\, dE_\lambda,$$

where the integral is of Riemann–Stieltjes type, in the sense that

$$\langle f(A)x, y\rangle = \int_{m-0}^{M} f(\lambda)\, d\langle E_\lambda x, y\rangle \text{ for all } x, y \in \mathscr{H}.$$

1.4 Löwner Order

If $A \in \mathbb{B}_{\rm sa}(\mathscr{H})$, by application of the functional calculus for A to the continuous functions $f_+(t) = \max\{t, 0\}$ and $f_-(t) = \max\{-t, 0\}$ with the properties $f = f_+ - f_-$, $f_+ f_- = 0$, $|f_\pm| \le |f|$, we obtain two positive operators $A_+, A_- \in \mathbb{B}(\mathscr{H})$ such that $A = A_+ - A_-$, $A_+A_- = 0 = A_-A_+$, $\|A_\pm\| \le \|A\|$. The decomposition $A = A_+ - A_-$ of a self-adjoint operator as the difference of two positive operators is called the *Jordan decomposition*, and A_+ and A_- are called the *positive part* and the *negative part* of A, respectively.

The following result plays an essential role in the theory of Hilbert space operators.

Theorem 1.4.1 *Let $A \in \mathbb{B}(\mathscr{H})$. The following assertions are equivalent:*

(a) *the operator A is positive;*
(b) *the operator A is of the form B^2 for some positive operator $B \in \mathbb{B}(\mathscr{H})$;*
(c) *the operator A is of the form B^*B for some $B \in \mathbb{B}(\mathscr{H})$;*
(d) *$\langle Ax, x\rangle \geq 0$ holds for every $x \in \mathscr{H}$.*

If A is a positive operator, then the operator B in part (b) of Theorem 1.4.1 is called the *positive square root* of A and is denoted by $A^{1/2}$.

If $A \in \mathbb{B}(\mathscr{H})$, then A^*A is positive. Hence, we can define the absolute value of $A \in \mathbb{B}(\mathscr{H})$ as $|A| = (A^*A)^{1/2}$.

Theorem 1.4.2 *Let $A, B \in \mathbb{B}_{\text{sa}}(\mathscr{H})$. Then the following assertions (i)-(ix) hold:*

(i) *if $A \geq 0$ and $t \in [0, \infty)$, then $tA \geq 0$;*
(ii) *if A and B are positive, then so is $A + B$;*
(iii) *if $-B \leq A \leq B$, then $\|A\| \leq \|B\|$;*
(iv) *if $0 \leq A \leq B$, then $X^*AX \leq X^*BX$ for all $X \in \mathbb{B}(\mathscr{H})$;*
(v) *$A > 0$ if and only if there is $m > 0$ such that $0 < m \leq A$;*
(vi) *if $0 < A \leq B$, then B is invertible and $0 < B^{-1} \leq A^{-1}$;*
(vii) *the set of positive operators is closed in $\mathbb{B}(\mathscr{H})$;*
(viii) *if $A, B \in \mathbb{B}(\mathscr{H})$ are positive and $AB = BA$, then $AB \geq 0$;*
(ix) *if $A, B \in \mathbb{B}(\mathscr{H})$ are positive, then $\text{sp}(AB) \subseteq [0, \infty)$;*

The following theorem is known as the *Löwner–Heinz theorem* [7, Theorem 1.8].

Theorem 1.4.3 (Löwner–Heinz theorem) *If $A, B \in \mathbb{B}(\mathscr{H})$ and $0 \leq A \leq B$, then $0 \leq A^t \leq B^t$ for all $t \in [0, 1]$.*

In the next result, we present the *Furuta inequality* [7, Theorem F, p. 202] in the literature.

Theorem 1.4.4 (Furuta inequality) *If $A, B \in \mathbb{B}(\mathscr{H})$ and $0 \leq A \leq B$, then for each $r \geq 0$*

$$(A^{r/2}B^pA^{r/2})^{1/q} \leq (A^{r/2}A^pA^{r/2})^{1/q}$$

and

$$(B^{r/2}B^pB^{r/2})^{1/q} \leq (B^{r/2}A^pB^{r/2})^{1/q}$$

hold for $p \geq 0$ and $q \geq 1$ with $(1 + r)q \geq p + r$.

The *polar decomposition* is a helpful tool in exploring operator inequalities; see, for example, [5, Theorem 2.3.4].

Theorem 1.4.5 (Polar decomposition) *Let $A \in \mathbb{B}(\mathscr{H})$. Then there is a unique partial isometry U with $\ker U = \ker A$ and $A = U|A|$. In addition, $|A| = U^*A$. When $A \in \mathbb{M}_n$, we can choose U to be a unitary.*

Finally, we assert the *Jensen inequality* for the inner product and the *Hölder–McCarthy inequality* [8, p. 20]. Both of them can be extended for positive linear maps.

Lemma 1.4.6 (Classical Jensen inequality) *Let $A \in \mathbb{B}(\mathscr{H})$ be self-adjoint with spectrum contained in an interval $[m, M]$, and let f be a continuous convex function on $[m, M]$. Then*

$$f(\langle Ax, x\rangle) \leq \langle f(A)x, x\rangle \tag{1.4.1}$$

for all unit vectors $x \in \mathscr{H}$. The inequality is reversed when f is a concave function.

It is worthy to mention that the classical Jensen inequality is used by von Neumann to prove that $A \mapsto \operatorname{tr}(f(A))$ is convex on Hermitian matrices with spectra in the domain of the convex function f, which establishes the concavity of the von Neumann entropy on density matrices.

Lemma 1.4.7 (Hölder–McCarthy inequality) *Let $A \in \mathbb{B}(\mathscr{H})$ be positive, and let $x \in \mathscr{H}$ with $\|x\| = 1$. Then*

$$\langle Ax, x\rangle^r \leq \langle A^r x, x\rangle, \quad r \geq 1.$$

The inequality is reversed when $0 \leq r \leq 1$.

1.5 Borel Functional Calculus

Let $A \in \mathbb{B}(\mathscr{H})$ be a normal operator. Then there is a unique map E, the so-called *spectral measure*, from the collection of all Borel sets of $\operatorname{sp}(A)$ to the set of projections in $\mathbb{B}(\mathscr{H})$ such that
(i) $E(\emptyset) = 0$, $E(\operatorname{sp}(A)) = I$, and $E(S_1 \cap S_2) = E(S_1)E(S_2)$ for all Borel sets $S_1, S_2 \subseteq \operatorname{sp}(A)$, and for each $x, y \in \mathscr{H}$, the function

$$E_{x,y} : S \mapsto \langle E(S)x, y\rangle$$

is a regular Borel measure on $\operatorname{sp}(A)$;
(ii) $A = \int z dE$, where z is the inclusion map of $\operatorname{sp}(A)$ into $\mathbb{C}$;
(iii) The map $f \mapsto f(A) = \int f dE$ is a $*$-homomorphism from the C^*-algebra of all bounded Borel measurable complex-valued functions on $\operatorname{sp}(A)$ into $\mathbb{B}(\mathscr{H})$. Here $\int f dE$ is the unique operator in $\mathbb{B}(\mathscr{H})$ satisfying

$$\left\langle \left(\int f dE \right) x, y \right\rangle = \int f dE_{x,y} \qquad (x, y \in \mathscr{H}).$$

This $*$-homomorphism is called the *Borel functional calculus*.

Note that for an unbounded normal operator A with the dense domain $D(A)$ in a Hilbert space $\mathscr{H}$, one can define the spectral measure as follows: Let $\lambda \in \mathbb{C}$ and $r \geq 0$. Then the range of $(|A - \lambda I| - rI)_+$ is closed in $\mathscr{H}$. Let $E(\lambda, r)$ denote the corresponding (bounded) projection. For a compact subset K of $\mathbb{C}$, we set

$$E(K) := \wedge_{s>0} \vee \{E(\lambda, s) : \lambda \in K\}.$$

For an arbitrary subset M of $\mathbb{C}$, we put

$$E(M) := \vee\{E(K) : K \text{ is a compact subset of } M\}.$$

Then the restriction of E to the set of all Borel subsets of $\mathbb{C}$ is the required spectral measure. In particular, for a self-adjoint operator A and $\lambda \in \mathbb{R}$, the projection $E(-\infty, \lambda)$ is called a *spectral projection* of A and is denoted by E_λ, or E_λ^A to avoid any ambiguity. Similarly, we may denote $E(M)$ by $E^A(M)$. This construction works for bounded and unbounded operators; see [9]. In addition, when A is self-adjoint, the projections E_λ are the same as constructed in the spectral representation Theorem 1.3.1.

In this book, we sometimes denote the spectral measure corresponding to a normal operator A by E^A.

1.6 Operator Matrices

Block matrices of operators are considered useful in investigating operator inequalities. They allow one to represent an operator by some operators with simple structures. Any bounded linear operator A on $\mathscr{H} = \mathscr{H}_1 \oplus \mathscr{H}_2$ can be represented by the block matrix $\begin{bmatrix} A_{11} & A_{12} \\ A_{21} & A_{22} \end{bmatrix}$, where $A_{ij} \in \mathbb{B}(\mathscr{H}_j, \mathscr{H}_i)$ is defined by $A_{ij} = \pi_i A \iota_j$. Here π_i is the natural projection from $\mathscr{H}$ onto $\mathscr{H}_i$ and ι_j is the natural embedding of $\mathscr{H}_j$ into $\mathscr{H}$ for $1 \leq i, j \leq 2$. One can define $n \times n$ operator matrices similarly.

Application of the functional calculus shows us that

$$f\left(\begin{bmatrix} A & 0 \\ 0 & B \end{bmatrix}\right) = \begin{bmatrix} f(A) & 0 \\ 0 & f(B) \end{bmatrix} \tag{1.6.1}$$

when f is a continuous function on the union of spectra of self-adjoint operators A and B. Furthermore, if $A, B \in \mathbb{B}(\mathscr{H})$, then

$$\|A \oplus B\| = \max\{\|A\|, \|B\|\}. \tag{1.6.2}$$

The following theorem is a crucial result in dealing with 2×2 block matrices of operators. Part (vi) goes back to Bourin and Lee [10].

Theorem 1.6.1 ([10–12]) *Let $A \in \mathbb{B}_{++}(\mathscr{H})$, $B \in \mathbb{B}_{+}(\mathscr{H})$ and $X \in \mathbb{B}(\mathscr{H})$. Then the following assertions are equivalent:*

(i) the block matrix $\begin{bmatrix} A & X \\ X^* & B \end{bmatrix}$ is positive;
(ii) there exists a contraction K such that $X = A^{1/2} K B^{1/2}$;
(iii) $B \geq X^* A^{-1} X$;
(iv) $|\langle x, Xy\rangle|^2 \leq \langle x, Ax\rangle \langle y, By\rangle$ for all $x, y \in \mathscr{H}$;
(v) there exists an operator $C \in \mathbb{B}(\mathscr{H})$ such that $X = C^* B^{1/2}$ and $C^* C \leq A$;
(vi) there exist two unitaries $U, V \in \mathbb{B}(\mathscr{H} \oplus \mathscr{H})$ such that

$$\begin{bmatrix} A & X \\ X^* & B \end{bmatrix} = U \begin{bmatrix} A & 0 \\ 0 & 0 \end{bmatrix} U^* + V \begin{bmatrix} 0 & 0 \\ 0 & B \end{bmatrix} V^*.$$

Remark 1.6.2 If $A \in \mathbb{B}_{+}(\mathscr{H})$ in Theorem 1.6.1, then the assertions (i), (ii), (iv), and (v) in Theorem 1.6.1 are mutually equivalent.

1.7 Tensor Product and Hadamard Product of Matrices

Let $A = [a_{ij}]$ and $B = [b_{pq}]$ be elements in $\mathbb{M}_m$ and $\mathbb{M}_n$, respectively. Let $(e_1, e_2, \ldots, e_m)$ and $(f_1, f_2, \ldots, f_n)$ be the standard orthonormal bases for $\mathbb{C}^m$ and $\mathbb{C}^n$, respectively. Then $a_{ij} = \langle Ae_j, e_i\rangle$ and $b_{pq} = \langle Bf_q, f_p\rangle$. The tensor product or Kronecker product of A and B is the matrix $A \otimes B$ represented relative to the basis elements $e_i \otimes f_p$ of the tensor product $\mathbb{C}^m \otimes \mathbb{C}^n$ in the following natural way:

$$\langle (A \otimes B)(e_j \otimes f_q), e_i \otimes f_p\rangle = \langle Ae_j \otimes Bf_q, e_i \otimes f_p\rangle = \langle Ae_j, e_i\rangle\langle Bf_q, f_p\rangle = a_{ij} b_{pq}.$$

Note that $A \otimes B$ is represented by a matrix such that the element in the (i, p)-th row and (j, q)-th column equals $a_{ij} b_{pq}$. Thus, by replacing each element a_{ij} of the $m \times m$ matrix $[a_{ij}]$ with the $n \times n$ matrix $a_{ij}[b_{pq}]$ we reach $A \otimes B$. If $A, C \in \mathbb{M}_m$, and $B, D \in \mathbb{M}_n$, then $(A \otimes B)(C \otimes D) = AC \otimes BD$.

Theorem 1.7.1 ([2]) *Let $A \in \mathbb{M}_n$ and $B \in \mathbb{M}_m$. If $\lambda_1, \lambda_2, \ldots, \lambda_n$ are the eigenvalues of A and $u_1, u_2, \ldots, u_n$, where $u_j \in \mathbb{C}^n$ for $j = 1, \ldots, n$, are the corresponding eigenvectors of A and if $\mu_1, \mu_2, \ldots, \mu_m$ are the eigenvalues of B and $v_1, v_2, \ldots, v_m$, where $v_j \in \mathbb{C}^m$ for $j = 1, \ldots, m$ are the corresponding eigenvectors of B, then $\mathrm{sp}(A \otimes B) = \{\lambda_i \mu_j : i = 1, \ldots, n,\ j = 1, \ldots, m\}$ (counting algebraic multiplicities) are the eigenvalues of $A \otimes B$ and $u_i \otimes v_j \in \mathbb{C}^{nm}$, $i = 1, \ldots, n$, $j = 1, \ldots, m$ are the corresponding eigenvectors of $A \otimes B$. Every eigenvalue of $A \otimes B$ can be expressed as a product of eigenvalues of A and B. In particular, $\mathrm{sp}(A \otimes B) = \mathrm{sp}(B \otimes A)$.*

Corollary 1.7.2 *If $A \in \mathbb{M}_n$ and $B \in \mathbb{M}_m$ are positive semidefinite (positive definite), then $A \otimes B$ is also positive semidefinite (positive definite).*

In the infinite-dimensional case, let $(e_n)_{n\in\mathbb{N}}$ and $(f_m)_{m\in\mathbb{N}}$ be orthonormal bases for the separable Hilbert spaces $\mathscr{H}$ and $\mathscr{K}$, respectively. Then $\mathscr{H}\otimes\mathscr{K}$ is defined as the completion of the algebraic tensor product consisting of all formal finite linear combinations $\sum \alpha_{nm} e_n \otimes f_m$ under the inner product

$$\langle \sum \alpha_{nm} e_n \otimes f_m, \sum \beta_{nm} e_n \otimes f_m \rangle = \sum \alpha_{nm} \bar{\beta}_{nm} .$$

Then $(e_n \otimes f_m)_{(n,m)\in\mathbb{N}^2}$ is an orthonormal basis for $\mathscr{H}\otimes\mathscr{K}$.
Also, if $A \in \mathbb{B}(\mathscr{H})$ and $B \in \mathbb{B}(\mathscr{K})$, then $A\otimes B$ is the unique operator on $\mathbb{B}(\mathscr{H} \otimes \mathscr{K})$ such that

$$(A\otimes B)(e_n \otimes f_m) = \sum_{k,l} \langle Ae_n, e_k\rangle\langle Bf_m, f_l\rangle e_k \otimes f_l \qquad (n, m \in \mathbb{N}) .$$

In addition, $\|A\otimes B\| = ||A\|\,\|B\|$.

Given an orthonormal basis (e_j) of a Hilbert space $\mathscr{H}$, the Hadamard product $A\circ B$ of two operators $A, B \in \mathbb{B}(\mathscr{H})$ is defined by $\langle (A\circ B)e_i, e_j\rangle = \langle Ae_i, e_j\rangle\langle Be_i, e_j\rangle$. In the case of matrices, the Hadamard product (Schur product) of $A = [a_{ij}]$ and $B = [b_{ij}]$ is $A\circ B = [a_{ij}b_{ij}]$. It differs from the usual matrix product. For instance, the usual matrix product is noncommutative, but the Hadamard product is commutative. The Hadamard product of two $n\times n$ matrices $A = [a_{ij}]$ and $B = [b_{ij}]$ is a principal submatrix of the tensor product $A\otimes B$. Recall that a *submatrix* of a given matrix is a matrix lying in the specified subsets of the rows and columns of a given matrix. For $C \in \mathbb{M}_n$ and index sets $r \subseteq \{1,\ldots,n\}$ and $s \subseteq \{1,\ldots,n\}$ we denote by $A(r, s)$ the submatrix that lies in the rows of A indexed by r and the columns indexed by s. If $s = r$, the submatrix is called a *principal submatrix* and is denoted by $A(r)$.

In the infinite-dimensional case, it is known that the Hadamard product can be presented by filtering the tensor product through a positive linear map. Let (e_n) be an orthonormal basis for the separable Hilbert space $\mathscr{H}$. To each operator $A \in \mathbb{B}(\mathscr{H})$, we can assign a doubly infinite matrix $[\alpha_{ij}]$, where $\alpha_{ij} = \langle Ae_j, e_i\rangle$. Let B be another operator with the corresponding matrix $[\beta_{ij}]$ relative to the basis (e_n). If $U : \mathscr{H} \to \mathscr{H}\otimes\mathscr{H}$ is the isometry defined by $Ue_j = e_j \otimes e_j$, then $A\circ B = U^*(A\otimes B)U \in \mathbb{B}(\mathscr{H})$ is an operator, whose matrix is $[\gamma_{ij}]$ in which $\gamma_{ij} = \alpha_{ij}\beta_{ij}$ for all i, j.

1.8 Exercises and Problems

Exercise 1.8.1 Show that if $A \in \mathbb{B}(\mathscr{H})$, then $(\ker A)^{\perp} = \overline{\operatorname{ran}(A^*)}$.

Exercise 1.8.2 Let (e_n) be an orthonormal basis for a separable Hilbert space $\mathscr{H}$. Let A be the *unilateral shift* defined by

$$A\left(\sum_{n=1}^{\infty}\lambda_n e_n\right)=\sum_{n=1}^{\infty}\lambda_n e_{n+1}.$$

Show that A^* is the *backward shift* defined by $A^*e_1 = 0$ and $A^*e_n = e_{n-1}$ for $n > 1$. Prove that A has no eigenvalues while every complex number λ with $|\lambda| < 1$ is an eigenvalue of A^*.

Exercise 1.8.3 Prove that an operator $A \in \mathbb{B}(\mathscr{H})$ is compact if and only if $\langle e_n, Ae_n\rangle \to 0$ as $n \to \infty$ for every orthonormal set (e_n) in $\mathscr{H}$.

Exercise 1.8.4 Show that if $A, B \in \mathbb{B}(\mathscr{H})$ with $AB = BA$, then $r(A + B) \leq r(A) + r(B)$ and $r(AB) \leq r(A)r(B)$.

Exercise 1.8.5 Show that if A is a compact operator, then $\mathrm{sp}(A)$ is countable. Moreover, each nonzero element of $\mathrm{sp}(A)$ is an eigenvalue of A and an isolated point of $\mathrm{sp}(A)$.

Exercise 1.8.6 Let $A \in \mathbb{B}_{\mathrm{sa}}(\mathscr{H})$. Show that

$$\inf_{\|x\|=1}\langle Ax, x\rangle = \sup\{\alpha \in \mathbb{R} : \alpha I \leq A\}$$

and

$$\sup_{\|x\|=1}\langle Ax, x\rangle = \inf\{\alpha \in \mathbb{R} : A \leq \alpha I\}.$$

Exercise 1.8.7

(i) Show that the C^*-algebra $(C(\Omega), \|\cdot\|_\infty)$ has a nontrivial projection if and only if Ω is a disconnected (compact Hausdorff) space.
(ii) Give an example of a noncommutative C^*-algebra without any nontrivial projections.

Exercise 1.8.8 Let $A \in \mathbb{B}(\mathscr{H})$ with the polar decomposition $A = U|A|$. Prove that

(i) the polar decomposition of A^* is $A^* = U^*|A^*|$;
(ii) $|A^*|^r = U|A|^rU^*$ for any positive number r;
(iii) if A is invertible, then there is a unique unitary $U \in \mathbb{B}(\mathscr{H})$ such that $A = U|A|$.

Exercise 1.8.9 Give an example of a positive block matrix $\begin{bmatrix} A & X \\ X^* & B\end{bmatrix}$ such that $\begin{bmatrix} A & X^* \\ X & B\end{bmatrix}$ is not positive.

Problem 1.8.10 Let $A \in \mathbb{B}(\mathscr{H})$ be such that $\|A\| < 1 - 2/n$ for some $n > 2$. Prove that there exist unitaries $U_1, \ldots, U_n$ such that

$$A = \frac{U_1 + \cdots + U_n}{n}.$$

Problem 1.8.11 (i) Show that

$$\mathscr{A} := \{f \in C([0, 1], \mathbb{M}_2) : f(0) \text{ and } f(1) \text{ are diagonal}\}$$

endowed with the involution $f \mapsto f^*$, where $f^*(t) = f(t)^*$ and the norm $\|f\| = \sup_{t\in[0,1]} \|f(t)\|$ in which $\|f(t)\|$ denotes the operator norm of the matrix $f(t) \in \mathbb{M}_2$, is a unital C^*-algebra.

(ii) Prove that $\mathscr{A}$ is generated by two projections

$$p(t) = \begin{bmatrix} 1 & 0 \\ 0 & 0 \end{bmatrix} \quad \text{and} \quad q(t) = \begin{bmatrix} t & \sqrt{t(1-t)} \\ \sqrt{t(1-t)} & 1-t \end{bmatrix}.$$

(iii) Let P and Q be projections in a unital C^*-algebra $\mathscr{B}$. Prove that there is a unital homomorphism $\phi : \mathscr{A} \to \mathscr{B}$ such that $\phi(p) = P$ and $\phi(q) = Q$.

1.9 Notes, Hints, and References

This chapter has provided concise and condensed information on functional analysis, operator theory, and matrix analysis. For those seeking more detailed explanations on the topics discussed here, notable references for further reading include works by Furuta [13] for Hilbert space operators, Murphy [5] for C^*-algebras, and Bhatia [14] for matrix analysis.

Let us provide some hints and references for the exercises and problems.

For Exercise 1.8.1, one can use $\langle Ax, y\rangle = \langle x, A^*y\rangle$. For Exercise 1.8.2, employ $\langle A^*e_n, e_m\rangle = \langle e_n, Ae_m\rangle$ to find A^*e_n. Show that if A had an eigenvalue μ with the corresponding eigenvector x (that is, $Ax = \mu x$), a contradiction would arise.

A good reference for Exercise 1.8.3 is [15, p. 59]. To solve Exercise 1.8.4, consider the commutative C^*-algebra generated by A, B, and I. Apply the Gelfand theorem to conclude that it should be of the form $C(\Omega)$ for some compact Hausdorff space Ω, and hence A and B correspond to two continuous functions. Now, use the fact that $r(h) = \|h\|_\infty$.

Exercise 1.8.5 is the same as [5, Theorem 1.4.11]. To solve Exercise 1.8.6, use the fact that $\alpha I \le A$ if and only if $\alpha \le \langle Ax, x\rangle$ for all unit vector x. A useful hint for Exercise 1.8.7 is that if $f \in C(\Omega)$ is a projection, then $f^2 = f = \overline{f}$, and hence the range of f is contained in $\{0, 1\}$. Now, use the fact that Ω is disconnected if and only if there exists a non-surjective continuous function f from Ω into the discrete topological space $\{0, 1\}$. The polar decomposition Theorem 1.4.5 can be utilized for Exercise 1.8.8. An example required in Exercise 1.8.9 is given in [16, p. 259]. Problem 1.8.10 is borrowed from [17, Proposition 3.2.23]. A suitable reference for Problem 1.8.11 is [18].

References

1. M.S. Moslehian, G.A. Muñoz-Fernández, A.M. Peralta, J.B. Seoane-Sepúlveda, Similarities and differences between real and complex Banach spaces: an overview and recent developments. Rev. R. Acad. Cienc. Exactas Fís. Nat. Ser. A Mat. RACSAM **116**, 88 (2022), 80 pp
2. R.A. Horn, C.R. Johnson, *Matrix Analysis*, 2nd edn. (Cambridge University Press, Cambridge, 2013)
3. P. Bhunia, S.S. Dragomir, M.S. Moslehian, K. Paul, *Lectures on Numerical Radius Inequalities, Infosys Science Foundation Series in Mathematical Sciences* (Springer, Cham, 2022)
4. S.S. Dragomir, *Inequalities for the numerical radius of linear operators in Hilbert spaces* SpringerBriefs in Mathematics. (Springer, Cham, 2013)
5. G.J. Murphy, *C^*-algebras and Operator Theory* (Academic, Boston, MA, 1990)
6. F. Zhang, *Matrix Theory*, 2nd edn. (Universitext, Springer, New York, Basic Results and Techniques, 2011)
7. T. Furuta, J. Mićić Hot, J. Pečarić, Y. Seo, *Mond–Pečarić Method in Operator Inequalities*, Monographs in Inequalities 1, Element, Zagreb (2005)
8. B. Simon, *Trace Ideals and Their Applications*, Mathematical Surveys and Monographs, vol. 120, 2nd edn. (American Mathematical Society, Providence, RI, 2005)
9. S.J. Bernau, The spectral theorem for unbounded normal operators. Pacific J. Math. **19**, 391–406 (1966)
10. J.-C. Bourin, E.Y. Lee, Unitary orbits of Hermitian operators with convex or concave functions. Bull. London Math. Soc. **44**, 1085–1102 (2012)
11. T. Ando, *Topics on Operator Inequalities* (Hokkaido University, Sapporo, 1978)
12. M.S. Moslehian, M. Kian, Q. Xu, Positivity of 2×2 block matrices of operators. Banach J. Math. Anal. **13**, 726–743 (2019)
13. T. Furuta, *Invitation to Linear Operators: From Matrices to Bounded Linear Operators on a Hilbert Space* (Taylor & Francis Group, London, 2001)
14. R. Bhatia, *Matrix Analysis* (Springer, New York, 1997)
15. J.R. Ringrose, *Compact Non-Self-Adjoint Operators* (Van Nostrand Reinhold Co., London, 1971)
16. M. Gumus, J. Liu, S. Raouafi, T.-Y. Tam, On positive partial transpose matrices. Electron. J. Linear Algebra **36**, 256–264 (2020)
17. G.K. Pedersen, *Analysis now Graduate Texts in Mathematics, 118* (Springer, New York, 1989)
18. I. Raeburn, A.M. Sinclair, The C^*-algebra generated by two projections. Math. Scand. **65**, 278–290 (1989)

Chapter 2
Unitarily Invariant Norms and Inequalities

In this chapter, we first introduce the notion of the trace of a matrix and its interesting properties. We then study the concept of the trace in a broader context. We explore the unitarily invariant norms on Hilbert space operators, which are symmetric gauge functions of singular values. Each of these norms is defined on a two-sided ideal of the algebra of bounded linear operators acting on a Hilbert space transforming the ideal into a Banach space. Examples of unitarily invariant norms include the operator norm, the Hilbert–Schmidt norm, the trace norm, and the Ky Fan norms. The study of unitarily invariant norms is traced back to von Neumann [1], and Fan and Hoffman [2]. We also present several elegant inequalities that are related to unitarily invariant norms, particularly the Hilbert–Schmidt and trace norms.

2.1 Trace of a Matrix

In this section, we focus on $n \times n$ complex matrices. We begin by discussing factorizations of matrices into specific types of matrices, which are valuable tools in matrix theory. We highlight three specific types of factorizations. The proofs for these factorizations can be found in textbooks on matrix theory, such as [3, 4]. In addition, it is notable to recall the *Cayley–Hamilton theorem*. This theorem states that if $p_A(\lambda) = \det(\lambda I - A)$ is the characteristic polynomial of a matrix $A \in \mathbb{M}_n$, then $p_A(A) = 0$.

Theorem 2.1.1 (Schur decomposition) *Let $A \in \mathbb{M}_n$ with eigenvalues $\lambda_1, \ldots, \lambda_n$. Then*

$$A = UTU^*,$$

where T is an upper-triangular matrix (that is, its (i, j)-entry is 0 if $i > j$) with $\lambda_1, \ldots, \lambda_n$ on its diagonal, and $U \in \mathbb{M}_n$ is a unitary matrix.

A. M. Bikchentaev et al., *Trace Inequalities*, Forum for Interdisciplinary Mathematics,
https://doi.org/10.1007/978-981-97-6520-1_2

Theorem 2.1.2 (Spectral decomposition) *Let $A \in \mathbb{M}_n$ with eigenvalues $\lambda_1, \ldots, \lambda_n$. Then A is normal if and only if A is unitarily diagonalizable; that is, there exists a unitary matrix U such that*

$$A = U\text{diag}(\lambda_1, \ldots, \lambda_n)U^*.$$

In particular, A is Hermitian (positive semidefinite, respectively) if and only if all λ_i's are real (nonnegative, respectively).

By a *singular value* of A, we mean an eigenvalue of $|A|$. We denote by $s(A) = (s_1(A), \ldots, s_n(A))$, the n-tuples of singular values of A, taken in nonincreasing order and with multiplicity.

Theorem 2.1.3 (Singular value decomposition) *Let $A \in \mathbb{M}_n$ with nonzero singular values $s_1, \ldots, s_k$. Then there exist unitaries $U, V \in \mathbb{M}_n$ such that*

$$A = U\text{diag}(s_1, \ldots, s_k, 0, \ldots, 0)V^*.$$

The *trace of a matrix* $A \in \mathbb{M}_n$ is denoted by tr A (or tr(A) if there is ambiguity) and is defined as the sum of its eigenvalues, which turns out to be the sum of the diagonal entries of A.

The *determinant* $\det(A)$ of a matrix $A \in \mathbb{M}_n$ is defined as the product of its eigenvalues. It is known that if A^{ij} is the $(n-1) \times (n-1)$ matrix obtained by removing the ith row and the jth column from the matrix $A = [a_{ij}]$, then

$$\det(A) = \sum_{i=1}^{n} (-1)^{i+j} a_{ij} \det(A^{ij})$$

for any $1 \le j \le n$.

The determinant has the multiplicative property $\det(AB) = \det(A)\det(B)$ for all matrices A and B. If 0 is an eigenvalue of AB, then $0 = \det(AB) = \det(A)\det(B) = \det(BA)$, and so 0 is an eigenvalue of BA. Also, if $\lambda \neq 0$ is an eigenvalue of AB with the eigenvector $x \neq 0$, then $Bx \neq 0$ and $BA(Bx) = B(ABx) = B(\lambda x) = \lambda Bx$. Hence, λ is also an eigenvalue of BA. This shows that for matrices, $\text{sp}(AB) = \text{sp}(BA)$. Therefore,

$$\text{tr}(AB) = \text{tr}(BA).$$

Moreover, $\text{tr}(\cdot)$ is a positive linear functional on $\mathbb{M}_n$. In addition, $\langle A, B\rangle = \text{tr}(B^*A)$ generates an inner product on $\mathbb{M}_n$, which is called the *Hilbert–Schmidt inner product*. Thus, the following Cauchy–Schwarz inequality holds:

$$|\,\text{tr}(AB^*)|^2 \le \text{tr}(A^*A)\,\text{tr}(B^*B). \tag{2.1.1}$$

If A is a matrix with real eigenvalues λ_i $(i = 1, \ldots, n)$, then by taking x_i to be λ_i and $y_i = 1$ in classical Cauchy–Schwarz inequality (1.1.4), we arrive to

$$\mathrm{tr}(A)^2 = \left(\sum_{i=1}^{n} \lambda_i\right)^2 \le n \sum_{i=1}^{n} \lambda_i^2 = n\,\mathrm{tr}(A^2).$$

Therefore, we have the following proposition.

Proposition 2.1.4 *If the eigenvalues of a matrix A are real, then*

$$\mathrm{tr}(A)^2 \le n\,\mathrm{tr}(A^2).$$

The eigenvalues of A are the zeros of the characteristic polynomial $\det(\lambda I - A)$. Expanding the determinant, we readily arrive at

$$\mathrm{tr}\,A = \sum_{i=1}^{n} a_{ii} = \sum_{i=1}^{n} \langle Ae_i, e_i\rangle.$$

It is immediately seen that $\mathrm{tr}(A \otimes B) = \mathrm{tr}\,A\,\mathrm{tr}\,B$. Moreover, one can define the *partial trace* tr_1 on $\mathbb{M}_n \otimes \mathbb{M}_m$ into $\mathbb{M}_m$ by $\mathrm{tr}_1(A \otimes B) = (\mathrm{tr}\,A)B$ as well as the *partial trace* tr_2 from $\mathbb{M}_n \otimes \mathbb{M}_m$ into $\mathbb{M}_n$ by $\mathrm{tr}_2(A \otimes B) = (\mathrm{tr}\,B)A$.

It follows from the arithmetic–geometric mean inequality (see [5, Chap. 3])

$$\sqrt[n]{a_1 \cdots a_n} \le \frac{a_1 + \cdots + a_n}{n} \qquad (a_1, \cdots, a_n \ge 0)$$

applied to the eigenvalues of a positive semidefinite matrix $A \in \mathbb{M}_n$ that

$$\det(A)^{\frac{1}{n}} \le \frac{\mathrm{tr}\,A}{n}. \tag{2.1.2}$$

In particular, equality holds if and only if all eigenvalues of A are the same number; that is, A is a scalar multiplication of the identity matrix.

Theorem 2.1.5 *Let $A \in \mathbb{M}_n$ be positive definite with* $\det(A) \le 1$. *Then*

$$\det(A)^{\frac{1}{n}} = \min_{B \ge 0, \det(B) \ge 1} \frac{\mathrm{tr}(BA)}{n}.$$

Proof Let $B \in \mathbb{M}_n$ be positive semidefinite with $\det(B) \ge 1$. If we apply inequality (2.1.2) to $B^{1/2}AB^{1/2}$, then

$$\det(A)^{\frac{1}{n}} \le \det(B^{1/2}AB^{1/2})^{\frac{1}{n}} \le \frac{\mathrm{tr}(B^{1/2}AB^{1/2})}{n} = \frac{\mathrm{tr}(BA)}{n}.$$

The equality occurs if we set $B = \det(A)^{1/n}A^{-1}$. □

Proposition 2.1.6 *Let $A \in \mathbb{M}_n$ be positive definite. Then*

$$\log(\det A) \le \mathrm{tr}\,A - n.$$

Proof Since $\log t \le t - 1$ for all $t > 0$, we have

$$\log(\det A) = \log\left(\prod_{i=1}^{n} \lambda_i\right) = \sum_{i=1}^{n} \log(\lambda_i) \le \sum_{i=1}^{n} (\lambda_i - 1) = \operatorname{tr} A - n.$$

Clearly, equality holds if and only if $A = I$. □

The eigenvalues of the product of two Hermitian matrices may not be real numbers. However, the trace of the product is real.

Theorem 2.1.7 *If* $A, B \in \mathbb{M}_n$ *are Hermitian, then* $\operatorname{tr}(AB)$ *is real.*

Proof (*First proof due to Coope* [6]) Since $\operatorname{sp}(A) \subseteq \mathbb{R}$ and $\operatorname{sp}(A + \lambda I) = \operatorname{sp}(A) + \lambda$, we can choose the real number λ such that $X = A + \lambda I \ge 0$. Hence,

$$\operatorname{tr}(AB) = \operatorname{tr}(XB) - \lambda \operatorname{tr} B = \operatorname{tr}\left(X^{1/2} B X^{1/2}\right) - \lambda \operatorname{tr} B \in \mathbb{R},$$

since $X^{1/2} B X^{1/2}$ and B are Hermitian, and hence have real traces.

(*Second proof*) Observe that

$$\overline{\operatorname{tr}(AB)} = \operatorname{tr}(AB)^* = \operatorname{tr}\left(B^* A^*\right) = \operatorname{tr}(BA) = \operatorname{tr}(AB).$$

So, $\operatorname{tr}(AB)$ is real. □

Theorem 2.1.8 *If* $A \ge 0$ *and* $B \ge 0$ *are in* $\mathbb{M}_n$, *then*

$$\operatorname{tr}(A^{\alpha} B^{1-\alpha}) \le \alpha \operatorname{tr} A + (1 - \alpha) \operatorname{tr} B$$

for all $0 \le \alpha \le 1$.

Proof Let $A = UCU^*$ and $B = VDV^*$ be the spectral decompositions of A and B, respectively, where $C = \operatorname{diag}(\lambda_1, \ldots, \lambda_n)$ and $D = \operatorname{diag}(\mu_1, \ldots, \mu_n)$ with $\lambda_i \ge 0$, $\mu_i \ge 0$, $1 \le i \le n$. Since $U^*V = (u_{ij})$ is a unitary matrix, it follows that

$$\begin{aligned}
\operatorname{tr}(A^{\alpha} B^{1-\alpha}) &= \operatorname{tr}(UC^{\alpha}U^*VD^{1-\alpha}V^*) \\
&= \operatorname{tr}((U^*V)^*C^{\alpha}(U^*V)D^{1-\alpha}) \\
&= \sum_{i,j=1}^{n} |u_{ij}|^2 \lambda_i^{\alpha} \mu_j^{1-\alpha} \\
&\le \sum_{i,j=1}^{n} |u_{ij}|^2 (\alpha\lambda_i + (1-\alpha)\mu_j) \\
&= \alpha \sum_{i=1}^{n} \lambda_i + (1-\alpha) \sum_{i=1}^{n} \mu_i \\
&= \alpha \operatorname{tr} A + (1-\alpha) \operatorname{tr} B,
\end{aligned}$$

where the inequality follows from the following *weighted arithmetic–geometric mean inequality*

$$\left(a_1^{w_1}\cdots a_n^{w_n}\right)^{\frac{1}{w}} \le \frac{w_1 a_1 + \cdots + w_n a_n}{w} \qquad (a_1, \cdots, a_n \ge 0),$$

in which $w_1, \ldots, w_n \ge 0$ are weights with $w = w_1 + \cdots + w_n > 0$. □

The next theorem reads as follows.

Theorem 2.1.9 *If $A \ge 0$ and $B \ge 0$ are in $\mathbb{M}_n$, then*

$$\operatorname{tr}\left((AB)^k\right) \le \|B\|^k \operatorname{tr}(A)^k$$

for all positive integers k.

Proof First note that if the eigenvalues of a positive semidefinite matrix C are $\lambda_1, \ldots, \lambda_n$, which are nonnegative, it follows from the spectral mapping theorem that

$$\operatorname{tr}(C^k) = \sum_{j=1}^{n} \lambda_j^k \le \left(\sum_{j=1}^{n} \lambda_j\right)^k = \operatorname{tr}(C)^k.$$

The matrices $(AB)^k$ and $(A^{1/2}BA^{1/2})^k$ have the same spectra and trace. Therefore,

$$\begin{aligned}\operatorname{tr}\left((AB)^k\right) &= \operatorname{tr}\left(\left(A^{1/2}BA^{1/2}\right)^k\right) \le \operatorname{tr}\left(A^{1/2}BA^{1/2}\right)^k \\ &\le \operatorname{tr}\left(A^{1/2}\|B\|A^{1/2}\right)^k \le \|B\|^k (\operatorname{tr} A)^k.\end{aligned}$$

□

From $\|B\| \le \operatorname{tr} B$ for a positive semidefinite matrix B, we get the following consequence; see also [7].

Corollary 2.1.10 *If $A \ge 0$ and $B \ge 0$ are in $\mathbb{M}_n$, then*

$$\operatorname{tr}\left((AB)^k\right) \le \operatorname{tr}(A)^k \operatorname{tr}(B)^k$$

for all positive integers k. In particular,

$$\operatorname{tr}(AB) \le \operatorname{tr} A \operatorname{tr} B. \tag{2.1.3}$$

Proposition 2.1.11 *A matrix $A \in \mathbb{M}_n$ is positive semidefinite if and only if $\operatorname{tr}(AB) \ge 0$ for all positive semidefinite matrices $B \in \mathbb{M}_n$.*

Proof If A is positive semidefinite, then $\operatorname{tr}(AB) = \operatorname{tr}(B^{1/2}AB^{1/2}) \ge 0$ for all positive semidefinite matrices B. Conversely, let x be a unit vector in $\mathbb{C}^n$ considered an $n \times 1$ column matrix. Then its adjoint x^* is a $1 \times n$ matrix. If we take $B = xx^*$, then

$$\langle Ax, x\rangle = \operatorname{tr}(x^*Ax) = \operatorname{tr}(Axx^*) = \operatorname{tr}(AB) \geq 0.$$

Hence, $A \geq 0$. □

For Hermitian matrices, one has the following assertion.

Theorem 2.1.12 *If A and B are Hermitian matrices in $\mathbb{M}_n$, then*

$$\operatorname{tr}\left((AB)^2\right) \leq \operatorname{tr}\left(A^2B^2\right).$$

Proof (*First proof*) First note that when A is Hermitian, A^2 is positive semidefinite. If A is Hermitian, we have

$$\operatorname{tr}\left((\mathrm{i}A)^2\right) = \operatorname{tr}(-A^2) = -\operatorname{tr}(A^2) \leq 0,$$

whence we conclude that the trace of the square of a skew-Hermitian matrix is non-positive. Since $AB - BA$ is skew-Hermitian, we therefore have $\operatorname{tr}\left((AB-BA)^2\right) \leq 0$. On the other hand,

$$\begin{aligned}\operatorname{tr}\left((AB-BA)^2\right) &= \operatorname{tr}\left((AB)^2 + (BA)^2 - AB^2A - BA^2B\right)\\ &= 2\operatorname{tr}\left((AB)^2\right) - 2\operatorname{tr}\left(A^2B^2\right),\end{aligned}$$

whence

$$\operatorname{tr}\left((AB)^2\right) \leq \operatorname{tr}\left(A^2B^2\right).$$

(*Second proof*) First assume that $D = \operatorname{diag}(\lambda_1, \ldots, \lambda_n)$ is a diagonal matrix with real entries and $B = [b_{ij}]$ is Hermitian. We have

$$\begin{aligned}\operatorname{tr}\left((DB)^2\right) &= \operatorname{tr}\left(\left[\lambda_i b_{ij}\right]^2\right) = \operatorname{tr}\left(\left[\sum_{k=1}^{n} \lambda_i b_{ik}\lambda_k b_{kj}\right]\right) = \sum_{i=1}^{n}\sum_{k=1}^{n} \lambda_i\lambda_k b_{ik}b_{ki}\\ &= \sum_{i=1}^{n}\sum_{k=1}^{n} \lambda_i\lambda_k |b_{ik}|^2 = \sum_{i=1}^{n} \lambda_i^2 |b_{ii}|^2 + \sum_{i<k} 2\lambda_i\lambda_k |b_{ik}|^2\end{aligned}$$

and

$$\begin{aligned}\operatorname{tr}\left(D^2B^2\right) &= \operatorname{tr}\left(\operatorname{diag}(\lambda_1^2, \ldots, \lambda_n^2)[b_{ij}]^2\right) = \operatorname{tr}\left(\left[\sum_{k=1}^{n} \lambda_i^2 b_{ik}b_{kj}\right]\right)\\ &= \sum_{i=1}^{n}\sum_{k=1}^{n} \lambda_i^2 b_{ik}b_{ki} = \sum_{i=1}^{n}\sum_{k=1}^{n} \lambda_i^2 |b_{ik}|^2 = \sum_{i=1}^{n} \lambda_i^2 |b_{ii}|^2 + \sum_{i<k} (\lambda_i^2 + \lambda_k^2)|b_{ik}|^2.\end{aligned}$$

Since $\sum_{i<k}(\lambda_i - \lambda_k)^2 \geq 0$, we arrive to

$$\operatorname{tr}\left((DB)^2\right) \leq \operatorname{tr}\left(D^2B^2\right).$$

Next, assume that $A \geq 0$ and $B \geq 0$ are given. We apply the spectral decomposition $A = UDU^*$ and the first part of this proof to get

$$\begin{aligned}\operatorname{tr}\left((AB)^2\right) &= \operatorname{tr}\left(UDU^*BUDU^*B\right) = \operatorname{tr}\left(D(U^*BU)D(U^*BU)\right)\\ &\leq \operatorname{tr}\left(D^2(U^*BU)^2\right) = \operatorname{tr}\left(D^2U^*B^2U\right) = \operatorname{tr}\left(UD^2U^*B^2\right) = \operatorname{tr}\left(A^2B^2\right).\end{aligned}$$

□

Proposition 2.1.13 *If* $\begin{bmatrix} A & X \\ X^* & C \end{bmatrix} \geq 0$, *then* $\operatorname{tr}(X^*X) \leq \operatorname{tr} A \operatorname{tr} C$.

Proof By Theorem 1.6.1 (ii), there exists a contraction K such that $X = A^{1/2}KC^{1/2}$. It follows from (2.1.3) that

$$\begin{aligned}\operatorname{tr}(X^*X) &= \operatorname{tr}(C^{1/2}K^*A^{1/2}A^{1/2}KC^{1/2}) = \operatorname{tr}(K^*AKC) \leq \operatorname{tr}(K^*AK)\operatorname{tr} C\\ &= \operatorname{tr}(A^{1/2}KK^*A^{1/2})\operatorname{tr} C \leq \|KK^*\| \operatorname{tr} A \operatorname{tr} C \leq \operatorname{tr} A \operatorname{tr} C,\end{aligned}$$

since K is a contraction. □

Proposition 2.1.14 *If* $\begin{bmatrix} A & X \\ X^* & C \end{bmatrix} \geq 0$, *then* $\begin{bmatrix} \operatorname{tr} A & \operatorname{tr} X \\ \operatorname{tr} X^* & \operatorname{tr} C \end{bmatrix} \geq 0$.

Proof Let $A = [a_{ij}]$, $C = [c_{ij}]$, and $X = [x_{ij}]$. We have

$$\begin{bmatrix} \operatorname{tr} A & \operatorname{tr} X \\ \operatorname{tr} X^* & \operatorname{tr} C \end{bmatrix} = \begin{bmatrix} \sum_{i=1}^n a_{ii} & \sum_{i=1}^n x_{ii} \\ \sum_{i=1}^n \overline{x_{ii}} & \sum_{i=1}^n c_{ii} \end{bmatrix} = \sum_{i=1}^n \begin{bmatrix} a_{ii} & x_{ii} \\ \overline{x_{ii}} & c_{ii} \end{bmatrix}.$$

Since $\begin{bmatrix} a_{ii} & x_{ii} \\ \overline{x_{ii}} & c_{ii} \end{bmatrix}$ is a principal submatrix of the positive semidefinite matrix $\begin{bmatrix} A & X \\ X^* & C \end{bmatrix}$ considered as an element of $\mathbb{M}_{2n}$, it is also positive semidefinite. In light of the fact that the sum of positive definite matrices is positive semidefinite, we get the desired result. □

Proposition 2.1.15 *Let* $A \in \mathbb{M}_n$. *Then* $A^n = 0$ *if and only if* $\operatorname{sp}(A) = \{0\}$.

Proof If $A^n = 0$, then we conclude from the spectral mapping theorem that $\operatorname{sp}(A) = \{0\}$.

Conversely, let $\operatorname{sp}(A) = \{0\}$. It follows from the Schur decomposition Theorem 2.1.1 that

$$A = U \begin{bmatrix} 0 & & * \\ & \ddots & \\ 0 & & 0 \end{bmatrix} U^*,$$

and so $A^n = 0$. Another proof for this part is to use the Cayley–Hamilton theorem and the fact that $\lambda^n = 0$ is the characteristic function of A. □

Now, we follow outlines of [3, Theorem 8.2].

Theorem 2.1.16 *A matrix* $A \in \mathbb{M}_n$ *is Hermitian if and only if* $\operatorname{tr}(A^2) = \operatorname{tr}(A^*A)$.

Proof ($\Longrightarrow$) It is clear.
($\Longleftarrow$) (*First proof*) We first show that if $A \in \mathbb{M}_n$ is Hermitian and $\operatorname{tr}(A^2) = 0$, then $A = 0$. To this end, we use the spectral decomposition to write $A = UDU^*$, where $D = \operatorname{diag}(\lambda_1, \ldots, \lambda_n)$. Since A is Hermitian, all its eigenvalues are real. We have $0 = \operatorname{tr}(A^2) = \operatorname{tr}(UDU^*UDU^*) = \operatorname{tr}(D^2U^*U) = \operatorname{tr}(D^2)$. Hence, $\sum_{j=1}^n \lambda_j^2 = 0$, and all of the numbers λ_j must be zero. Thus, $D = 0$ and $A = 0$. Next, we consider the Cartesian decomposition of an arbitrary matrix A with $\operatorname{tr}(A^2) = \operatorname{tr}(A^*A)(\geq 0)$. Considering the imaginary part $B = \frac{A-A^*}{2\mathrm{i}}$, we have

$$\operatorname{tr}(B^2) = \operatorname{tr}\left(\frac{A^2 - AA^* - A^*A + A^{*2}}{-4}\right) = 0,$$

since $\operatorname{tr} C^* = \overline{\operatorname{tr} C}$ for all matrix C. Therefore, using the first part of the proof, we conclude that $B = 0$. Hence, A is Hermitian.

(*Second proof*) By the Schur decomposition Theorem 2.1.1, there exists an upper-triangular matrix $T = [t_{ij}]$ with $\lambda_1, \ldots, \lambda_n$ on its diagonal such that $A = UTU^*$. Hence,

$$\operatorname{tr}(A^2) = \operatorname{tr}(UTU^*UTU^*) = \operatorname{tr}(UT^2U^*) = \operatorname{tr}(T^2).$$

Similarly, $\operatorname{tr}(A^*A) = \operatorname{tr}(T^*T)$. A straightforward computation on $\operatorname{tr}(T^2) = \operatorname{tr}(TT^*)$ shows that

$$\sum_{j=1}^n \lambda_j^2 = \sum_{j=1}^n |\lambda_j|^2 + \sum_{j=1}^n \sum_{i<j} |t_{ij}|^2.$$

If we write $\lambda_j = \alpha_j + \mathrm{i}\beta_j$ with $\alpha_j, \beta_j \in \mathbb{R}$, then we conclude from the above equality that $\beta_j = 0$ for all j and $t_{ij} = 0$ for all $i < j$. Thus, λ_j's are real. Since T is upper-triangular, we infer that T is diagonal with real entries. Hence, $A = UTU^*$ is Hermitian.

(*Third proof*) If $X = [x_{ij}]$ and $\operatorname{tr}(X^*X) = \sum_{1\leq i,j\leq n} |x_{ij}|^2 = 0$, then $X = 0$. Thus, if $\operatorname{tr}(A^2) = \operatorname{tr}(A^*A)$, then $\operatorname{tr}(A - A^*)^2 = 0$. Hence, $A - A^* = 0$, and so A is Hermitian. □

Remark 2.1.17 When A is a matrix with real entries, we can prove the theorem above. To achieve our aim, let $\operatorname{tr}(A^2) = \operatorname{tr}(A^*A)$ and $B = \mathrm{i}(A - A^*)$. Then $\operatorname{tr}(B^*B) = \operatorname{tr}(AA^* - A^2 - A^{*2} + A^*A) = 0$ since $\operatorname{tr}(A^2)^t = \operatorname{tr}(A^2)$. Hence, $B = 0$, which ensures that A is Hermitian.

Let $A = [a_{ij}]$. The Schur decomposition Theorem 2.1.1 entails that there exists an upper-triangular matrix $T = [t_{ij}]$ with $\lambda_1, \ldots, \lambda_n$ on its diagonal such that $A = UTU^*$. We have

$$\sum_{i,j=1}^{n} |a_{ij}|^2 = \operatorname{tr}(A^*A) = \operatorname{tr}(T^*T) = \sum_{j=1}^{n} |\lambda_j|^2 + \sum_{i<j} |t_{ij}|^2.$$

Equality holds if and only if $t_{ij} = 0,\ i < j$, and this occurs if and only if T is diagonal, and this holds if and only if A is normal. Hence, we arrive at the following result [3, Theorem 9.5].

Theorem 2.1.18 (Schur inequality) *If $A \in \mathbb{M}_n$ with the eigenvalues $\lambda_1, \ldots, \lambda_n$, then*

$$\sum_{j=1}^{n} |\lambda_j|^2 \le \sum_{i,j=1}^{n} |a_{ij}|^2.$$

Equality holds if and only if A is normal.

Theorem 2.1.19 *If $A, B \in \mathbb{M}_n$ are positive definite, then*

$$\operatorname{tr}\left((A^{-1} - B^{-1})(A - B)\right) \le 0.$$

Proof Apply the functional calculus to a positive definite matrix X. It follows from the inequality $2 - t - t^{-1} \le 0$ $(t > 0)$ that $2 - X - X^{-1} \le 0$, and hence $\operatorname{tr}(2 - X - X^{-1}) \le 0$. Employing the cyclicity property of the trace and taking $X = A^{-1/2}BA^{-1/2}$, we get

$$\begin{aligned} \operatorname{tr}\left((A^{-1} - B^{-1})(A - B)\right) &= \operatorname{tr}\left(I - A^{-1}B - B^{-1}A + I\right) \\ &= \operatorname{tr}\left(2I - A^{-1/2}BA^{-1/2} - A^{1/2}B^{-1}A^{1/2}\right) \le 0. \end{aligned}$$

□

The next assertion is due to Parker [8, Theorem 9].

Lemma 2.1.20 *If $A \in \mathbb{M}_n$, then there exists a unitary $U \in \mathbb{M}_n$ such that UAU^* has all its main diagonal entries equal to $n^{-1}\operatorname{tr} A$.*

Proof The numerical range $\Sigma = \{\langle Ax, x\rangle : \|x\| = 1, x \in \mathbb{C}^n\}$ is a closed convex set that includes the diagonal entries $\langle Ae_i, e_i\rangle$ of A. Hence, $t_0 = n^{-1}\operatorname{tr} A \in \Sigma$.

Suppose that u is a unit vector such that $t_0 = \langle Au, u\rangle$ and that U is a unitary matrix such that $Uu = e_1$. Then the first diagonal entry of UAU^* is $\langle UAU^*e_1, e_1\rangle = \langle AU^*e_1, U^*e_1\rangle = t_0$. We now apply induction. Let V be a unitary matrix such that $VAV^* = \begin{bmatrix} A_0 & A_1 \\ A_2 & A_3 \end{bmatrix}$ in which A_0 is a matrix with all k diagonal entries equal to t_0. Since $\operatorname{tr}(VAV^*) = \operatorname{tr} A$, we have $\operatorname{tr}(A_3) = (n-k)t_0$ and there exists a unitary matrix W_3 such that $W_3A_3W_3^*$ has its first diagonal entry equal to t_0. If we define $W = \begin{bmatrix} I_k & 0 \\ 0 & W_3 \end{bmatrix}$, then $WVAV^*W^*$ has its first $k+1$ diagonal entries equal to t_0. Thus, the induction is completed. □

Now we are ready to prove the next interesting result due to Shoda.

Theorem 2.1.21 *If $A \in \mathbb{M}_n$ with zero trace, then there exist matrices X and Y so that $A = XY - YX$ in which X is Hermitian and Y has trace zero.*

Proof It follows from Lemma 2.1.20 that there exists a unitary matrix U so that $UAU^* = B = [b_{ij}]$ has zero diagonal. Pick distinct real numbers $d_1, \ldots, d_n$ and set $D = \mathrm{diag}(d_1, \ldots, d_n)$ and $Y_1 = [y_{ij}]$ with

$$y_{ij} = \begin{cases} \frac{b_{ij}}{d_i - d_j} & \text{if } i \neq j, \\ 0 & \text{if } i = j. \end{cases}$$

One can readily verify that $B = DY_1 - Y_1 D$, and hence $A = XY - YX$, where $X = U^* DU$ and $Y = U^* Y_1 U$. Moreover, X is Hermitian and $\mathrm{tr}\, Y = 0$. □

The next result is significant.

Theorem 2.1.22 (Schur product theorem) *If $A, B \in \mathbb{M}_n$ are positive definite, then so is $A \circ B$.*

Proof Let $A = [a_{ij}]$, $B = [b_{ij}]$, and x be the column matrix with entries $x_1, \ldots, x_n$. We have

$$\begin{aligned} \langle (A \circ B)x, x \rangle &= \sum_{i=1}^{n} \sum_{j=1}^{n} a_{ij} b_{ij} x_j \overline{x_i} \\ &= \sum_{i,j=1}^{n} (\overline{x_i} a_{ij})(b_{ij} x_j) \\ &= \mathrm{tr}\left(\left(\mathrm{diag}(\overline{x_1}, \ldots, \overline{x_n})[a_{ij}] \right) \left([b_{ij}] \mathrm{diag}(x_1, \ldots, x_n) \right)^t \right) \\ &= \mathrm{tr}\left(\mathrm{diag}(\overline{x_1}, \ldots, \overline{x_n})[a_{ij}] \mathrm{diag}(x_1, \ldots, x_n)[b_{ji}] \right) \\ &= \mathrm{tr}\left([b_{ji}]^{1/2} \mathrm{diag}(\overline{x_1}, \ldots, \overline{x_n})[a_{ij}] \mathrm{diag}(x_1, \ldots, x_n)[b_{ji}]^{1/2} \right) \geq 0, \end{aligned}$$

since the matrix $[b_{ji}]^{1/2} \mathrm{diag}(\overline{x_1}, \ldots, \overline{x_n})[a_{ij}] \mathrm{diag}(x_1, \ldots, x_n)[b_{ji}]^{1/2}$ is positive semidefinite. □

Theorem 2.1.23 *If $A \in \mathbb{M}_n$, then*

$$\mathrm{tr}\, |A| = \max_{\|X\| \leq 1} |\, \mathrm{tr}(A \circ X)|. \tag{2.1.4}$$

Proof It follows from Theorem 1.6.1 that if $A \in \mathbb{M}_n$ and X is a contraction, then

$$\begin{bmatrix} |A| & A^* \\ A & |A^*| \end{bmatrix} \geq 0 \quad \text{and} \quad \begin{bmatrix} I & X^* \\ X & I \end{bmatrix} \geq 0.$$

It follows from Schur product Theorem 2.1.22 that

$$\begin{bmatrix} |A| & (A \circ X)^* \\ A \circ X & |A^*| \end{bmatrix} = \begin{bmatrix} |A| \circ I & A^* \circ X^* \\ A \circ X & |A^*| \circ I \end{bmatrix} \geq 0.$$

Hence,

$$\begin{bmatrix} \operatorname{tr}|A| & \overline{\operatorname{tr}(A \circ X)} \\ \operatorname{tr}(A \circ X) & \operatorname{tr}|A^*| \end{bmatrix} \geq 0,$$

so that

$$|\operatorname{tr}(A \circ X)| \leq (\operatorname{tr}|A| \operatorname{tr}|A^*|)^{1/2} = \operatorname{tr}|A|.$$

The equality occurs if we take $X = I$. □

If we pick $X = I$ in (2.1.4), then we have

Corollary 2.1.24 *If $A \in \mathbb{M}_n$, then*

$$|\operatorname{tr} A| \leq \operatorname{tr}|A|. \tag{2.1.5}$$

The following trace inequality was first proved by von Neumann [1]. The presented proof is due to Grigorieff [9].

Theorem 2.1.25 (von Neumann trace inequality) *Let $A, B \in \mathbb{M}_n$. Then*

$$|\operatorname{tr}(AB)| \leq \sum_{j=1}^{n} s_j(A)s_j(B). \tag{2.1.6}$$

Proof Consider the singular value decomposition of matrices A and B as follows:

$$A = \sum_{j=1}^{n} s_j(A)x_j \otimes x_j'^* \quad \text{and} \quad B = \sum_{j=1}^{n} s_j(B)y_j \otimes y_j'^*,$$

where (x_j), (x_j'), (y_j) and (y_j') are orthonormal sets in $\mathbb{C}^n$. In addition, for any orthonormal basis (z_j), we have

$$\operatorname{tr}(AB) = \sum_{j=1}^{n} \langle ABz_j, z_j \rangle.$$

Both sides of (2.1.6) are positively homogeneous with respect to $s(A)$ and $s(B)$, respectively. Therefore, it is sufficient to consider the case where $s_1(A) \leq 1$ and $s_1(B) \leq 1$. In addition, the left-hand side of (2.1.6) is a convex function of the vectors $f_j := (1, \ldots, 1, 0, \ldots, 0)$ with j components equal to 1, for $j = 0, 1, \ldots, n$. Therefore, it is sufficient to prove (2.1.6) for the case $s(A) = f_r$ and $s(B) = f_s$ for

$r, s = 0, \ldots, n$. By the cyclicity property of $\operatorname{tr}(\cdot)$, we can assume that $r \leq s$. Taking $z_j = y_j$, we get

$$|\operatorname{tr}(AB)| = \left|\sum_{j=1}^{n}\sum_{k=1}^{r}\sum_{l=1}^{t}\langle y_l', x_k\rangle\langle z_j, y_l\rangle\langle x_k', z_j\rangle\right|$$
$$= \left|\sum_{k=1}^{r}\sum_{l=1}^{t}\langle y_l', x_k\rangle\langle x_k', y_l\rangle\right| \leq \sum_{k=1}^{r}\|x_k\|\,\|x_k'\| = r.$$

In this case, we have $\langle s(A), s(B)\rangle = \sum_{j=1}^{n} s_j(A)s_j(B) = r$. □

The next result reads as follows. Its proof slightly differs from that of [10, Theorem 3].

Theorem 2.1.26 *If $A, B \in \mathbb{M}_n$ are positive semidefinite, then*

$$\operatorname{tr}(A - B) \leq \operatorname{Re}\ \operatorname{tr}(A - UBV) \leq \operatorname{tr}(A + B)$$

for all contractions $U, V \in \mathbb{M}_n$.

Proof First note that when U is a contraction, we have $U^*U \leq 1$ and $UU^* \leq 1$. Therefore, the column and row vectors of U have a norm of at most one.

We first apply the spectral decomposition $A = UDU^*$ with $D = \operatorname{diag}(\lambda_1, \ldots, \lambda_n)$ to prove that $\operatorname{Re}\ \operatorname{tr}(A - VAW) = \operatorname{Re}\ \operatorname{tr}(D - U^*VUDU^*WU) \geq 0$ for all contractions $V, W \in \mathbb{M}_n$. Hence, it suffices to show that $\operatorname{Re}\ \operatorname{tr}(D - VDW) \geq 0$ for all contractions $V = [v_{ij}]$ and $W = [w_{ij}]$. One has

$$\operatorname{Re}\ \operatorname{tr}(D \pm VDW) = \operatorname{Re}\sum_{i=1}^{n}\left(\lambda_i \pm \sum_{k=1}^{n} v_{ik}\lambda_k w_{ki}\right)$$
$$= \operatorname{Re}\sum_{i=1}^{n}\left(\lambda_i \pm \sum_{k=1}^{n} v_{ki}\lambda_i w_{ik}\right)$$
$$= \sum_{i=1}^{n}\lambda_i\left(1 \pm \operatorname{Re}\sum_{k=1}^{n} v_{ki}w_{ik}\right);$$

the value $\sum_{k=1}^{n} v_{ki}w_{ik}$ is the inner product of the ith columns of V and W^*, which have norms not greater than 1. Hence, the Cauchy–Schwarz inequality (1.1.4) yields that

$$\left|\operatorname{Re}\sum_{k=1}^{n} v_{ki}w_{ik}\right| \leq 1.$$

Thus, $\operatorname{Re}\ \operatorname{tr}(D \pm VDW) \geq 0$. Next, applying the polar decomposition $A - UBV = P|A - UBV|$, we have

$$
\begin{aligned}
\operatorname{tr}(A-B) &= \operatorname{Re}\operatorname{tr}(A-B) \\
&= \operatorname{Re}\operatorname{tr}(A-UBV) - \operatorname{Re}\operatorname{tr}(B-UBV) \\
&\le \operatorname{Re}\operatorname{tr}(A-UBV) \qquad \text{(by the first part of the proof)} \\
&\le |\operatorname{tr}(A-UBV)| \\
&\le \operatorname{tr}|A-UBV| \qquad \text{(by Corollary 2.1.24)} \\
&= \operatorname{Re}\operatorname{tr}(P^*A - P^*UBV) \\
&= \operatorname{tr}(A+B) - \operatorname{Re}\operatorname{tr}(A-P^*A) - \operatorname{Re}\operatorname{tr}(B+P^*UBV) \\
&\le \operatorname{tr}(A+B) \qquad \text{(by the first part of the proof).}
\end{aligned}
$$

□

2.2 Majorization

We first provide materials for generalizing the notion of trace from matrices to operators on separable Hilbert spaces.

Let $x = (x_1, x_2, \ldots, x_n) \in \mathbb{R}^n$ and $x^{\downarrow} = (x_1^{\downarrow}, x_2^{\downarrow}, \ldots, x_n^{\downarrow})$ and $x^{\uparrow} = (x_1^{\uparrow}, x_2^{\uparrow}, \ldots, x_n^{\uparrow})$ be the vectors obtained by rearranging the coordinates of x in the decreasing and the increasing orders, respectively.

Definition 2.2.1 Let $x, y \in \mathbb{R}^n$. If

$$
\sum_{j=1}^{k} x_j^{\downarrow} \le \sum_{j=1}^{k} y_j^{\downarrow} \quad \text{for all} \quad 1 \le k \le n, \tag{2.2.1}
$$

then x is said to be *weakly majorized* by y in the sense of Hardy–Littlewood–Pólya [11], denoted by $x \prec_w y$. If in addition to (2.2.1),

$$
\sum_{j=1}^{n} x_j^{\downarrow} = \sum_{j=1}^{n} y_j^{\downarrow}
$$

holds, then x is said to be *majorized* by y and it is denoted by $x \prec y$.

Example 2.2.2 In $\mathbb{R}^n$,

$$
\begin{aligned}
\left(\frac{1}{n}, \frac{1}{n}, \ldots, \frac{1}{n}\right) &\prec \left(\frac{1}{n-1}, \frac{1}{n-1}, \ldots, \frac{1}{n-1}, 0\right) \prec \cdots \\
&\prec \left(\frac{1}{2}, \frac{1}{2}, 0, \ldots, 0\right) \prec (1, 0, \ldots, 0).
\end{aligned}
$$

Definition 2.2.3 Let $x, y \in \mathbb{R}_+^n$, where $\mathbb{R}_+^n$ is the set of all $(x_1, \ldots, x_n)$ such that $x_i \ge 0$ for all $1 \le i \le n$. If

$$\prod_{j=1}^{k} x_j^{\downarrow} \leq \prod_{j=1}^{k} y_j^{\downarrow} \quad \text{for all} \quad 1 \leq k \leq n, \tag{2.2.2}$$

then x is said to be *weakly log-majorized* by y and we denote this by $x \prec_{w\log} y$. If in addition to (2.2.2),

$$\prod_{j=1}^{n} x_j^{\downarrow} = \prod_{j=1}^{n} y_j^{\downarrow}$$

holds, then x is said to be *log-majorized* by y and we denote this by $x \prec_{log} y$.

Weak log-majorization is stronger than weak majorization as the following theorem [12, p. 64] shows.

Theorem 2.2.4 *Let $x, y \in \mathbb{R}_+^n$. If $\prod_{j=1}^{k} x_j^{\downarrow} \leq \prod_{j=1}^{k} y_j^{\downarrow}$ for all $1 \leq k \leq n$, then $\sum_{j=1}^{k} x_j^{\downarrow} \leq \sum_{j=1}^{k} y_j^{\downarrow}$ for all $1 \leq k \leq n$.*

For a Hermitian matrix $A \in \mathbb{M}_n$, we denote by $\lambda_j^{\downarrow}(A)$ the eigenvalues of A arranged in the decreasing order with their multiplicities counted. The *minimax theorem* states that if $A \in \mathbb{M}_n$ is Hermitian and S is a subspace of $\mathbb{C}^n$, then

$$\lambda_k^{\downarrow}(A) = \max_{\dim S = k} \min_{x \in S, \|x\|=1} \langle Ax, x \rangle = \min_{\dim S = n-k+1} \max_{x \in S, \|x\|=1} \langle Ax, x \rangle . \tag{2.2.3}$$

In addition, the *Ky Fan maximum principle* says that

$$\sum_{j=1}^{k} \lambda_j^{\downarrow}(A) = \max \sum_{j=1}^{k} \langle Ax_j, x_j \rangle, \tag{2.2.4}$$

where maximum is taken over all orthonormal k-tuples of vectors $(x_1, \ldots, x_n)$ in $\mathbb{C}^n$.

For a Hermitian matrix A, the notation $\lambda(A)$ stands for $(\lambda_1^{\downarrow}(A), \lambda_2^{\downarrow}(A), \ldots, \lambda_n^{\downarrow}(A))$. For any Hermitian matrices A and B, the eigenvalue inequality $\lambda(A) \leq \lambda(B)$ means that $\lambda_j^{\downarrow}(A) \leq \lambda_j^{\downarrow}(B)$ for all $1 \leq j \leq n$. As a matter of fact, $\lambda(A) \leq \lambda(B)$ holds if and only if $A \leq U^* B U$ for some unitary matrix U. It is known that the three types of orders defined above satisfy

$$A \leq B \Rightarrow \lambda(A) \leq \lambda(B) \Rightarrow \lambda(A) \prec_w \lambda(B) .$$

The first above implication is called the *Weyl monotonicity principal*, which follows from the minimax theorem.

Theorem 2.2.5 *If A and B are Hermitian matrices such that $0 \leq A \leq B$ and f is a continuous increasing real function on $[0, \infty)$, then*

$$\operatorname{tr} f(A) \le \operatorname{tr} f(B).$$

A generalization of the above inequality is given in Exercise [13].

Proof It follows from $0 \le A \le B$ that $\lambda_j(A) \le \lambda_j(B)$ for all $1 \le j \le n$. Since f is increasing, we have

$$\lambda_j(f(A)) = f(\lambda_j(A)) \le f(\lambda_j(B)) = \lambda_j(f(B)),$$

which leads to $\operatorname{tr} f(A) \le \operatorname{tr} f(B)$. □

For Hermitian matrices A and B, the weak majorization $A \prec_w B$ means that $\lambda(A) \prec_w \lambda(B)$. Furthermore, the majorization $A \prec B$ means that $\lambda(A) \prec \lambda(B)$. For $A, B \ge 0$, we write $A \prec_{w\log} B$ and refer to it as the weak log-majorization if $\lambda(A) \prec_{w\log} \lambda(B)$. Moreover, the log-majorization $A \prec_{\log} B$ means that $\lambda(A) \prec_{\log} \lambda(B)$. When $A, B > 0$, the log-majorization $A \prec_{\log} B$ is equivalent to $\log A \prec \log B$. It is known that if A and B are positive semidefinite, then

$$A \prec_{w\log} B \Longrightarrow A \prec_w B \Longrightarrow |||A||| \le |||B||| \tag{2.2.5}$$

for any unitarily invariant norm $|||\cdot|||$ (see the next section); see [14, 15] for the theory of majorization for matrices.

2.3 Unitarily Invariant Norms

Let $A \in \mathbb{K}(\mathscr{H})$. As in the finite-dimensional case, a *singular value* of A is defined as an eigenvalue of $|A|$. We denote by $s(A) = (s_k(A))_{k\in\mathbb{N}}$, the sequence of singular values of A, taken in nonincreasing order and with multiplicity. If $\operatorname{rank}(A) = n < \infty$, we take $s_k(A) = 0$ for $k > n$. The *spectral norm* $s_1(A)$ of a matrix $A \in \mathbb{M}_n$ coincides with its operator norm since

$$\|A\| = \|A^*A\|^{\frac{1}{2}} = r(A^*A)^{\frac{1}{2}} = \lambda_1(A^*A)^{\frac{1}{2}} = \lambda_1(|A|^2)^{\frac{1}{2}} = \lambda_1(|A|) = s_1(A).$$

The *singular value decomposition* says that if $A \in \mathbb{K}(\mathscr{H})$, then $A = \sum_{n=1}^{\infty} s_n(A) f_n \otimes e_n^*$ for some orthonormal sets (e_n) and (f_n).

Let c_0 be the space of all complex sequences converging to zero. For $a \in c_0$, denote $\lfloor a \rfloor = (|a_n|)_{n\in\mathbb{N}} \in c_0$. By a *symmetric gauge function* (or *symmetric norm*) we mean a real-valued function g defined on the space $c_F \subseteq c_0$ of all sequences with a finite number of nonzero entries, which satisfies (i) g is a norm on c_F, (ii) $g(a) = g(\lfloor a \rfloor)$ for every $a \in c_F$ (or, equivalently, g is monotone in the sense that if $|a_n| \le |b_n|$ for all n, then $g(a) \le g(b)$), and (iii) g is invariant under permutations. We say that g is a *normalized gauge function* if $g(e_1) = 1$. For $a = (a_i) \in c_0$, let us set

$$g(a) := \sup_{n\in\mathbb{N}} g(a_1, \ldots, a_n, 0, \ldots),$$

whose values can be infinity.

A *unitarily invariant norm* on $\mathbb{K}(\mathscr{H})$ is a map $|||\cdot|||$ on $\mathbb{K}(\mathscr{H})$ given by $|||A||| = g(s(A))$, $A \in \mathbb{K}(\mathscr{H})$ in which g is a symmetric norm. The set

$$\mathcal{C}_{|||\cdot|||} = \{A \in \mathbb{K}(\mathscr{H}) : |||A||| < \infty\}$$

is a (two-sided) ideal of $\mathbb{B}(\mathscr{H})$, which is self-adjoint, meaning it is closed under the adjoint operation. It is called the *norm ideal* associated with $|||\cdot|||$. Furthermore, $(\mathcal{C}_{|||\cdot|||}, |||\cdot|||)$ is a Banach space. Generally, when we consider $|||A|||$ we are assuming that the operator A belongs to $\mathcal{C}_{|||\cdot|||}$.

Typical and important examples are as follows: An operator $A \in \mathbb{K}(\mathscr{H})$ is said to be in the *Schatten p-class* denoted by $\mathcal{C}_p$ $(1 \le p < \infty)$, if $\sum_j s_j(A)^p < \infty$. The Schatten p-norm of A is defined by $\|A\|_p = \left(\sum_j s_j(A)^p\right)^{\frac{1}{p}}$, which is a typical example of a unitarily invariant norm. In particular, $\mathcal{C}_1$ and $\mathcal{C}_2$ are called the *trace class* and the *Hilbert–Schmidt class*, respectively. It is reasonable to let $\mathcal{C}_\infty$ denote $\mathbb{K}(\mathscr{H})$ because $\lim_{p\to\infty} \|A\|_p = \|A\|$.

For compact operators $A, B \in \mathbb{K}(\mathscr{H})$, set

$$|||A \oplus B||| = g(s_1(A), s_1(B), s_2(A), s_2(B), \ldots).$$

Since $s_j(A) = s_j(A^*)$, we have

$$\left|\left|\left|A \oplus A^*\right|\right|\right| = |||A \oplus A||| \,. \tag{2.3.1}$$

In addition,

$$\|A \oplus B\|_p = (\|A\|_p^p + \|B\|_p^p)^{1/p} \tag{2.3.2}$$

for all $1 \le p \le \infty$; see (1.6.2).

Other examples of unitarily invariant norms are the Ky Fan norms $\|A\|_{(k)} := \sum_{j=1}^k s_j(A)$, $k \in \mathbb{N}$. It is known [14] that

$$\begin{aligned}\|A\|_{(k)} &= \max\{\mathrm{tr}(B^*A) : B^*B \ \text{ is a projection of rank } k\} \\ &= \max\left\{\sum_{i=1}^{k} |(VAU)_{ii}| : U \text{ and } V \text{ are unitary operators}\right\},\end{aligned} \tag{2.3.3}$$

in which $(VAU)_{ii}$ stands for the (i, i) entry of VAU with respect to some orthonormal basis.

Some basic properties of a unitarily invariant norm are as follows:

(i) $|||UAV||| = |||A|||$, where U and V are unitary operators and $A \in \mathcal{C}_{|||\cdot|||}$, since

$$\begin{aligned}|||UAV||| &= g(s(UAV)) = g(\lambda(V^*A^*AV)^{\frac{1}{2}})\\ &= g(\lambda(A^*A)^{\frac{1}{2}}) = g(s(A)) = |||A|||\,.\end{aligned}$$

(ii) If $B \in \mathcal{C}_{|||\cdot|||}$, then $|||B||| = |||\,|B|\,||| = |||B^*|||$ since

$$\begin{aligned}|||B||| &= g(s(B)) = g(\lambda(|B|) = g(s(|B|)) = |||\,|B|\,||| = g(\lambda(B^*B)^{\frac{1}{2}})\\ &= g(\lambda(BB^*)^{\frac{1}{2}}) = g(\lambda(|B^*|)) = \left|\left|\left|B^*\right|\right|\right|\,.\end{aligned}$$

(iii) For every $B \in \mathcal{C}_{|||\cdot|||}$ and $A, C \in \mathbb{B}(\mathscr{H})$ it holds that

$$|||ABC||| \le \|A\|\; |||B|||\; \|C\| \tag{2.3.4}$$

since

$$\begin{aligned}s_j(AB) &= \lambda_j(|AB|) = \lambda_j(B^*A^*AB)^{1/2} \le \lambda_j(B^*\|A^*A\|B)^{1/2}\\ &= \|A\|\lambda_j(B^*B)^{1/2} = \|A\|s_j(B)\,,\end{aligned}$$

whence

$$s_j(ABC) \le \|A\|s_j(B)\|C\|. \tag{2.3.5}$$

It follows from the monotonicity of g that

$$|||AB||| = g(s(AB)) \le g(\|A\|s(B)) = \|A\|g(s(B)) = \|A\|\; |||B|||\,.$$

Therefore,

$$\begin{aligned}|||ABC||| &\le \|A\|\; |||BC||| = \|A\|\; \left|\left|\left|C^*B^*\right|\right|\right| \le \|A\|\; \|C^*\|\; \left|\left|\left|B^*\right|\right|\right|\\ &= \|A\|\; |||B|||\; \|C\|\,.\end{aligned}$$

A unitarily invariant norm $\|\cdot\|_Q$ is called a *Q-norm* if there exists a unitarily invariant norm $\|\cdot\|_{\widehat{Q}}$ such that $\|BB^*\|_{\widehat{Q}} = \|B\|_Q^2$.

The following theorem is a key tool for establishing inequalities involving unitarily invariant norms; see [12, Theorem IV.2.2] and [16].

Theorem 2.3.1 (Fan dominance theorem) *Let* $A, B \in \mathbb{K}(\mathscr{H})$. *Then* $\|A\|_{(k)} \le \|B\|_{(k)}$ *for all* $1 \le k \le n$ *if and only if* $|||A||| \le |||B|||$ *for all unitarily invariant norms* $|||\cdot|||$.

The Fan dominance principle yields the equivalence of the following three inequalities on operators X, Y

$$|||X||| \le |||Y||| \text{ for all unitarily invariant norms;} \tag{2.3.6}$$

$$|||X \oplus 0||| \le |||Y \oplus 0||| \text{ for all unitarily invariant norms;} \tag{2.3.7}$$

$$|||X \oplus X||| \le |||Y \oplus Y||| \text{ for all unitarily invariant norms.} \tag{2.3.8}$$

Remark 2.3.2 From now on, we assume that a unitarily invariant norm $|||\cdot|||$ is given on a two-sided ideal of bounded linear operators that act on a separable Hilbert space.

When we say that $|||A||| \le |||B|||$ for $A \in \mathbb{M}_n$ and $B \in \mathbb{M}_{2n}$ we mean that $|||A \oplus 0||| \le |||B|||$. In addition, $s_j(A \oplus 0) = s_j(A)$ for $j = 1, \ldots, n$ and $s_j(A \oplus 0) = 0$ for $j = n+1, \ldots, 2n$.

In the infinite-dimensional setting, $\mathbb{M}_n$ and $\mathbb{M}_{2n}$ can be replaced with $\mathbb{B}(\mathscr{H})$ and $\mathbb{B}(\mathscr{H} \oplus \mathscr{H})$, respectively.

Furthermore,

$$|||X^*X||| = |||XX^*|||, \tag{2.3.9}$$

since positive operators XX^* and X^*X satisfy (1.1.7). To see this for matrices, note that

$$\begin{bmatrix} I & X \\ 0 & I \end{bmatrix}^{-1} \begin{bmatrix} XY & 0 \\ Y & 0 \end{bmatrix} \begin{bmatrix} I & X \\ 0 & I \end{bmatrix} = \begin{bmatrix} 0 & 0 \\ Y & YX \end{bmatrix}.$$

Hence, $\begin{bmatrix} XY & 0 \\ Y & 0 \end{bmatrix}$ and $\begin{bmatrix} 0 & 0 \\ Y & YX \end{bmatrix}$ are similar and so have the same eigenvalues. Thus, XY and YX have the same eigenvalues.

It follows from the definition of unitary invariance that

$$\left|\left|\left|\begin{bmatrix} X & 0 \\ 0 & Y \end{bmatrix}\right|\right|\right| = \left|\left|\left|\begin{bmatrix} X & 0 \\ 0 & Y \end{bmatrix}\begin{bmatrix} 0 & I \\ I & 0 \end{bmatrix}\right|\right|\right| = \left|\left|\left|\begin{bmatrix} 0 & X \\ Y & 0 \end{bmatrix}\right|\right|\right|. \tag{2.3.10}$$

In addition, $U = \frac{1}{\sqrt{2}}\begin{bmatrix} I & I \\ I & -I \end{bmatrix}$ is a unitary matrix. Hence, the unitary invariance of $|||\cdot|||$ yields that

$$\begin{aligned} \left|\left|\left|\begin{bmatrix} X & Y \\ Y & X \end{bmatrix}\right|\right|\right| &= \left|\left|\left| U\begin{bmatrix} X+Y & 0 \\ 0 & X-Y \end{bmatrix}U^* \right|\right|\right| \\ &= \left|\left|\left|\begin{bmatrix} X+Y & 0 \\ 0 & X-Y \end{bmatrix}\right|\right|\right|. \end{aligned} \tag{2.3.11}$$

The matrix $C(A) = \begin{bmatrix} A_{11} & 0 \\ 0 & A_{22} \end{bmatrix}$ is called a *pinching of the matrix* $A = \begin{bmatrix} A_{11} & A_{12} \\ A_{21} & A_{22} \end{bmatrix}$. We frequently use the following pinching inequality:

$$|||C(A)||| = \left|\left|\left|(A + UAU^*)/2\right|\right|\right| \leq |||A||| , \qquad (2.3.12)$$

where $U = \begin{bmatrix} I & 0 \\ 0 & -I \end{bmatrix}$.

There are other equivalent approaches in order to define the Hilbert–Schmidt and trace norms:
For $A \in \mathbb{B}_+(\mathscr{H})$, define the trace of A to be

$$\operatorname{tr} A = \sum_j \langle Ae_j, e_j \rangle \in [0, \infty].$$

Then

$$\begin{aligned}
\operatorname{tr}(A^*A) &= \sum_i \langle A^*Ae_i, e_i \rangle = \sum_i \langle Ae_i, Ae_i \rangle \\
&= \sum_i \langle \sum_j \langle Ae_i, e_j \rangle e_j, Ae_i \rangle = \sum_i \sum_j \langle Ae_i, e_j \rangle \langle e_j, Ae_i \rangle \\
&= \sum_i \sum_j |\langle Ae_i, e_j \rangle|^2 = \sum_j \sum_i |\langle Ae_i, e_j \rangle|^2 \\
&= \sum_j \sum_i \langle e_j, Ae_i \rangle \langle Ae_i, e_j \rangle = \sum_j \sum_i \langle A^*e_j, e_i \rangle \langle e_i, A^*e_j \rangle \\
&= \sum_j \langle \sum_i \langle A^*e_j, e_i \rangle e_i, A^*e_j \rangle = \sum_j \langle A^*e_j, A^*e_j \rangle \\
&= \sum_j \langle AA^*e_j, e_j \rangle = \operatorname{tr}(AA^*) .
\end{aligned}$$

Therefore, if U is a unitary operator, then

$$\operatorname{tr}(UAU^*) = \operatorname{tr}(UA^{1/2}A^{1/2}U^*) = \operatorname{tr}(A^{1/2}U^*UA^{1/2}) = \operatorname{tr} A .$$

Since for two orthonormal bases $(e_j : j \in J)$ and $(f_j : j \in J)$ the operator U defined by $Ue_j = f_j$ $(j \in J)$ is a unitary operator, we conclude that the definition of trace is independent of the choice of basis.

Thus, $\mathcal{C}_1$ is the norm-closed linear span of $\{A \in \mathbb{K}(\mathscr{H}) : A \geq 0, \operatorname{tr} A < \infty\}$, and so

$$\mathcal{C}_1 = \{A \in \mathbb{B}(\mathscr{H}) : \|A\|_1 := \operatorname{tr}(|A|) < \infty\}.$$

In addition,

$$\mathcal{C}_2 = \{A \in \mathbb{K}(\mathscr{H}) : \|A\|_2^2 := \operatorname{tr}(A^*A) < \infty\} = \{A \in \mathbb{B}(\mathscr{H}) : A^*A \in \mathcal{C}_1\}.$$

Because of

$$\operatorname{tr}(sA_1 + A_2) = s\ \operatorname{tr}(A_1) + \operatorname{tr}(A_2)\ (s \geq 0,\ A_1 \geq 0,\ A_2 \geq 0)$$

and that any operator $A \in \mathbb{B}(\mathscr{H})$ is a linear combination of positive operators as $A = \sum_{k=1}^{4} \mathrm{i}^k A_k$, where $A_k \geq 0$, we can extend the trace to a linear functional on $\mathcal{C}_1$. The trace of a matrix is the sum of its eigenvalues. The infinite-dimensional version of this result is known as the *Lidskii theorem* stating that the trace of a trace class operator is the sum of its eigenvalues; see [17, 18].

Thus,

$$A \in \mathcal{C}_1 \Longleftrightarrow \|A\|_1 = \sum_j \langle |A| e_j, e_j \rangle < \infty$$

and

$$A \in \mathcal{C}_2 \Longleftrightarrow \|A\|_2 = \left(\sum_j \|Ae_j\|^2 \right)^{1/2} < \infty.$$

These relations are independent of the choice of basis.

When $\mathscr{H} = \mathbb{M}_n$, we can use the standard basis for $\mathbb{C}^n$ to get

$$\operatorname{tr} A = \sum_{i=1}^{n} \langle Ae_i, e_i \rangle = \sum_{i=1}^{n} a_{ii}$$

and

$$\|A\|_2^2 = \operatorname{tr}(A^*A) = \operatorname{tr}\left(\left[\sum_{k=1}^{n} \overline{a_{ki}} a_{kj} \right] \right) = \sum_{1 \leq i,j \leq n} |a_{ij}|^2,$$

where $A = [a_{ij}]$.

It follows from (1.2.1) that if $A, B \in \mathcal{C}_2$, then $B^*A \in \mathcal{C}_1$ and

$$\langle A, B \rangle_{tr} = \operatorname{tr}(B^*A) \qquad (A, B \in \mathcal{C}_2)$$

gives rise to an inner product on $\mathcal{C}_2$. Note that

$$\|A\| = \|A^*A\|^{1/2} \leq \operatorname{tr}(A^*A)^{1/2} = \langle A, A \rangle_{tr}^{1/2} = \|A\|_2\,.$$

The relation $\|A\|_2^2 = \operatorname{tr}(A^*A)$ is significant. For example, we can show that if $A, B \in \mathcal{C}_2$ are normal operators and $X \in \mathbb{B}(\mathscr{H})$, then $\|AX - XB\|_2 = \|A^*X - XB^*\|_2$. In fact, using the cyclicity property, we have

$$\|AX - XB\|_2^2 = \operatorname{tr}(((AX - XB)^*(AX - XB))$$
$$= \operatorname{tr}(X^*A^*AX + XBB^*X^*) - \operatorname{tr}(A^*XBX^* + AXB^*X^*)$$

and the second term is unchanged if we replace A and B with A^* and B^*, respectively. The first term is also not changed since A and B are normal.

We present the basic properties of the trace class and Hilbert–Schmidt operators.

Theorem 2.3.3 *If $A, B \in \mathcal{C}_2$ and $X \in \mathbb{B}(\mathscr{H})$, then*

(i) $\|\lambda A\|_2 = |\lambda| \, \|A\|_2$ *for all* $\lambda \in \mathbb{C}$;
(ii) $\|A + B\|_2 \leq \|A\|_2 + \|B\|_2$;
(iii) $\|A\| \leq \|A\|_2 = \|A^*\|_2$;
(iv) $\|AX\|_2 \leq \|A\|_2 \, \|X\|$ *and* $\|XB\|_2 \leq \|X\| \, \|B\|_2$;
(v) $\|x \otimes y^*\|_2 = \|x\| \, \|y\|$ *for all* $x, y \in \mathscr{H}$.

Proof Let (e_j) be an orthonormal basis of $\mathscr{H}$.

(i) It is obvious.
(ii) It follows from

$$\left(\sum_{j=1}^{n} \|Ae_j + Be_j\|^2\right)^{1/2} \leq \left(\sum_{j=1}^{n} (\|Ae_j\| + \|Be_j\|)^2\right)^{1/2}$$
$$\leq \left(\sum_{j=1}^{n} \|Ae_j\|^2\right)^{1/2} + \left(\sum_{j=1}^{n} \|Be_j\|^2\right)^{1/2}.$$

(iii) For each unit vector $x \in \mathscr{H}$, by Zorn's lemma, we can extend $\{x\}$ to an orthonormal basis (f_j) of $\mathscr{H}$. Hence,

$$\|Ax\|^2 \leq \sum_j \|Af_j\|^2 = \|A\|_2^2,$$

whence $\|A\| \leq \|A\|_2$. In addition, $\|A^*\|_2 = \sqrt{\operatorname{tr}(AA^*)} = \sqrt{\operatorname{tr}(A^*A)} = \|A\|_2$.

(iv) We only prove the second inequality. The first one can be proved from the second one and (iii). We have

$$\|XB\|_2 = \left(\sum_j \|XBe_j\|^2\right)^{1/2} \leq \|X\| \left(\sum_j \|Be_j\|^2\right)^{1/2} = \|X\| \, \|B\|_2.$$

(v) We have

$$\|x \otimes y^*\|_2^2 = \sum_j \|(x \otimes y^*)e_j\|^2 = \sum_j \|(\langle e_j, y\rangle x\|^2 = \|x\|^2 \sum_j |\langle y, e_j\rangle|^2 = \|x\|^2 \, \|y\|^2,$$

where we apply the Parseval equality.

□

Theorem 2.3.4 *If* $A, B \in \mathcal{C}_2$, *then* $\operatorname{tr}(AB) = \operatorname{tr}(BA)$.

Proof By the polarization identity (1.1.3) we get

$$\begin{aligned}\operatorname{tr}(AB) &= \sum_j \langle ABe_j, e_j\rangle = \sum_j \langle Be_j, A^*e_j\rangle = \sum_j \frac{1}{4}\sum_{k=0}^{3} \mathrm{i}^k \|Be_j + \mathrm{i}^k A^* e_j\|^2 \\ &= \frac{1}{4}\sum_{k=0}^{3} \mathrm{i}^k \sum_j \|(B + \mathrm{i}^k A^*)e_j\|^2 = \frac{1}{4}\sum_{k=0}^{3} \mathrm{i}^k \|B + \mathrm{i}^k A^*\|_2^2 \qquad (2.3.13) \\ &= \frac{1}{4}\sum_{k=0}^{3} \mathrm{i}^k \|(B + \mathrm{i}^k A^*)^*\|_2^2 = \frac{1}{4}\sum_{k=0}^{3} \mathrm{i}^k \|A + \mathrm{i}^k B^*\|_2^2 = \operatorname{tr}(BA).\end{aligned}$$

□

Theorem 2.3.5 *If* $A \in \mathbb{K}(\mathscr{H})$, *then the following assertions are equivalent:*

(i) $A \in \mathcal{C}_1$;
(ii) $|A|^{1/2} \in \mathcal{C}_2$;
(iii) A *is the product of two Hilbert–Schmidt operators.*

Proof (i) $\Rightarrow$ (ii): $\|\,|A|^{1/2}\|_2^2 = \operatorname{tr}(|A|) < \infty$.
(ii) $\Rightarrow$ (iii): By the polar decomposition $A = U|A|$, we have $A = U|A| = (U|A|^{1/2})|A|^{1/2}$. Moreover, $U|A|^{1/2}$ is a Hilbert–Schmidt operator, by Theorem 2.3.3(iv).
(iii) $\Rightarrow$ (i): Again, application of the polar decomposition of A shows that $|A| = B_1B_2$ is a product of two Hilbert–Schmidt operators. Via (2.3.13) we arrive at

$$\sum_j \langle |A|e_j, e_j\rangle = \frac{1}{4}\sum_{k=0}^{3} \mathrm{i}^k \|B_2 + \mathrm{i}^k B_1^*\|_2^2 < \infty.$$

□

Corollary 2.3.6 *If* $A \in \mathcal{C}_1$ *and* $B \in \mathbb{B}(\mathscr{H})$, *then* $\operatorname{tr}(AB) = \operatorname{tr}(BA)$.

Proof It follows from Theorem 2.3.5 (iii) that $A = C_1C_2$, where $C_1, C_2 \in \mathcal{C}_2$. In light of Theorem 2.3.4 that

$$\operatorname{tr}(AB) = \operatorname{tr}(C_1(C_2B)) = \operatorname{tr}(C_2(BC_1)) = \operatorname{tr}((BC_1)C_2) = \operatorname{tr}(BA).$$

□

Lemma 2.3.7 *If* $A, B \in \mathbb{B}(\mathscr{H})$, *then* $|AB| \leq \|A\|\,|B|$.

Proof Clearly,

$$|AB| = (B^*A^*AB)^{1/2} \le (\|A\|^2 B^*B)^{1/2} = \|A\|\,|B|.$$

□

Theorem 2.3.8 *If $A, B \in \mathcal{C}_1$ and $X \in \mathbb{B}(\mathscr{H})$, then*

(i) $\|\lambda A\|_1 = |\lambda|\,\|A\|_1$ *for all* $\lambda \in \mathbb{C}$*;*
(ii) $\|A + B\|_1 \le \|A\|_1 + \|B\|_1$*;*
(iii) $\|A\| \le \|A\|_1 = \|A^*\|_1$*;*
(iv) $\|AX\|_1 \le \|A\|_1\,\|X\|$ *and* $\|XB\|_1 \le \|X\|\,\|B\|_1$*;*
(v) $\|x \otimes y^*\|_1 = \|x\|\,\|y\|$ *for all* $x, y \in \mathscr{H}$*.*

Proof (i) It is obvious.
(ii) It can be proved by applying the polar decomposition.
(iii) Using the polar decomposition $A = U|A|$ (1.4.5), we reach

$$\|A^*\|_1 = \operatorname{tr}|A^*| = \operatorname{tr}(U|A|U^*) = \operatorname{tr}(AU^*) = \operatorname{tr}(U^*A) = \operatorname{tr}|A| = \|A\|_1.$$

In addition, from Theorem 2.3.3 (iii) we get

$$\|A\|_1 = \operatorname{tr}|A| = \|\,|A|^{1/2}\|_2^2 \ge \|\,|A|^{1/2}\|^2 = \|A\|.$$

(iv) By Lemma 2.3.7 we obtain

$$\|XB\|_1 = \sum_j \langle |XB|e_j, e_j\rangle \le \sum_j \langle \|X\|\,|B|e_j, e_j\rangle = \|X\|\,\|B\|_1.$$

The other inequality can be proved through item (iii).
(v) We have $|x \otimes y^*|^2 = (y \otimes x^*)(x \otimes y^*) = \langle x, x\rangle (y \otimes y^*)$. Taking the square root (see Theorem 1.4.3) yields that $|x \otimes y^*| = \frac{\|x\|}{\|y\|}(y \otimes y^*)$. Hence,

$$\begin{aligned}\|x \otimes y^*\|_1 &= \sum_j \langle \frac{\|x\|}{\|y\|}(y \otimes y^*)e_j, e_j\rangle = \frac{\|x\|}{\|y\|}\sum_j \langle e_j, y\rangle\langle y, e_j\rangle \\ &= \frac{\|x\|}{\|y\|}\sum_j |\langle y, e_j\rangle|^2 = \|x\|\,\|y\|,\end{aligned}$$

where we use the Parseval identity.

□

2.4 Basic Inequalities Related to Unitarily Invariant Norms

We present some fundamental theorems involving unitarily invariant norms. Such inequalities can be specifically restated for the trace class norm.

Proposition 2.4.1 *Let $A \in \mathcal{C}_2$. If* $\operatorname{Re} A = |A|$, *then A is positive.*

Proof From the assumption, we derive that $(A + A^*)^2 = 4|A|^2 = 4A^*A$. Hence, $A^2 + A^{*2} = 3A^*A - AA^*$. Therefore,

$$\begin{aligned}\|A - A^*\|_2^2 &= \operatorname{tr}\left((A - A^*)^*(A - A^*)\right) = \operatorname{tr}(A^*A - A^{*2} - A^2 + AA^*)\\ &= \operatorname{tr}(A^*A - 3A^*A + AA^* + AA^*) = 0.\end{aligned}$$

Thus, $A^* = A$, whence $A = \operatorname{Re} A = |A| \geq 0$. □

Theorem 2.4.2 *If $A, B \in \mathcal{C}_2$ are self-adjoint operators, then*

$$\operatorname{tr}\left((AB)^2\right) \leq \operatorname{tr}\left(A^2B^2\right).$$

The equality holds if and only if A and B commute.

Proof We have

$$\begin{aligned}0 &\leq \|AB - BA\|_2^2 = \operatorname{tr}\left((AB - BA)^*(AB - BA)\right)\\ &= \operatorname{tr}(BA^2B + AB^2A - BABA - ABAB) = 2\operatorname{tr}(A^2B^2) - 2\operatorname{tr}\left((AB)^2\right).\end{aligned}$$

Thus, $\operatorname{tr}\left((AB)^2\right) \leq \operatorname{tr}\left(A^2B^2\right)$. Since $\|\cdot\|_2$ is a norm, the equality $\operatorname{tr}\left((AB)^2\right) = \operatorname{tr}\left(A^2B^2\right)$ holds if and only if $AB = BA$. □

Let $A, B \in \mathbb{B}(\mathscr{H})$ be positive and one of them is a trace class operator. Taking the trace, we deduce from

$$2(AB + BA) = (A + B)^2 - (A - B)^2 \leq (A + B)^2$$

that

$$\operatorname{tr}(AB) \leq \operatorname{tr}\left(\left(\frac{A + B}{2}\right)^2\right).$$

This is a *trace arithmetic–geometric mean inequality*. We state the result of [19, p. 205] as an arithmetic–geometric mean inequality involving the Hilbert–Schmidt norm $\|\cdot\|_2$.

Theorem 2.4.3 *Let $A, B \in \mathcal{C}_2$ be positive operators. Then*

$$\|AB\|_2 \leq \left\| \left(\frac{A+B}{2} \right)^2 \right\|_2 . \tag{2.4.1}$$

Furthermore, equality occurs if and only if $A = B$.

Proof First note that for any operator $C \in \mathcal{C}_2$, we have

$$\sum_{i,j} |\langle Ce_i, f_j \rangle|^2 = \sum_{i,j} |\langle U^* Ce_i, e_j \rangle|^2 = \|U^* C\|_2^2 = \|C\|_2^2, \tag{2.4.2}$$

where (e_j) and (f_j) are two arbitrary orthonormal bases of $\mathscr{H}$ and U is the unitary which maps e_j to f_j for all j.

Since A and B are diagonalizable, there are two orthonormal bases (e_j) and (f_j) of $\mathscr{H}$ such that $Ae_j = \alpha_j e_j$, and $Bf_j = \beta_j f_j$ with nonnegative α_js and β_js. Then

$$\begin{aligned} \|A^2 + B^2 + 2AB\|_2^2 &= \sum_{i,j} |\langle (A^2 + B^2 + 2AB)e_i, f_j \rangle|^2 \\ &= \sum_{i,j} |\langle A^2 e_i, f_j \rangle + \langle e_i, B^2 f_j \rangle + 2\langle Ae_i, Bf_j \rangle|^2 \\ &= \sum_{i,j} (\alpha_i^2 + \beta_j^2 + 2\alpha_i \beta_j)^2 |\langle e_i, f_j \rangle|^2 \\ &\geq \sum_{i,j} \left((\alpha_i^2 + \beta_j^2 - 2\alpha_i \beta_j)^2 + 16\alpha_i^2 \beta_j^2 \right) |\langle e_i, f_j \rangle|^2 \\ &= \|A^2 + B^2 - 2AB\|_2^2 + 16\|AB\|_2^2. \end{aligned} \tag{2.4.3}$$

Since

$$\|T\|_2^2 = \operatorname{tr}((\operatorname{Re} T - \mathrm{i}\operatorname{Im} T)(\operatorname{Re} T + \mathrm{i}\operatorname{Im} T)) = \|\operatorname{Re} T\|_2^2 + \|\operatorname{Im} T\|_2^2,$$

for all $T \in \mathcal{C}_2$, we obtain from (2.4.3) that

$$\begin{aligned} &\|\operatorname{Re}(A^2 + B^2 + 2AB)\|_2^2 + \|\operatorname{Im}(A^2 + B^2 + 2AB)\|_2^2 \\ \geq &\|\operatorname{Re}(A^2 + B^2 - 2AB)\|_2^2 + \|\operatorname{Im}(A^2 + B^2 - 2AB)\|_2^2 + 16\|AB\|_2^2 . \end{aligned}$$

It follows from

$$\operatorname{Re}(A^2 + B^2 \pm 2AB) = (A \pm B)^2$$

and

$$\operatorname{Im}(A^2 + B^2 \pm 2AB) = \pm \frac{1}{\mathrm{i}}(AB - BA)$$

that

$$\|(A+B)^2\|_2^2 \geq \|(A-B)^2\|_2^2 + 16\|AB\|_2^2,$$

whence

$$4\|AB\|_2 \leq \|(A+B)^2\|_2, \tag{2.4.4}$$

and equality occurs if and only if $A = B$. □

An extension of the above theorem to arbitrary unitarily invariant norms applying the block matrices is given in [19].

The proof of the following McCarthy's inequalities [20] can be inferred from that of [12, Theorem IV.2.13] as an application of the pinching inequality (2.3.12).

Theorem 2.4.4 *Let $A, B \in \mathcal{C}_p$ be positive operators. The following equivalent inequalities hold for all $p \geq 1$:*

$$2^{1-p}\|A+B\|_p^p \leq \|A\|_p^p + \|B\|_p^p \leq \|A+B\|_p^p;$$

$$2^{1-p}\,\mathrm{tr}\left((A+B)^p\right) \leq \mathrm{tr}(A^p + B^p) \leq \mathrm{tr}\left((A+B)^p\right).$$

Proof Since $(A+B) \oplus (A+B) = (A \oplus B) + (B \oplus A)$, we have

$$\begin{aligned} 2^{1-p}\|A+B\|_p^p &= 2^{-p}\|(A+B) \oplus (A+B)\|_p^p \\ &= 2^{-p}\|(A \oplus B) + (B \oplus A)\|_p^p \\ &\leq \|A \oplus B\|_p^p = \|A\|_p^p + \|B\|_p^p. \end{aligned}$$

To prove the right inequality, we set $X = \begin{bmatrix} A^{1/2} & B^{1/2} \\ 0 & 0 \end{bmatrix}$. Then

$$\begin{aligned} \|A\|_p^p + \|B\|_p^p &= \left\|\begin{bmatrix} A & 0 \\ 0 & B \end{bmatrix}\right\|_p^p \\ &\leq \left\|\begin{bmatrix} A & A^{1/2}B^{1/2} \\ B^{1/2}A^{1/2} & B \end{bmatrix}\right\|_p^p \qquad \text{(by the pinching inequality (2.3.12))} \\ &= \|X^*X\|_p^p = \|XX^*\|_p^p \qquad \text{(by (2.3.9))} \\ &= \left\|\begin{bmatrix} A+B & 0 \\ 0 & 0 \end{bmatrix}\right\|_p^p = \|A+B\|_p^p. \end{aligned}$$

The other inequality is a restatement of the first one in terms of the trace functional. □

Next, we intend to prove that

$$\mathrm{tr}\,|A-B| \leq \mathrm{tr}\,|A+B|$$

for some Hermitian matrices. We prove a general assertion. To this end, we first state a lemma; see [21, Lemma 2.1].

Lemma 2.4.5 *If $X, Y \in \mathcal{C}_{\|\cdot\|}$ are self-adjoint and $\pm Y \leq X$, then $|||Y||| \leq |||X|||$ for all unitarily invariant norms.*

Proof Let (f_j) be an orthonormal basis such that $Yf_j = \lambda_j f_j$ with $|\lambda_1| \geq \cdots \geq \cdots$. For any $k \in \mathbb{N}$, one has

$$\sum_{j=1}^{k} s_j(Y) = \sum_{j=1}^{k} |\lambda_j| = \sum_{j=1}^{k} |\langle Yf_j, f_j\rangle| \leq \sum_{j=1}^{k} \langle Xf_j, f_j\rangle \leq \sum_{j=1}^{k} s_j(X),$$

where the last inequality follows from Ky Fan maximum principle (2.2.4). The required assertion is deduced from Fan dominance Theorem 2.3.1. □

Applying the above lemma to the operators $A - B$ and $A + B$ and noting that $\pm(A - B) \leq A + B$ for positive operators A and B give the following result.

Corollary 2.4.6 *If $A, B \in \mathcal{C}_{\|\cdot\|}$ are positive, then*

$$|||A - B||| \leq |||A + B||| .$$

To achieve a unitarily invariant norm inequality from which we obtain a version of the arithmetic–geometric mean inequality, we need the following lemma [22, Lemma 1]. In its proof we use the following inequality [16, Theorem II.4.4]:

$$|\mathrm{Re}(\lambda(X))| \prec \lambda(|\mathrm{Re}\, X|). \tag{2.4.5}$$

Lemma 2.4.7 *Let $A, B \in \mathcal{C}_{\|\cdot\|}$. If AB is self-adjoint, then*

$$|||\mathrm{Re}\,(BA)||| \geq |||AB||| . \tag{2.4.6}$$

Proof It can be concluded that if $X \in \mathcal{C}_{\|\cdot\|}$, then

$$\begin{aligned} |||\mathrm{Re}\, X||| &= g(s(\mathrm{Re}\, X)) \\ &= g(\lambda(|\mathrm{Re}\, X|)) \\ &\geq g(|\mathrm{Re}\, \lambda(X)|) \qquad \text{(by (2.4.5))} \\ &= g(|\mathrm{Re}\, \lambda_1(X)|, |\mathrm{Re}\, \lambda_2(X)|, \ldots), \end{aligned} \tag{2.4.7}$$

where g is the symmetric gauge function associated with $|||\cdot|||$. Applying (2.4.7) to the operator BA, we get

$$\begin{aligned} |||\mathrm{Re}(BA)||| &\geq g(|\mathrm{Re}\, \lambda_1(BA)|, |\mathrm{Re}\, \lambda_2(BA)|, \ldots) \\ &= g(|\mathrm{Re}\, \lambda_1(AB)|, |\mathrm{Re}\, \lambda_2(AB)|, \ldots) \end{aligned}$$

$$\begin{aligned}&= g(|\lambda_1(AB)|, |\lambda_2(AB)|, \ldots)\\ &= g(s_1(AB), s_2(AB), \ldots)\\ &= |||AB|||\,.\end{aligned}$$

□

Utilizing appropriate 4×4 block matrices, one can prove the following assertion [23].

Theorem 2.4.8 *If $A_1, A_2, B_1, B_2 \in \mathcal{C}_{|||\cdot|||}$ and $X, Y \in \mathbb{B}(\mathscr{H})$, then*

$$2\,|||A_1XA_2^* + B_1YB_2^*||| \le \left|\left|\left|\begin{bmatrix} A_1^*A_1X + XA_2^*A_2 & A_1^*B_1Y + XA_2^*B_2 \\ B_1^*A_1X + YB_2^*A_2 & B_1^*B_1Y + YB_2^*B_2 \end{bmatrix}\right|\right|\right| \tag{2.4.8}$$

for all unitarily invariant norms.

Proof Consider the following block $4n \times 4n$ matrices T and R:

$$T = \begin{bmatrix} A_1 & B_1 & 0 & 0 \\ 0 & 0 & 0 & 0 \\ 0 & 0 & A_2 & B_2 \\ 0 & 0 & 0 & 0 \end{bmatrix} \quad \text{and} \quad R = \begin{bmatrix} 0 & 0 & X & 0 \\ 0 & 0 & 0 & Y \\ X^* & 0 & 0 & 0 \\ 0 & Y^* & 0 & 0 \end{bmatrix}.$$

Then

$$TRT^* = \begin{bmatrix} 0 & 0 & A_1XA_2^* + B_1YB_2^* & 0 \\ 0 & 0 & 0 & 0 \\ A_2X^*A_1^* + B_2^*B_1^* & 0 & 0 & 0 \\ 0 & 0 & 0 & 0 \end{bmatrix},$$

$$T^*TR = \begin{bmatrix} 0 & 0 & A_1^*A_1X & A_1^*B_1Y \\ 0 & 0 & B_1^*A_1X & B_1^*B_1Y \\ A_2^*A_2X & A_2^*B_2Y & 0 & 0 \\ B_2^*A_2X^* & B_2^*B_2Y^* & 0 & 0 \end{bmatrix},$$

and

$$\begin{aligned}&\mathrm{Re}(T^*TR)\\ &= \frac{1}{2}\begin{bmatrix} 0 & 0 & A_1^*A_1X + XA_2^*A_2 & A_1^*B_1Y + XA_2^*B_2 \\ 0 & 0 & B_1^*A_1X + YB_2^*A_2 & B_1^*B_1Y + YB_2^*B_2 \\ A_2^*A_2X + X^*A_1^*A_1 & A_2^*B_2Y^* + X^*A_1^*B_1 & 0 & 0 \\ B_2^*A_2X^* + Y^*B_1^*A_1 & B_2^*B_2Y^* + Y^*B_1^*B_1 & 0 & 0 \end{bmatrix}.\end{aligned}$$

Since TRT^* is self-adjoint, it follows from Lemma 2.4.7 that $|||TRT^*||| \le |||\mathrm{Re}(T^*TR)|||$. Now the desired inequality is concluded from (2.3.10), (2.3.1), and the equivalence of inequalities (2.3.6) and (2.3.8). □

Several inequalities can be derived from inequality (2.4.8).

If $X = Y = I, A, B \in \mathcal{C}_{|||\cdot|||}$ are positive operators and $A_1 = A_2 = A^{\frac{1}{2}}, B_1 = B_2 = B^{\frac{1}{2}}$, then (2.4.8) and the triangle inequality entails that

$$|||A + B||| \le |||A \oplus B||| + \left|\left|\left|A^{\frac{1}{2}} B^{\frac{1}{2}} \oplus A^{\frac{1}{2}} B^{\frac{1}{2}}\right|\right|\right|. \tag{2.4.9}$$

If $A_i, B_i \in \mathcal{C}_{|||\cdot|||}$ have orthogonal ranges for $i = 1, 2$, or equivalently if $A_i^* B_i = 0$ for $i = 1, 2$, then

$$2\left|\left|\left|A_1 X A_2^* + B_1 Y B_2^*\right|\right|\right| \le \left|\left|\left|(A_1^* A_1 X + X A_2^* A_2) \oplus (B_1^* B_1 Y + Y B_2^* B_2)\right|\right|\right|. \tag{2.4.10}$$

In particular, if $X = I$ and $A_i = 0$ or $B_i = 0$ for $i = 1, 2$, then the equivalence of inequalities (2.3.7) and (2.3.6) yields an operator arithmetic–geometric mean inequality stating that if $A, B \in \mathcal{C}_{|||\cdot|||}$, then

$$2\left|\left|\left|AB^*\right|\right|\right| \le \left|\left|\left|A^* A + B^* B\right|\right|\right| \tag{2.4.11}$$

for all unitarily invariant norms.

If $X = I, Y = -I, A, B \in \mathcal{C}_{|||\cdot|||}$ are positive, $A_1 = A_2 = A^{\frac{1}{2}}$ and $B_1 = B_2 = B^{\frac{1}{2}}$, then (2.4.10) implies that

$$|||A - B||| \le |||A \oplus B|||.$$

An interesting consequence of the previous inequality is

$$|||\mathcal{O}(M)||| \le \frac{|||M \oplus M|||}{2},$$

where $M = \begin{bmatrix} A & X \\ X^* & C \end{bmatrix}$ is a positive operator matrix and $\mathcal{O}(M) = \begin{bmatrix} 0 & X \\ X^* & 0 \end{bmatrix}$ is its off-diagonal matrix. To verify this norm inequality, take the unitary operator $U = \begin{bmatrix} I & 0 \\ 0 & -I \end{bmatrix}$ and consider the equality $\mathcal{O}(M) = (M - UMU^*)/2$. Now the unitary invariance of $|||\cdot|||$ gives the desired inequality.

If $X = I, Y = \mathrm{i}I, A, B \in \mathcal{C}_{|||\cdot|||}$ are positive, $A_1 = A_2 = A^{\frac{1}{2}}$ and $B_1 = B_2 = B^{\frac{1}{2}}$, then

$$|||A + \mathrm{i}B||| \le \left|\left|\left|\begin{bmatrix} A & \left(\frac{1+\mathrm{i}}{2}\right) A^{\frac{1}{2}} B^{\frac{1}{2}} \\ \left(\frac{1+\mathrm{i}}{2}\right) B^{\frac{1}{2}} A^{\frac{1}{2}} & iB \end{bmatrix}\right|\right|\right|. \tag{2.4.12}$$

From inequalities (2.3.1), (2.4.12) and the triangle inequality, we obtain the following inequality:

$$|||A + \mathrm{i}B||| \le |||A \oplus B||| + \frac{1}{\sqrt{2}}\left|\left|\left|A^{\frac{1}{2}} B^{\frac{1}{2}} \oplus A^{\frac{1}{2}} B^{\frac{1}{2}}\right|\right|\right|.$$

The following result is an operator version of the arithmetic–geometric mean inequality [22] that can be proved by the same method.

Theorem 2.4.9 *Let $A, B \in \mathcal{C}_{|||\cdot|||}$ and $X \in \mathbb{B}(\mathscr{H})$. Then*

$$\left|\left|\left|AA^*X + XBB^*\right|\right|\right| \geq 2\left|\left|\left|A^*XB\right|\right|\right|. \tag{2.4.13}$$

The following theorem gives a unitarily invariant norm inequality for operators; see [24].

Theorem 2.4.10 *Let $A, B \in \mathcal{C}_{|||\cdot|||}$. Then*

$$2\left|\left|\left|A^*B + B^*A\right|\right|\right| \leq \left|\left|\left|\big((A+B)^*(A+B)\big) \oplus \big((A-B)^*(A-B)\big)\right|\right|\right|.$$

Proof Put $T = \begin{bmatrix} A & 0 \\ B & 0 \end{bmatrix}$ and $S = \begin{bmatrix} B & 0 \\ A & 0 \end{bmatrix}$. It follows from (2.4.11) that $2\,|||T^*S||| \leq |||TT^* + SS^*|||$. Therefore,

$$2\left|\left|\left|\begin{bmatrix} A^*B + B^*A & 0 \\ 0 & 0 \end{bmatrix}\right|\right|\right| \leq \left|\left|\left|\begin{bmatrix} AA^* + BB^* & AB^* + BA^* \\ BA^* + AB^* & AA^* + BB^* \end{bmatrix}\right|\right|\right|.$$

We derive from (2.3.11) that

$$\begin{aligned} &2\left|\left|\left|\begin{bmatrix} A^*B + B^*A & 0 \\ 0 & 0 \end{bmatrix}\right|\right|\right| \\ &\leq \left|\left|\left|\begin{bmatrix} AA^* + BB^* + AB^* + BA^* & 0 \\ 0 & AA^* + BB^* - AB^* - BA^* \end{bmatrix}\right|\right|\right| \\ &= \left|\left|\left|\begin{bmatrix} (A+B)(A+B)^* & 0 \\ 0 & (A-B)(A-B)^* \end{bmatrix}\right|\right|\right| \\ &= \left|\left|\left|\begin{bmatrix} (A+B)^*(A+B) & 0 \\ 0 & (A-B)^*(A-B) \end{bmatrix}\right|\right|\right| \quad \text{(by (2.3.9))}. \end{aligned}$$

□

For positive operators A and B, Davidson and Power [25] showed that

$$\|A + B\| \leq \max\{\|A\|, \|B\|\} + \|AB\|^{1/2}.$$

The next result is an extension of the above inequality to all unitarily invariant norms [26, Theorem 1].

Theorem 2.4.11 *Let $A, B \in \mathcal{C}_{|||\cdot|||}$ be positive operators. If f and g are nonnegative functions on $[0, \infty)$ satisfying the relation $f(t)g(t) = t$ for all $t \in [0, \infty)$, then*

$$\begin{aligned} &|||A + B||| \\ &\leq |||A \oplus B||| + \frac{1}{2}\,|||(f(A)g(B) + g(A)f(B)) \oplus (f(A)g(B) + g(A)f(B))|||. \end{aligned} \tag{2.4.14}$$

Proof On $\mathscr{H} \oplus \mathscr{H}$, let $X = \begin{bmatrix} f(A) & f(B) \\ 0 & 0 \end{bmatrix}$ and $Y = \begin{bmatrix} g(A) & 0 \\ g(B) & 0 \end{bmatrix}$. Since $f(A)g(A) = A$ and $f(B)g(B) = B$, it follows that

$$XY = \begin{bmatrix} A+B & 0 \\ 0 & 0 \end{bmatrix} \quad \text{and} \quad YX = \begin{bmatrix} A & g(A)f(B) \\ g(B)f(A) & B \end{bmatrix}.$$

Since XY is self-adjoint, it follows from Lemma 2.4.7 that $|||XY||| \le |||\mathrm{Re}(YX)|||$. This implies that

$$|||A+B||| \le \left|\left|\left| \begin{bmatrix} A & \frac{1}{2}(f(A)g(B)+g(A)f(B)) \\ \frac{1}{2}(g(B^*)f(A^*)+f(B^*)g(A^*)) & B \end{bmatrix} \right|\right|\right|.$$

The triangle inequality now yields

$$\begin{aligned} &|||A+B||| \\ &\le \left|\left|\left| \begin{bmatrix} A & 0 \\ 0 & B \end{bmatrix} \right|\right|\right| + \left|\left|\left| \begin{bmatrix} 0 & \frac{1}{2}(f(A)g(B)+g(A)f(B)) \\ \frac{1}{2}(g(B^*)f(A^*)+f(B^*)g(A^*)) & 0 \end{bmatrix} \right|\right|\right| \\ &= |||A \oplus B||| + \left|\left|\left| \left(\frac{1}{2}(f(A)g(B)+g(A)f(B))\right) \oplus \left(\frac{1}{2}(f(A)g(B)+g(A)f(B))\right) \right|\right|\right|, \\ &\qquad\qquad\qquad\qquad \text{(by (2.3.10) and (2.3.1))} \end{aligned}$$

as required. □

Observe that some special cases follow from (2.4.14) if $f(t) = t^r$ and $g(t) = t^{1-r}$, where r is a real number with $0 < r < 1$. In particular, if $r = \frac{1}{2}$, we have the following corollary.

Corollary 2.4.12 *If $A, B \in \mathcal{C}_{|||\cdot|||}$, then*

$$|||A+B||| \le |||A \oplus B||| + \left|\left|\left| A^{\frac{1}{2}}B^{\frac{1}{2}} \oplus A^{\frac{1}{2}}B^{\frac{1}{2}} \right|\right|\right|. \tag{2.4.15}$$

Now applying (2.4.15) to the usual operator norm and the Schatten p-norms, we have the following two corollaries.

Corollary 2.4.13 *If $A, B \in \mathbb{B}_+(\mathscr{H})$, then*

$$\|A+B\| \le \max\{\|A\|, \|B\|\} + \|A^{\frac{1}{2}}B^{\frac{1}{2}}\|. \tag{2.4.16}$$

Corollary 2.4.14 *If $A, B \in \mathcal{C}_p$ are positive operators and $1 \le p < \infty$, then*

$$\|A+B\|_p \le (\|A\|_p^p + \|B\|_p^p)^{\frac{1}{p}} + 2^{\frac{1}{p}}\|A^{\frac{1}{2}}B^{\frac{1}{2}}\|_p. \tag{2.4.17}$$

The next three theorems and their corollaries are mainly established in [27]. The first extends the Powers–Størmer inequality for the operator norm.

Theorem 2.4.15 *If $A, B \in \mathbb{B}(\mathscr{H})$ are such that $A + B$ is self-adjoint and $A + B \geq \pm X$, where $X \in \mathbb{B}(\mathscr{H})$ is self-adjoint, then $\|AX + XB\| \geq \|X\|^2$.*

Proof Since X is self-adjoint, $\|X\|$ or $-\|X\|$ is in $\mathrm{sp}(X)$ as well as the numerical range $W(X)$ of X is the closed convex hull of $\mathrm{sp}(X)$. Hence, $\|X\|$ or $-\|X\|$ is in $W(X)$, and so there exists a sequence (f_n) of unit vectors in $\mathscr{H}$ such that $\langle Xf_n, f_n\rangle \to t$ as $n \to \infty$, where $|t| = \|X\|$. For any $n \in \mathbb{N}$ one has

$$\|Xf_n - tf_n\|^2 = \|Xf_n\|^2 + t^2 - 2t\langle Xf_n, f_n\rangle \leq 2t^2 - 2t\langle Xf_n, f_n\rangle.$$

Therefore, $Xf_n - tf_n \to 0$ as $n \to \infty$. Now

$$\begin{aligned}\|AX + XB\| \geq& |\langle (AX + XB)f_n, f_n\rangle| \\ =& |\langle A(X-t)f_n, f_n\rangle + \langle Bf_n, (X-t)f_n\rangle + t\langle (A+B)f_n, f_n\rangle| \\ \geq& |t|\,|\langle (A+B)f_n, f_n\rangle| - |\langle A(X-t)f_n, f_n\rangle + \langle Bf_n, (X-t)f_n\rangle| \\ \geq& |t|\,|\langle Xf_n, f_n\rangle| - |\langle A(X-t)f_n, f_n\rangle + \langle Bf_n, (X-t)f_n\rangle| \\ \geq& |t|\,|\langle Xf_n, f_n\rangle| - \|A\|\,|\langle (X-t)f_n, f_n\rangle| - \|B\|\,|\langle f_n, (X-t)f_n\rangle|.\end{aligned}$$

Letting $n \to \infty$, we get $\|AX + XB\| \geq \|X\|^2$. □

Corollary 2.4.16 *If $A, B \in \mathbb{B}(\mathscr{H})$ are such that $A + B$ is self-adjoint and $A + B \geq \pm X$, where $X \in \mathbb{B}(\mathscr{H})$ is self-adjoint and $AX + XB = 0$, then $X = 0$.*

The corresponding inequality for a p-norm reads as follows.

Theorem 2.4.17 *Suppose that A and B are operators such that $A + B$ is self-adjoint and $A + B \geq \pm X$, where $X \in \mathbb{B}(\mathscr{H})$ is self-adjoint. Then,*

$$\|AX + XB\|_p \geq \|X\|_{2p}^2$$

for $1 \leq p < \infty$.

Proof Assume that $AX + XB \in \mathcal{C}_p$; otherwise $\|AX + XB\|_p = \infty$ and the inequality trivially is valid. Hence, $AX + XB$ is compact. If $\pi : \mathbb{B}(\mathscr{H}) \to \mathbb{B}(\mathscr{H})/\mathbb{K}(\mathscr{H})$ is the quotient map of $\mathbb{B}(\mathscr{H})$ onto the Calkin algebra $\mathbb{B}(\mathscr{H})/\mathbb{K}(\mathscr{H})$, we have $\pi(A)\pi(X) + \pi(X)\pi(B) = 0$ and $\pi(A) + \pi(B) \geq \pm\pi(X)$, since π is a $*$-homomorphism. Corollary 2.4.16 entails that $\pi(X) = 0$, in other words, X is compact. It is known that a compact self-adjoint operator is diagonalizable [28, Theorem 2.4.4], hence $Xe_n = t_ne_n$, where (e_n) is an orthonormal basis for $\mathscr{H}$ and $t_n \in \mathbb{R}$. Therefore,

$$\begin{aligned}\|AX + XB\|_p^p \geq& \sum_n |\langle (AX + XB)e_n, e_n\rangle|^p \\ =& \sum_n |\langle AXe_n, e_n\rangle + \langle Be_n, Xe_n\rangle|^p\end{aligned}$$

$$
\begin{aligned}
&= \sum_n |t_n \langle (A+B)e_n, e_n\rangle|^p \\
&\geq \sum_n |t_n|^p |\langle X e_n, e_n\rangle|^p \\
&= \sum_n |t_n|^{2p} \\
&= \|X\|_{2p}^{2p}.
\end{aligned}
$$

□

A consequence of Theorems 2.4.15 and 2.4.17 with $X = A - B$ is the Powers–Størmer inequality [29, Lemma 4.1].

Corollary 2.4.18 (Powers–Størmer inequality) *If A and B are positive operators, then $\|A - B\|_{2p}^2 \leq \|A^2 - B^2\|_p$ for $1 \leq p \leq \infty$.*

Theorem 2.4.17 can be generalized further by removing the restriction on X. To accomplish this, we first recall the following lemma, which can be proved easily by taking the square root of $T^*T = \begin{bmatrix} B^*B & 0 \\ 0 & A^*A \end{bmatrix}$.

Lemma 2.4.19 *If $T = \begin{bmatrix} 0 & A \\ B & 0 \end{bmatrix}$ is defined on $\mathscr{H} \oplus \mathscr{H}$, then $|T| = \begin{bmatrix} |B| & 0 \\ 0 & |A| \end{bmatrix}$ and $\|T\|_p^p = \|A\|_p^p + \|B\|_p^p$ for $1 \leq p < \infty$.*

Theorem 2.4.20 *Let $A, B,$ and X be operators with $A + B \geq |X|$ and $A + B \geq |X^*|$. Then $\|AX + XB\|_p^p + \|AX^* + X^*B\|_p^p \geq 2\|X\|_{2p}^{2p}$ for $1 \leq p < \infty$, and $\max\{\|AX + XB\|, \|AX^* + X^*B\|\} \geq \|X\|^2$.*

Proof Let $\tilde{A} = \begin{bmatrix} A & 0 \\ 0 & A \end{bmatrix}$, $\tilde{B} = \begin{bmatrix} B & 0 \\ 0 & B \end{bmatrix}$ and $\tilde{X} = \begin{bmatrix} 0 & X \\ X^* & 0 \end{bmatrix}$. Then $\tilde{X}$ is self-adjoint. In light of Lemma 2.4.19 we have $|\tilde{X}| = \begin{bmatrix} |X^*| & 0 \\ 0 & |X| \end{bmatrix}$. From $A + B \geq |X|$ and $A + B \geq |X^*|$, we obtain that $\tilde{A} + \tilde{B} \geq |\tilde{X}|$. Since $\tilde{X}$ is self-adjoint, it follows that $\tilde{A} + \tilde{B} \geq |\tilde{X}| \geq \pm\tilde{X}$. Applying Theorems 2.4.15 and 2.4.17 to the operators $\tilde{A}$, $\tilde{B}$ and $\tilde{X}$, we get

$$\|\tilde{A}\tilde{X} + \tilde{X}\tilde{B}\|_p \geq \|\tilde{X}\|_{2p}^2$$

for $1 \leq p \leq \infty$. Since

$$\tilde{A}\tilde{X} + \tilde{X}\tilde{B} = \begin{bmatrix} 0 & AX + XB \\ AX^* + X^*B & 0 \end{bmatrix},$$

the theorem can be deduced from Lemma 2.4.19. □

Corollary 2.4.21 *If A and X are operators with $A + A^* \geq |X|$ and $A + A^* \geq |X^*|$, then $\|AX + XA^*\|_p \geq \|X\|_{2p}^2$ for $1 \leq p \leq \infty$.*

Proof This follows from Theorem 2.4.20 applied to A and A^* with the observation that $\|AX + XA^*\|_p = \|AX^* + X^*A^*\|_p$ for $1 \leq p \leq \infty$. □

We investigate some Hilbert–Schmidt norm inequalities related to Lipschitz functions of normal operators. The next three results appeared in [30].

Theorem 2.4.22 *Let A be a normal operator and f be a function defined on* $\mathrm{sp}(A)$. *If $|f(z) - f(w)| \leq k|z - w|$ for all $z, w \in \mathrm{sp}(A)$ and some positive constant k, then*

$$\|f(A)X - Xf(A)\|_2 \leq k\|AX - XA\|_2$$

for all $X \in \mathbb{B}(\mathscr{H})$, where $\mathscr{H}$ is separable.

Proof Voiculescu's perturbation result [31] states that for given $\varepsilon > 0$, there is a diagonal operator D_ε and a Hilbert–Schmidt operator K_ε with $\|K_\varepsilon\|_2 < \varepsilon$ such that $A = D_\varepsilon + K_\varepsilon$. If $D_\varepsilon e_n = \lambda_n e_n$ and $X = [x_{ij}]$ is the corresponding matrix of X, relative to the basis (e_n), then the (i, j)-entry for $D_\varepsilon X - XD_\varepsilon$ is $(\lambda_i - \lambda_j)x_{ij}$. In a similar way, the (i, j)-entry for $f(D_\varepsilon)X - Xf(D_\varepsilon)$ is $(f(\lambda_i) - f(\lambda_j))x_{ij}$. Since

$$\|D_\varepsilon X - XD_\varepsilon\|_2^2 = \sum_{i,j} |(\lambda_i - \lambda_j)x_{ij}|^2$$

and

$$\|f(D_\varepsilon)X - Xf(D_\varepsilon)\|_2^2 = \sum_{i,j} |(f(\lambda_i) - f(\lambda_j))x_{ij}|^2,$$

it follows that

$$\|f(D_\varepsilon)X - Xf(D_\varepsilon)\|_2 \leq k\|D_\varepsilon X - XD_\varepsilon\|_2. \tag{2.4.18}$$

Next, let $A = \int_{\mathrm{sp}(A)} z dE(z)$ be the spectral representation of A. Then

$$\begin{aligned}\|((f(A) - f(D_\varepsilon))e_n\|^2 =& \int_{\mathrm{sp}(A)} |f(z) - f(\lambda_n)|^2 d\|E(z)e_n\|^2 \\ \leq& k^2 \int_{\mathrm{sp}(A)} |z - \lambda_n|^2 d\|E(z)e_n\|^2 \\ =& k^2\|(A - D_\varepsilon)e_n\|^2.\end{aligned}$$

Consequently,

$$\|f(A) - f(D_\varepsilon)\|_2 \leq k\|A - D_\varepsilon\|_2 = k\|K_\varepsilon\|_2 < \varepsilon k.$$

Thus, $f(A) = f(D_\varepsilon) + C_\varepsilon$ with $\|C_\varepsilon\|_2 \to 0$ as $\varepsilon \to 0$. It follows from

$$| \, \|AX - XA\|_2 - \|D_\varepsilon X - XD_\varepsilon\|_2| \leq \|K_\varepsilon X - XK_\varepsilon\|_2 \leq 2\|K_\varepsilon\|_2\|X\| < 2\|X\|\varepsilon$$

that

$$\lim_{\varepsilon \to 0} \|D_\varepsilon X - XD_\varepsilon\|_2 = \|AX - XA\|_2.$$

Similarly, we have

$$\lim_{\varepsilon \to 0} \|f(D_\varepsilon)X - Xf(D_\varepsilon)\|_2 = \|f(A)X - Xf(A)\|_2.$$

The required result now follows by letting $\varepsilon \to 0$ in (2.4.18). □

Corollary 2.4.23 (Fuglede's theorem modulo $\mathcal{C}_2$) *Let A be a normal operator. Then $\|AX - XA\|_2 = \|A^*X - XA^*\|_2$ for all $X \in \mathbb{B}(\mathscr{H})$.*

Proof Apply Theorem 2.4.22 to the function $f(z) = \bar{z}$.

The Berberian trick allows us to extend Theorem 2.4.22 as follows. □

Corollary 2.4.24 *Let A and B be normal operators and f be a function defined on the union of the spectra of A and B. If $|f(z) - f(w)| \leq k|z - w|$ for all $z, w \in \operatorname{sp}(A) \cup \operatorname{sp}(B)$ and some positive constant k, then $\|f(A)X - Xf(B)\|_2 \leq k\|AX - XB\|_2$ for all $X \in \mathbb{B}(\mathscr{H})$. In particular, $\|f(A) - f(B)\|_2 \leq k\|A - B\|_2$.*

Proof Define operators C and Y on the space $\mathscr{H} \oplus \mathscr{H}$ by

$$C := \begin{bmatrix} A & 0 \\ 0 & B \end{bmatrix} \quad \text{and} \quad Y := \begin{bmatrix} 0 & X \\ 0 & 0 \end{bmatrix}.$$

Then

$$\|AX - XB\|_2 = \|CY - YC\|_2$$

and

$$\|f(A)X - Xf(B)\|_2 = \|f(C)Y - Yf(C)\|_2.$$

Applying Theorem 2.4.22 to C and Y, we get the required result. □

Via the triangle inequality and the submultiplicativity of the operator norm, we have

$$\|AB - BA\| \leq 2\|A\| \, \|B\|$$

for all $A, B \in \mathbb{B}(\mathscr{H})$. Now we extend it to unitarily invariant norms restricted to positive operators. To this end, we follow arguments in [27, Lemma 3, Theorem 2]. We need the following theorem.

Theorem 2.4.25 *Let $A \in \mathcal{C}_{|||\cdot|||}$ and $P \in \mathbb{B}(\mathscr{H})$ be a projection. Then*

$$|||PA - AP||| \leq |||A|||.$$

Proof (*First proof*) We have $PA - AP = \frac{1}{2}(S_P A - AS_P)$, where $S_P = 2P - I$ is a unitary operator. Thus,

$$\begin{aligned} |||PA - AP||| &= \frac{1}{2}\,|||S_P A - AS_P||| \le \frac{1}{2}(|||S_P A||| + |||AS_P|||) \\ &\le \frac{1}{2}(||S_P||\;|||A||| + ||S_P||\;|||A|||) = |||A|||\,. \end{aligned}$$

(*Second proof*) We consider the orthogonal decomposition $\mathscr{H} = \operatorname{ran}(P) \oplus \ker P$. Then $P = \begin{bmatrix} I' & 0 \\ 0 & 0 \end{bmatrix}$, where I' denotes the identity operator on $\operatorname{ran}(P)$. Consider $A = \begin{bmatrix} A_{11} & A_{12} \\ A_{21} & A_{22} \end{bmatrix}$, represented as a block matrix of operators concerning to this decomposition. We have

$$PA - AP = \begin{bmatrix} 0 & A_{12} \\ -A_{21} & 0 \end{bmatrix}.$$

Assume that I'' is the identity operator on $\ker P$. Evidently, $U = \begin{bmatrix} I' & 0 \\ 0 & I'' \end{bmatrix}$ a unitary operator and

$$\begin{bmatrix} 0 & A_{12} \\ -A_{21} & 0 \end{bmatrix} = \frac{1}{2}(UA - AU).$$

Thus,

$$\begin{aligned} |||PA - AP||| &= \left|\left|\left| \begin{bmatrix} 0 & A_{12} \\ -A_{21} & 0 \end{bmatrix} \right|\right|\right| \\ &= \frac{1}{2}\,|||UA - AU||| \\ &\le \frac{1}{2}(||U||\;|||A||| + |||A|||\;||U||) \\ &\le |||A|||\,. \end{aligned}$$

□

The previous result can be extended. To achieve it, we follow the proof presented in [32], which is based on the ideas of Stampfli [33]. This result also extends the inequality $|||AX - XB||| \le (\|A\| + \|B\|)\,|||X|||$, deduced from (2.3.4), for positive operators A and B.

Theorem 2.4.26 *Suppose that $A, B \in \mathbb{B}(\mathscr{H})$ are positive operators and $X \in \mathcal{C}_{|||\cdot|||}$. Then*

$$|||AX - XB||| \le \max\{\|A\|, \|B\|\}\,|||X|||\,. \tag{2.4.19}$$

Proof Let $z \in \mathbb{C}$. We have

$$|||AX - XA||| = |||(A - z)X - X(A - z)||| \leq 2\|A - z\| \, |||X||| .$$

Hence

$$|||AX - XA||| \leq 2\Delta(A, \|\cdot\|) \, |||X||| ,$$

where $\Delta(A, \|\cdot\|) = \inf_{z\in\mathbb{C}} \|A - z\|$ is the distance from scalar operators. Since A is positive, $\Delta(A, \|\cdot\|) \leq \frac{1}{2}\|A\|$; see [33]. Hence

$$|||AX - XA||| \leq \|A\| \, |||X||| . \tag{2.4.20}$$

Now, set $C = \begin{bmatrix} A & 0 \\ 0 & B \end{bmatrix} \geq 0$ and $Z = \begin{bmatrix} 0 & X \\ 0 & 0 \end{bmatrix}$. Then $\|C\| = \max\{\|A\|, \|B\|\}$. Also, (2.3.10) yields that

$$|||Z||| = \left|\left|\left| \begin{bmatrix} 0 & X \\ 0 & 0 \end{bmatrix} \right|\right|\right| = \left|\left|\left| \begin{bmatrix} X & 0 \\ 0 & 0 \end{bmatrix} \right|\right|\right| .$$

Thus,

$$\begin{aligned} \left|\left|\left| \begin{bmatrix} AX - XB & 0 \\ 0 & 0 \end{bmatrix} \right|\right|\right| &= |||CZ - ZC||| \\ &\leq \|C\| \, |||Z||| \qquad \text{(by inequality (2.4.20))} \\ &= \max\{\|A\|, \|B\|\} \, |||X \oplus 0||| . \end{aligned}$$

□

Jocić [34] established the following norm inequality for self-adjoint operators $A, B \in \mathbb{B}(\mathscr{H})$, $\alpha \in (0, 1)$, and $r \geq 1$:

$$\left|\left|\left| \, |\alpha A + (1-\alpha)B|^r \, \right|\right|\right| \leq \left|\left|\left| \alpha|A|^r + (1-\alpha)|B|^r \right|\right|\right| .$$

It can be extended as follows; see [35].

Theorem 2.4.27 *For each $i = 1, \ldots, n$, let $A_i, X_i \in \mathbb{B}(\mathscr{H})$ be such that $X_i \in \mathcal{C}_{|||\cdot|||}$ is a contraction, let $\alpha_i \in (0, 1)$ such that $\sum_{i=1}^n \alpha_i = 1$, and let $r \geq 2$. Then*

$$\left|\left|\left| \left| \sum_{i=1}^n \alpha_i A_i X_i \right|^r \right|\right|\right| \leq \left|\left|\left| \left| \sum_{i=1}^n \alpha_i X_i^* |A_i|^r X_i \right| \right|\right|\right| \tag{2.4.21}$$

for every unitarily invariant norm.

Proof Let y be any unit vector in $\mathscr{H}$. Then

$$\begin{aligned}
\left\langle \left(\sum_{i=1}^{n} \alpha_i X_i^* |A_i|^r X_i \right) y, y \right\rangle &= \sum_{i=1}^{n} \alpha_i \langle (|A_i|^2)^{\frac{r}{2}} X_i y, X_i y \rangle \\
&\geq \sum_{i=1}^{n} \alpha_i \langle |A_i|^2 X_i y, X_i y \rangle^{\frac{r}{2}} \\
&\qquad \text{(by Lemma 1.4.7 and } \|X_i y\| \leq 1) \\
&= \sum_{i=1}^{n} \alpha_i \, \langle A_i X_i y, A_i X_i y \rangle^{\frac{r}{2}} \\
&= \sum_{i=1}^{n} \alpha_i \, \|A_i X_i y\|^r \\
&\geq \left(\sum_{i=1}^{n} \alpha_i \|A_i X_i y\| \right)^r \quad \text{(by the convexity of } f(t) = t^r) \\
&\geq \left\| \sum_{i=1}^{n} \alpha_i A_i X_i y \right\|^r \quad \text{(by the triangle inequality)} \\
&= \left\langle \sum_{i=1}^{n} \alpha_i A_i X_i y, \sum_{i=1}^{n} \alpha_i A_i X_i y \right\rangle^{\frac{r}{2}} \\
&= \left\langle \left| \sum_{i=1}^{n} \alpha_i A_i X_i \right|^2 y, y \right\rangle^{\frac{r}{2}} \\
&= \left\langle \left| \sum_{i=1}^{n} \alpha_i A_i X_i \right| y, \left| \sum_{i=1}^{n} \alpha_i A_i X_i \right| y \right\rangle^{\frac{r}{2}} \\
&= \left\| \left| \sum_{i=1}^{n} \alpha_i A_i X_i \right| y \right\|^r \\
&\geq \left\langle \left| \sum_{i=1}^{n} \alpha_i A_i X_i \right| y, y \right\rangle^r . \\
&\qquad \text{(by the Cauchy–Schwarz inequality).} \qquad (2.4.22)
\end{aligned}$$

For the operator norm, inequality (2.4.21) follows immediately from (2.4.22). For other unitarily invariant norms, we need to assume that $\sum_{i=1}^{n} \alpha_i X_i^* |A_i|^r X_i$ is compact. In this case, it follows from (2.4.22) and Exercise 1.8.3 that $|\sum_{i=1}^{n} \alpha_i A_i X_i|$ is compact, and so $\sum_{i=1}^{n} \alpha_i A_i X_i$ is compact. Now, by (2.4.22) and the minimax theorem (2.2.3), we get

$$s_j \left(\left| \sum_{i=1}^{n} \alpha_i A_i X_i \right|^r \right) \leq s_j \left(\sum_{i=1}^{n} \alpha_i X_i^* |A_i|^r X_i \right) \tag{2.4.23}$$

for $j = 1, 2, \ldots$ that yields inequality (2.4.21), by Theorem 2.3.1. □

Our next result appeared in [36] and can be regarded as an extension of (2.3.5).

Theorem 2.4.28 *Let $A_i, B_i \in \mathbb{B}(\mathscr{H})$, and let $X_i \in \mathcal{C}_{|||\cdot|||}$ for $i = 1, \ldots, n$. Then*

$$\left|\left|\left|\sum_{i=1}^{n} A_i X_i B_i\right|\right|\right| \leq \left(\sum_{i=1}^{n} \|A_i\| \, \|B_i\|\right) |||\oplus_{i=1}^{n} X_i||| .$$

Proof It suffices to show that

$$s_j\left(\sum_{i=1}^{n} A_i X_i B_i\right) \leq \left(\sum_{i=1}^{n} \|A_i\| \, \|B_i\|\right) s_j\left(\oplus_{i=1}^{n} X_i\right)$$

for all $j = 1, 2, \ldots$. By (2.3.5) we get

$$\begin{aligned}
s_j\left(\sum_{i=1}^{n} A_i X_i B_i\right) &= s_j\left(\begin{bmatrix} \sum_{i=1}^{n} A_i X_i B_i & 0 \cdots 0 \\ 0 & 0 \cdots 0 \\ \vdots & \vdots \ddots \vdots \\ 0 & 0 \cdots 0 \end{bmatrix}\right) \\
&= s_j\left(\begin{bmatrix} A_1 & A_2 & \cdots & A_n \\ 0 & 0 & \cdots & 0 \\ \vdots & \vdots & \ddots & \vdots \\ 0 & 0 & \cdots & 0 \end{bmatrix}\begin{bmatrix} X_1 & 0 & \cdots & 0 \\ 0 & X_2 & \cdots & 0 \\ \vdots & \vdots & \ddots & \vdots \\ 0 & 0 & \cdots & X_n \end{bmatrix}\begin{bmatrix} B_1 & 0 \cdots 0 \\ B_2 & 0 \cdots 0 \\ \vdots & \vdots \ddots \vdots \\ B_n & 0 \cdots 0 \end{bmatrix}\right) \\
&\leq \left\|\begin{bmatrix} A_1 & A_2 & \cdots & A_n \\ 0 & 0 & \cdots & 0 \\ \vdots & \vdots & \ddots & \vdots \\ 0 & 0 & \cdots & 0 \end{bmatrix}\right\| \left\|\begin{bmatrix} B_1 & 0 \cdots 0 \\ B_2 & 0 \cdots 0 \\ \vdots & \vdots \ddots \vdots \\ B_n & 0 \cdots 0 \end{bmatrix}\right\| s_j\left(\oplus_{i=1}^{n} X_i\right) \\
&\leq \left(\sum_{i=1}^{n} \|A_i\|^2\right)^{1/2} \left(\sum_{i=1}^{n} \|B_i\|^2\right)^{1/2} s_j\left(\oplus_{i=1}^{n} X_i\right).
\end{aligned}$$

Let $t_i > 0,\ 1 = 1, \ldots, n$ be arbitrary. We replace A_i and B_i with $\sqrt{t_i}A_i$ and $\frac{1}{\sqrt{t_i}}B_i$ in the above inequality to get

$$s_j\left(\sum_{i=1}^{n} A_i X_i B_i\right) \leq \left(\sum_{i=1}^{n} t_i \|A_i\|^2\right)^{1/2} \left(\sum_{i=1}^{n} \frac{1}{t_i}\|B_i\|^2\right)^{1/2} s_j\left(\oplus_{i=1}^{n} X_i\right).$$

Taking infimum over all $t_1, \ldots, t_n > 0$ and noting that

$$\min_{t_1,\ldots,t_n>0} \left(\sum_{i=1}^{n} t_i \|A_i\|^2\right)^{1/2} \left(\sum_{i=1}^{n} \frac{1}{t_i}\|B_i\|^2\right)^{1/2} = \sum_{i=1}^{n} \|A_i\| \, \|B_i\|,$$

we arrive at the required inequality. □

The following theorem, due to Jocić [37, Theorem 2.1], extends [16, Theorem 2.5.1], and [18, Theorem 1.19]. It can be considered as a variant of Theorem 2.4.28. The associated ideals $\mathcal{C}_{(k)}$ consist of all compact operators as every Ky Fan k-norm is equivalent to the norm in $\mathbb{K}(\mathscr{H})$.

Theorem 2.4.29 *If $\sum_{n=1}^{\infty} A_n^* A_n \leq I$, $\sum_{n=1}^{\infty} A_n A_n^* \leq I$, and $\sum_{n=1}^{\infty} B_n B_n^* \leq I$, $\sum_{n=1}^{\infty} B_n^* B_n \leq I$ for some operator families $\{A_n\}_{n=1}^{\infty}$ and $\{B_n\}_{n=1}^{\infty}$, then $\sum_{n=1}^{\infty} A_n Y B_n \in \mathcal{C}_{|||\cdot|||}$ whenever $Y \in \mathcal{C}_{|||\cdot|||}$ for any unitarily invariant norm $|||\cdot|||$. Moreover,*

$$\left|\left|\left| \sum_{n=1}^{\infty} A_n Y B_n \right|\right|\right| \leq |||Y||| \,. \tag{2.4.24}$$

Proof For arbitrary x and y in $\mathscr{H}$, a straightforward calculation gives

$$\begin{aligned}
\left|\left\langle\left(\sum_{n=1}^{\infty} A_n Y B_n\right) x, y\right\rangle\right| &= \left|\sum_{n=1}^{\infty} \langle Y B_n x, A_n^* y\rangle\right| \\
&\leq \sum_{n=1}^{\infty} \|Y\| \|B_n x\| \|A_n^* y\| \\
&\qquad \text{(by the Cauchy–Schwarz inequality)} \\
&\leq \|Y\| \left(\sum_{n=1}^{\infty} \|B_n x\|^2\right)^{\frac{1}{2}} \left(\sum_{n=1}^{\infty} \|A_n^* y\|^2\right)^{\frac{1}{2}} \\
&\qquad \text{(by the Cauchy–Schwarz inequality in } \mathbb{C}^n) \\
&= \|Y\| \left\langle \sum_{n=1}^{\infty} B_n^* B_n x, x\right\rangle^{\frac{1}{2}} \left\langle \sum_{n=1}^{\infty} A_n A_n^* y, y\right\rangle^{\frac{1}{2}} \\
&= \|Y\| \left\| \left(\sum_{n=1}^{\infty} B_n^* B_n\right)^{\frac{1}{2}} x\right\| \left\| \left(\sum_{n=1}^{\infty} A_n A_n^*\right)^{\frac{1}{2}} y\right\| \\
&\leq \|Y\| \, \|x\| \, \|y\|,
\end{aligned}$$

from which we conclude that

$$\|\sum_{n=1}^{\infty} A_n Y B_n\| \leq \|Y\|. \tag{2.4.25}$$

Therefore, for all $N = 1, 2, \ldots$, for $Y \in \mathcal{C}_1$ and for all $W \in \mathbb{B}(\mathscr{H})$ we have

$$\left|\operatorname{tr}\left(\sum_{n=1}^{N} A_n Y B_n W^*\right)\right| = \left|\operatorname{tr}\left(Y\left(\sum_{n=1}^{N} A_n^* W B_n^*\right)^*\right)\right| \le \left\|Y\sum_{n=1}^{N} A_n^* W B_n^*\right\|_1$$
$$\le \|Y\|_1 \left\|\sum_{n=1}^{N} A_n^* W B_n^*\right\| \le \|Y\|_1 \|W\| \qquad \text{by (2.4.25).}$$

Hence,

$$\sup_{\|W\|=1}\left|\operatorname{tr}\left(\sum_{n=1}^{N} A_n Y B_n W^*\right)\right| = \left\|\sum_{n=1}^{N} A_n Y B_n\right\|_1 \le \|Y\|_1. \tag{2.4.26}$$

If $Y \in \mathbb{K}(\mathscr{H})$, let $Y = \sum_{n=1}^{\infty} s_n(Y) f_n \otimes e_n^*$ be a singular value decomposition for some orthonormal sets (e_n) and (f_n). For all $k \ge 2$ we set

$$Z := \sum_{n=1}^{k-1}(s_n(Y) - s_{n+1}(Y)) \sum_{j=1}^{n} f_j \otimes e_j^*$$

and

$$V := s_k(Y)\sum_{n=1}^{k} f_n \otimes e_n^* + \sum_{n=k+1}^{\infty} s_n(Y) f_n \otimes e_n^*.$$

We observe that

$$\begin{aligned} Z &= \sum_{n=1}^{k-1}\sum_{j=1}^{n}(s_n(Y) - s_{n+1}(Y)) f_j \otimes e_j^* \\ &= \sum_{j=1}^{k}(s_j(Y) - s_k(Y)) f_j \otimes e_j^* \\ &= \sum_{n=1}^{k} s_n(Y) f_n \otimes e_n^* - s_k(Y)\sum_{n=1}^{k} f_n \otimes e_n^* = Y - V. \end{aligned}$$

We can also note that $s_1(V) = \cdots = s_k(V) = s_k(Y)$, due to orthogonality of the systems (e_n) and (f_n). For each Ky Fan k-norm $|||\cdot|||$ we have

$$\left|\left|\left|\sum_{n=1}^{N} A_n Y B_n\right|\right|\right|_{(k)} \le \left|\left|\left|\sum_{n=1}^{N} A_n Z B_n\right|\right|\right|_{(k)} + \left|\left|\left|\sum_{n=1}^{N} A_n V B_n\right|\right|\right|_{(k)}$$
$$\le \|Z\|_1 + k\left\|\sum_{n=1}^{N} A_n V B_n\right\| \tag{2.4.27}$$

$$\leq \sum_{n=1}^{k-1} (s_n(Y) - s_{n+1}(Y)) \sum_{j=1}^{n} \|f_j \otimes e_j^*\| + k\|V\| \tag{2.4.28}$$

$$\leq \sum_{n=1}^{k-1} n(s_n(Y) - s_{n+1}(Y)) + ks_k(Y)$$

$$= \sum_{n=1}^{k} s_k(Y) = \|\!|Y|\!\|_k \,, \tag{2.4.29}$$

with (2.4.27) following from (2.4.26) and (2.4.28) from (2.4.25).

Moreover, if Y is in $\mathbb{K}(\mathscr{H})$, then also $\sum_{n=1}^{\infty} A_n Y B_n \in \mathbb{K}(\mathscr{H})$. Indeed, elementary operators $R_N(Y) = \sum_{n=1}^{N} A_n Y B_n$ acting on $\mathcal{C}_{(k)}$ represent a bounded family, because $\|\!|R_N(Y)|\!\|_k \leq \|\!|Y|\!\|_k$ for all $Y \in \mathbb{K}(\mathscr{H})$ by (2.4.29). Also, for one-dimensional operators $f \otimes g^*$ and $M > N$ we have

$$\begin{aligned}\left|\!\left|\!\left| R_M(f \otimes g^*) - R_N(f \otimes g^*) \right|\!\right|\!\right|_k &\leq \left\| \sum_{n=N+1}^{M} B_n^* f \otimes (A_n g)^* \right\|_1 \\ &\leq \sum_{n=N+1}^{M} \|B_n^* f\| \|A_n g\| \\ &\leq \left\| \left(\sum_{n=N+1}^{M} A_n A_n^* \right)^{\frac{1}{2}} g \right\| \left\| \left(\sum_{n=N+1}^{M} B_n B_n^* \right)^{\frac{1}{2}} f \right\| ,\end{aligned}$$

which tends to 0 as $M, N \to \infty$. Therefore, $R_N(Y)$ converges to a compact operator in $\mathcal{C}_{(k)}$ for all finite-rank operators Y. By the uniform boundedness principle, the same is true for all $Y \in \mathcal{C}_{(k)}$, due to its separability. Hence, (2.4.24) holds for all Ky Fan k-norms. It follows from the Ky Fan dominance property that (2.4.24) is valid for all unitarily invariant norms. □

Van Hemmen and Ando [38] posed the following question: "If A and B are positive bounded operators such that $A - B$ is trace class, when is it true that the difference of their square roots is trace class?" They presented the following answer.

Theorem 2.4.30 *Let A and B be positive operators. If $A^{1/2} + B^{1/2} \geq M$ for some scalar $M > 0$, then*

$$M\|A^{1/2} - B^{1/2}\|_1 \leq \|A - B\|_1 .$$

Proof If $A - B \notin \mathcal{C}_1$, then the inequality trivially holds. We can therefore assume that $A - B \in \mathcal{C}_1$. Hence, $A - B$ is a compact operator. If $\pi : \mathbb{B}(\mathscr{H}) \to \mathbb{B}(\mathscr{H})/\mathbb{K}(\mathscr{H})$ is the quotient map of $\mathbb{B}(\mathscr{H})$ onto the Calkin algebra $\mathbb{B}(\mathscr{H})/\mathbb{K}(\mathscr{H})$, then $\pi(A) = \pi(B)$. Taking the positive square root we arrive to $\pi(A)^{1/2} = \pi(B)^{1/2}$. Since π is a $*$-homomorphism, we get $\pi(A^{1/2}) = \pi(B^{1/2})$. Thus, $A^{1/2} - B^{1/2}$ is a compact operator. Let us set

$$X = A^{1/2} - B^{1/2} \quad \text{and} \quad Y = A^{1/2} + B^{1/2} \geq M.$$

Since X is a compact self-adjoint operator, it is diagonalizable [28, Theorem 2.4.4]. Hence, $\mathscr{H}$ has a orthonormal basis (e_j) consisting of eigenvectors of X such that $Xe_i = \lambda_i e_i$ for some scalar λ_i and all i. Furthermore, $|X|e_i = |\lambda_i|e_i$. Clearly, $A - B = (XY + YX)/2$. We have

$$\begin{aligned}
\operatorname{tr}|A - B| &= \sum_i \frac{1}{2}\langle e_i, |XY + YX|e_i\rangle \\
&\geq \sum_i \frac{1}{2}|\langle e_i, (XY + YX)e_i\rangle| \\
&= \sum_i |\lambda_i|\langle e_i, Ye_i\rangle \\
&\geq M \sum_i |\lambda_i| \\
&= M \sum_i \langle e_i, |X|e_i\rangle \\
&= M \operatorname{tr}|A^{1/2} - B^{1/2}|,
\end{aligned}$$

which gives the required inequality. □

The singular value form of the arithmetic–geometric inequality (2.4.11) for matrices $A, B \in \mathbb{M}_n$ is

$$2s_j(AB^*) \leq s_j(A^*A + B^*B) \qquad (j = 1, \ldots, n); \tag{2.4.30}$$

see [39]. If X and Y are positive definite matrices, then

$$\begin{aligned}
2 \operatorname{tr} X = 2\sum_{j=1}^{n} s_j(X) &= 2\sum_{j=1}^{n} s_j\left(Y^{1/2}(XY^{-1/2})^*\right) \\
&\leq \sum_{j=1}^{n} s_j(Y + Y^{-1/2}X^2Y^{-1/2}) = \operatorname{tr}(Y + X^2Y^{-1}).
\end{aligned}$$

Therefore,

$$\operatorname{tr}(X - X^2Y^{-1}) \leq \operatorname{tr}(Y - X). \tag{2.4.31}$$

Utilizing this inequality, Zhan [40] established a trace inequality as follows.

Theorem 2.4.31 *Let $A_0 \in \mathbb{M}_n$ be positive definite and $A_1, \ldots A_k \in \mathbb{M}_n$ be positive semidefinite. Then*

$$\operatorname{tr}\left(\sum_{j=1}^{k}\left(\sum_{i=0}^{j} A_i\right)^{-2} A_j\right) \le \operatorname{tr}(A_0^{-1}).$$

Proof Picking $X = \left(\sum_{i=0}^{j} A_i\right)^{-1}$ and $Y = \left(\sum_{i=0}^{j-1} A_i\right)^{-1}$ in (2.4.31), one gets

$$\begin{aligned}
&\operatorname{tr}\left(\left(\sum_{i=0}^{j} A_i\right)^{-1} - \left(\sum_{i=0}^{j} A_i\right)^{-2}\left(\sum_{i=0}^{j-1} A_i\right)\right)\\
&\le \operatorname{tr}\left(\left(\sum_{i=0}^{j-1} A_i\right)^{-1} - \left(\sum_{i=0}^{j} A_i\right)^{-1}\right)
\end{aligned}$$

whence

$$\operatorname{tr}\left(\left(\sum_{i=0}^{j} A_i\right)^{-2} A_j\right) \le \operatorname{tr}\left(\left(\sum_{i=0}^{j-1} A_i\right)^{-1} - \left(\sum_{i=0}^{j} A_i\right)^{-1}\right)$$

for all $1 \le j \le k$, since summing these terms over all $1 \le j \le k$ we reach the following inequality:

$$\operatorname{tr}\left(\sum_{j=1}^{k}\left(\sum_{i=0}^{j} A_i\right)^{-2} A_j\right) \le \operatorname{tr}(A_0^{-1}) - \operatorname{tr}\left(\sum_{i=0}^{k} A_i\right)^{-1} \le \operatorname{tr}(A_0^{-1}).$$

□

2.5 Exercises and Problems

Exercise 2.5.1 Let $A, B \in \mathbb{B}(\mathscr{H})$. Show that the following statements are equivalent:

(i) $\operatorname{ran}(B) \subseteq \operatorname{ran}(A)$;
(ii) there is an operator $C \in \mathbb{B}(\mathscr{H})$ such that $AC = B$;
(iii) $BB^* \le \lambda AA^*$ for some $\lambda > 0$.

If one of these conditions holds, then establish that there is a unique operator $D \in \mathbb{B}(\mathscr{H})$, called the Douglas (or reduced) solution, such that $AD = B$ and $\operatorname{ran}(D) \subseteq \operatorname{ran}(A^*)$. Moreover, $\|D\|^2 = \inf\{\lambda > 0 : BB^* \le \lambda AA^*\}$.

Exercise 2.5.2 With the notation in the singular value decomposition $A = UDV^*$, set

$$A^\dagger := V\mathrm{diag}(1/s_1, \ldots, 1/s_k, 0, \ldots, 0)U^*.$$

It is called the *Moore–Penrose inverse* of A. Prove that it is a unique square matrix with the following properties:

(i) $AA^\dagger A = A$;
(ii) $A^\dagger AA^\dagger = A^\dagger$;
(iii) $AA^\dagger$ is Hermitian;
(iv) $A^\dagger A$ is Hermitian.

Moreover, if A is invertible, then $A^\dagger = A^{-1}$.

Exercise 2.5.3 Prove that an operator $A \in \mathbb{B}(\mathscr{H})$ has the Moore–Penrose inverse if and only if its range is closed.

Exercise 2.5.4 Prove if $A, B, X \in \mathbb{M}_n$ and $X \in \mathbb{M}_n$ is positive definite, then

$$|\operatorname{tr}(AB^*)|^2 \leq \operatorname{tr}(AXA^*)\operatorname{tr}(BX^{-1}B^*).$$

Exercise 2.5.5 Prove that

$$\left|\operatorname{tr}(XAB) - \operatorname{tr}(XA)\operatorname{tr}(XB)\right| \leq 4rs,$$

where $A, B \in \mathbb{M}_n$ and their numerical ranges lie in the circular disks of radii r and s, respectively, and X is a positive matrix of trace one. Show that if A and B are normal, then 4 can be replaced with 1.

Exercise 2.5.6 Show that $M = \{X \in \mathbb{M}_n : \operatorname{tr} X = 0\}$ is a subspace of the Hilbert space $\mathbb{M}_n$ equipped with the inner product $\langle A, B\rangle = \operatorname{tr}(B^*A)$ and $M^\perp = \mathbb{C}I$.

Exercise 2.5.7 Prove $A \in \mathbb{M}_n$ is zero matrix if and only if $\operatorname{tr}(AB) = 0$ for all Hermitian matrices $B \in \mathbb{M}_n$.

Exercise 2.5.8 Let $A, B \in \mathbb{M}_2$. Show that if $[A, B] := AB - BA$, then $[A, B]^2$ commutes with every matrix $C \in \mathbb{M}_2$.

Exercise 2.5.9 Introduce a class of matrices $A, B, C \in \mathbb{M}_n$ for which $\operatorname{tr}(ABC) = \operatorname{tr}(BAC)$.

Exercise 2.5.10 If $A \in \mathbb{M}_n$, do $\operatorname{tr}(\operatorname{Re} A) \leq \operatorname{tr}|A|$ and $\operatorname{tr}(\operatorname{Im} A) \leq \operatorname{tr}|A|$ hold?

Exercise 2.5.11 If $A, B \in \mathbb{M}_n$ are Hermitian with positive traces, show that

$$\frac{\operatorname{tr}\left((A+B)^2\right)}{\operatorname{tr}(A+B)} \leq \frac{\operatorname{tr}(A^2)}{\operatorname{tr} A} + \frac{\operatorname{tr}(B^2)}{\operatorname{tr} B}.$$

Exercise 2.5.12 Under what conditions on a real-valued function f defined on $[0, \infty)$ does the inequality

$$\operatorname{tr}(f(A+B)) \leq \operatorname{tr}(f(A)) + \operatorname{tr}(f(B))$$

hold for all positive semidefinite $n \times n$ matrices A and B?

Exercise 2.5.13 Let f be a real-valued function defined on $[0, \infty)$ with $f(0) = 0$. Prove that if

$$\operatorname{tr}(f(A+B)) \leq \operatorname{tr}(f(A)) + \operatorname{tr}(f(B))$$

holds for all positive semidefinite $n \times n$ matrices A and B, then f is concave.

Exercise 2.5.14 Show that Theorem 2.1.7 cannot be generalized to the product of three Hermitian matrices.

Exercise 2.5.15 Show that inequality (2.1.3) cannot be extended to the product of three positive semidefinite matrices.

Exercise 2.5.16 If A and B are Hermitian matrices in $\mathbb{M}_n$ and k is an arbitrary integer, then does the inequality $\operatorname{tr}\left((AB)^k\right) \leq \operatorname{tr}(A)^k \operatorname{tr}(B)^k$ hold?

Exercise 2.5.17 If $\begin{bmatrix} A & X \\ X^* & B \end{bmatrix} \geq 0$, then does $\det(X^*X) \leq \det(A)\det(B)$?

Exercise 2.5.18 If $\begin{bmatrix} A & X \\ X^* & B \end{bmatrix} \geq 0$, then prove that

$$|\operatorname{tr}(AB) - \operatorname{tr}(X^*X)| \leq \operatorname{tr} A \operatorname{tr} B - \operatorname{tr} X^* \operatorname{tr} X.$$

Exercise 2.5.19 Apply the Schur inequality to

$$\begin{bmatrix} 0 & a_1^{1/2} & 0 & \dots & 0 \\ 0 & 0 & a_2^{1/2} & \dots & 0 \\ \vdots & \vdots & \vdots & \dots & \vdots \\ 0 & 0 & 0 & \dots & a_{n-1}^{1/2} \\ a_n^{1/2} & 0 & 0 & \dots & 0 \end{bmatrix}$$

to get the usual arithmetic–geometric inequality for nonnegative real numbers $a_1, \dots, a_n$.

Exercise 2.5.20 Show that equality occurs in (2.1.5) if and only if $A = \lambda B$ for some $B \geq 0$ and $\lambda \in \mathbb{C}$ of modulus 1.

Exercise 2.5.21 Let $p > 1$ and $\frac{1}{p} + \frac{1}{q} = 1$. If $A \in \mathbb{M}_n$, then

$$|\operatorname{tr} A| = n^{1/q} \min\{\|A - C\|_p : \operatorname{tr} C = 0\}.$$

Exercise 2.5.22 Suppose that $\mathscr{H}$ is a Hilbert space. It follows from Theorem 2.3.8 that for $A \in \mathcal{C}_1$, the functional $\operatorname{tr}(A\cdot) : \mathbb{K}(\mathscr{H}) \to \mathbb{C}$ is bounded and the map from $\mathcal{C}_1$ into $\mathbb{K}(\mathscr{H})'$ taking A to $\operatorname{tr}(A\cdot)$ is linear and norm-decreasing, that is, $\| \operatorname{tr}(A\cdot)\| \leq \|A\|_1$. Show that this map is an isometric linear isomorphism and one can identify $\mathbb{K}(\mathscr{H})'$ with $\mathcal{C}_1$.

Exercise 2.5.23 Suppose that $\mathscr{H}$ is a Hilbert space. It follows from Theorem 2.3.8 that for $B \in \mathbb{B}(\mathscr{H})$, the linear functional $\operatorname{tr}(\cdot B) : \mathcal{C}_1 \to \mathbb{C}$ is bounded and the map from $\mathbb{B}(\mathscr{H})$ into $\mathcal{C}_1'$ taking B to $\operatorname{tr}(\cdot B)$ is linear and norm-decreasing. Show that this map is an isometric linear isomorphism and one can identify $\mathcal{C}_1'$ with $\mathbb{B}(\mathscr{H})$.

Exercise 2.5.24 Show that $\operatorname{tr}(\cdot)$ is a normal positive linear functional in the sense that for any bounded increasing sequence (A_n) of positive operators in $\mathcal{C}_1$, it holds that

$$\operatorname{tr}(\sup_n A_n) = \sup_n \operatorname{tr}(A_n).$$

Exercise 2.5.25 Is there a trace class operator $X \in \mathbb{B}(\mathscr{H})$ such that $\operatorname{tr}(X^n) = n$ for all $n \in \mathbb{N}$.

Exercise 2.5.26 Give examples showing that $\|T\|_p^2$ and $\|\operatorname{Re} T\|_p^2 + \|\operatorname{Im} T\|_p^2$, are not comparable for any $1 \leq p \leq \infty$ except $p = 2$.

Exercise 2.5.27 Prove that if $\begin{bmatrix} X & Y \\ Y^* & Z \end{bmatrix} \geq 0$, then

$$\left\| \begin{bmatrix} X & Y \\ Y^* & Z \end{bmatrix} \right\|_p \geq \left\| \begin{bmatrix} \|X\|_p & \|Y\|_p \\ \|Y\|_p & \|Z\|_p \end{bmatrix} \right\|_p$$

when $1 \leq p \leq 2$, and the reverse inequality holds for $p \geq 2$.

Problem 2.5.28 Let $A \in \mathbb{M}_n \otimes \mathbb{M}_m$ be a positive matrix. Prove that

$$\left(\operatorname{tr}_2\left((\operatorname{tr}_1 A)^p\right)\right)^{\frac{1}{p}} \leq \operatorname{tr}_1\left(\left(\operatorname{tr}_2(A^p)\right)^{\frac{1}{p}}\right)$$

for all $p \geq 1$.

Problem 2.5.29 For every unitarily invariant norm and every $A, B \in \mathbb{M}_n$, it holds that

$$|||AB||| \leq |||\, |A|^p \,|||^{1/p} \; |||\, |B|^{1/q} \,|||^{1/q}$$

for all $p > 1$ and $\frac{1}{p} + \frac{1}{q} = 1$. In particular,

$$\|AB\|_1 \leq \|A\|_p \, \|B\|_q.$$

Problem 2.5.30 For positive matrices A and B, show that the inequality

$$\operatorname{tr}(A+B) - \operatorname{tr}|A-B| \le 2 \operatorname{tr}(A^s B^{1-s})$$

holds for every $s \in [0, 1]$.

Problem 2.5.31 If A and B are positive operators and z is any complex number, then show that

$$|||A - |z|B||| \le |||A + zB||| \le |||A + |z|B||| .$$

Problem 2.5.32 Let $A, B \in \mathbb{B}(\mathscr{H})$ be such that $A \ge B \ge 0$, let $p > 1$, and let $\alpha \ge \max\{-1, -p/2\}$. Show that

(i) There exists a partial isometry U such that

$$A^{\alpha/2} B^p A^{\alpha/2} \le U^* A^{p+\alpha} U.$$

(ii) If f is a continuous increasing function on $[0, \infty)$ with $f(0) = 0$, then

$$\operatorname{tr} f(A^{\alpha/2} B^p A^{\alpha/2}) \le \operatorname{tr} f(A^{p+\alpha}).$$

In the above statements, the invertibility of A is assumed when $\alpha < 0$.

Problem 2.5.33 Let f be a nonzero linear functional on $\mathbb{M}_n$. Show that following statements are equivalent.

- f is a constant multiple of the trace;
- $f(AB - BA) = 0$ for all A and B;
- $f(CAC^{-1}) = f(A)$ for all A and all invertible matrices C;
- $|f(A)| \le cr(A)$ for some constant c and for all A;
- $\limsup_{k\to\infty} |f(A^k)|^{1/k} = r(A)$ for all A;
- $f(P) \ne 0$ for every nonzero idempotent P;
- $\sup\{|f(P)| : P^2 = P\}$ is finite;
- f is constant on idempotents of the same rank;
- If $A^2 = 0$, then $f(A) = 0$;
- $f(A) = f(A^2) = \cdots = f(A^n) = 0$ implies A is nilpotent.

2.6 Notes, Hints, and References

Excellent sources for studying operator ideals are the books by Pietsch [17], Simon [18], and Gohberg and Kreĭn [16]. Hints and references for the exercises and problems in this section are as follows.

Exercise 2.5.1 is known as the *Douglas majorization theorem*, which plays an essential role in operator theory; see also [41]. Exercise 2.5.2 is devoted to introducing the Moore–Penrose inverse. To prove it use straightforward computations to verify the mentioned properties. In particular, one can check that $AA^\dagger = UU^*$ and $A^\dagger A = VV^*$ are projections onto the ranges of A and the transpose of A, respectively.

To solve Exercise 2.5.3, assume that A has the Moore–Penrose inverse $A^\dagger$. Show that $AA^\dagger$ is a projection and $AA^\dagger(\mathscr{H}) = \operatorname{ran}(A)$, and conclude that the range of A is closed. For the converse statement, consider the projection P onto $\operatorname{ran}(A)$. Show that the Douglas solution to $AX = P$ is the Moore–Penrose inverse of A.

Consider the Cauchy–Schwarz inequality for the trace functional (2.1.1) and replace A and B with $AX^{1/2}$ and $BX^{-1/2}$, respectively, to solve Exercise 2.5.4. A solution to Exercise 2.5.5 can be found in [42]. For Exercise 2.5.6, note that $\langle I, X\rangle = \operatorname{tr}(I \cdot X^*) = \overline{\operatorname{tr} X} = 0$ for every $X \in M$. Thus, $I \perp M$. Since tr is a linear functional on the Hilbert space $\mathbb{M}_n$, its kernel (that is, M) has codimension 1. Therefore, $M^\perp = \mathbb{C}I$.

To establish Exercise 2.5.7, use the Cartesian decomposition to conclude that $\operatorname{tr}(AB) = 0$ for every matrix B, and then take B to be A^*. To solve Exercise 2.5.8, use the trace of $[A, B]$. For Exercise 2.5.9, consider the class of Hermitian matrices with real entries. For Exercise 2.5.10, note that

$$\operatorname{tr}(\operatorname{Re} A) = \operatorname{Re}(\operatorname{tr} A) \le |\operatorname{tr} A| \le \operatorname{tr}|A|.$$

A solution to Exercise 2.5.11 can be found in [3]. The papers [43, 44] provide interesting insights into Exercises 2.5.12 and 2.5.13.

A counterexample for both Exercises 2.5.14 and 2.5.15 can be found in [6]. For an answer to the question posed in Exercise 2.5.16 consult [45]. For Exercise 2.5.17, explore whether $\begin{bmatrix} \det(A) & \det(X) \\ \det(X^*) & \det(B) \end{bmatrix} \ge 0$? Exercise 2.5.18 is solved in [46] and [47, Theorem 2.1]. For Exercise 2.5.19; see [48]. A solution to Exercise 2.5.20 can be found in [3, Theorem 9.6]. Exercise 2.5.21 appeared in [49, Theorem 2]. Exercises 2.5.22 and 2.5.23 are extensively discussed in [28, Theorem 4.2.1] and [28, Theorem 4.2.3]. In Exercise 2.5.24, note that $\sup_n A_n$ is the limit of $\{A_n\}$ in strong operator topology and $\operatorname{tr} A = \sum_i \langle Ae_i, e_i\rangle$, where (e_i) is an arbitrary orthonormal basis in the Hilbert space $\mathscr{H}$; see also Example 8.1.2 (i).

In Exercise 2.5.25, one can explore the problem first for square matrices A with $\operatorname{tr} A = 0$ satisfying $A^m = A^{m+1}$ for some m; see [3, Problem 8.2.8]. A discussion on Exercise 2.5.26 can be found in [50]. Exercise 2.5.27 is significant; see [51, 52]. What do we say if we remove the condition of positivity of $\begin{bmatrix} X & Y \\ Y^* & Z \end{bmatrix}$ and replace this matrix with an arbitrary matrix $\begin{bmatrix} X & Y \\ W & Z \end{bmatrix}$? For Problem 2.5.28 dealing with partial traces, the reader may consult [53]. Problem 2.5.29 refers to the Hölder inequality for unitarily invariant norms; [12, Corollary IV.2.6] provides a proof for it.

The Chernoff bounds are used to analyze the distance measure between two probability distributions in quantum information theory. These measures play a crucial role in quantifying the difference between quantum states. In [54, Theorem 1], the inequality presented in Problem 2.5.30 concerning the trace distance to the quantum Chernoff bound is demonstrated. For a generalization; see [55]. Problem 2.5.31 is the main result of [56] and Problem 2.5.32 is discussed in [13].

Problem 2.5.32 appeared in [57, Theorem2]. To prove it employ the Shoda theorem (Theorem 2.1.21) stating that the commutators form a subspace of codimension one in $\mathbb{M}_n$, which is the kernel of the trace.

References

1. J. von Neumann, Some matrix-inequalities and metrization of matric-space. Mitt. Forsch.-Inst. Math. Mech. Univ. Tomsk **1**, 286–299 (1937)
2. K. Fan, A.J. Hoffman, Some metric inequalities in the space of matrices. Proc. Amer. Math. Soc. **6**, 111–116 (1955)
3. F. Zhang, *Matrix Theory: Basic Results and Techniques*, 2nd edn. (Universitext, Springer, New York, 2011)
4. R.A. Horn, C.R. Johnson, *Matrix Analysis*, 2nd edn. (Cambridge University Press, Cambridge, 2013)
5. D.J.H. Garling, *Inequalities: A Journey into Linear Analysis* (Cambridge University Press, Cambridge, 2007)
6. I.D. Coope, On matrix trace inequalities and related topics for products of Hermitian matrices. J. Math. Anal. Appl. **188**, 999–1001 (1994)
7. X.M. Yang, Some trace inequalities for operators. J. Austral. Math. Soc. Ser. A **58**, 281–286 (1995)
8. W.W. Parker, Sets of complex numbers associated with a matrix. Duke Math. J. **15**, 711–715 (1948)
9. R.D. Grigorieff, A note on von Neumann's trace inequality. Math. Nachr. **151**, 327–328 (1991)
10. B.-Y. Wang, B.-Y. Xi, F. Zhang, Some inequalities for sum and product of positive semidefinite matrices. Linear Algebra Appl. **293**, 39–49 (1999)
11. G.H. Hardy, J.E. Littlewood, G. Pólya, *Inequalities* (Cambridge University Press, Cambridge, 1988)
12. R. Bhatia, *Matrix Analysis* (Springer, New York, 1997)
13. H. Kosaki, On some trace inequality. Proc. Centre Math. Anal. Austral. Nat. Univ. 129–134 (1991)
14. T. Ando, Majorizations and inequalities in matrix theory. Linear Algebra Appl. **199**, 17–67 (1994)
15. A.W. Marshall, I. Olkin, *Inequalities: Theory of Majorization and Its Applications/* Mathematics in Science and Engineering, vol. 143 (Academic, 1979)
16. I.C. Gohberg, M.G. Kreĭn, *Introduction to the Theory of Linear Nonselfadjoint Operators, Translations of Mathematical Monographs*, vol. 18 (American Mathematical Society, Providence, R. I., 1969)
17. A. Pietsch, *Operator ideals, Mathematische Monographien [Mathematical Monographs], 16* (VEB Deutscher Verlag der Wissenschaften, Berlin, 1978), p.451
18. B. Simon, *Trace Ideals and Their Applications, Second Edition, Mathematical Surveys and Monographs*, vol. 120 (American Mathematical Society, Providence, RI, 2005)
19. R. Bhatia, F. Kittaneh, Notes on matrix arithmetic-geometric mean inequalities. Linear Algebra Appl. **308**, 203–211 (2000)

20. C.A. McCarthy, c_p. Isr. J. Math. **5**, 249–271 (1967)
21. R. Bhatia, F. Kittaneh, The matrix arithmetic geometric mean inequality revisited. Linear Algebra Appl. **428**, 2177–2191 (2008)
22. F. Kittaneh, A note on the arithmetic-geometric-mean inequality for matrices. Linear Algebra Appl. **171**, 1–8 (1992)
23. F. Kittaneh, Norm inequalities for sums of positive operators. J. Operator Theory **48**, 95–103 (2002)
24. F. Kittaneh, Some norm inequalities for operators. Canad. Math. Bull. **42**, 87–96 (1999)
25. K.R. Davidson, S.C. Power, Best approximation in C^* -algebras. J. Reine Angew. Math. **368**, 43–62 (1986)
26. F. Kittaneh, Norm inequalities for certain operator sums. J. Funct. Anal. **143**, 337–348 (1997)
27. F. Kittaneh, Inequalities for the Schatten p-Norm. IV. Commun. Math. Phys. **106**, 581–585 (1986)
28. G.J. Murphy, *C^* -algebras and Operator Theory* (Academic, Boston, MA, 1990)
29. R.T. Powers, E. Størmer, Free states of the canonical anticommutation relations. Commun. Math. Phys. **16**, 1–33 (1970)
30. F. Kittaneh, On Lipschitz functions of normal operators. Proc. Amer. Math. Soc. **94**, 416–418 (1985)
31. D. Voiculescu, Some results on norm-ideal perturbations of Hilbert space operators. J. Operator Theory **2**, 3–37 (1979)
32. R. Bhatia, F. Kittaneh, Commutators, pinchings, and spectral variation. Oper. Matrices **2**, 143–151 (2008)
33. J.G. Stampfli, The norm of a derivation. Pacific J. Math. **33**, 737–747 (1970)
34. D.R. Jocić, Norm inequalities for self-adjoint derivations. J. Funct. Anal. **145**, 24–34 (1997)
35. O. Hirzallah, F. Kittaneh, Norm inequalities for weighted power means of operators. Linear Algebra Appl. **341**, 181–193 (2002)
36. O. Hirzallah, F. Kittaneh, Singular values, norms, and commutators. Linear Algebra Appl. **432**, 1322–1336 (2010)
37. D.R. Jocić, Cauchy-Schwarz and means inequalities for elementary operators into norm ideals. Proc. Amer. Math. Soc. **126**, 2705–2711 (1998)
38. J.L. van Hemmen, T. Ando, An inequality for trace ideals. Commun. Math. Phys. **76**, 143–148 (1980)
39. R. Bhatia, F. Kittaneh, On the singular values of a product of operators. SIAM J. Matrix Anal. Appl. **11**, 272–277 (1990)
40. X. Zhan, On some matrix inequalities. Linear Algebra Appl. **376**, 299–303 (2004)
41. V. Manuilov, M.S. Moslehian, Q. Xu, Douglas factorization theorem revisited. Proc. Amer. Math. Soc. **148**, 1139–1151 (2020)
42. P.F. Renaud, A matrix formulation of Grüss inequality. Linear Algebra Appl. **335**, 95–100 (2001)
43. O.E. Tikhonov, Subadditivity inequalities in von Neumann algebras and characterization of tracial functionals. Positivity **9**, 259–264 (2005)
44. O.E. Tikhonov, On matrix-subadditive functions and a relevant trace inequality. Linear Multilinear Algebra **44**, 25–28 (1998)
45. P.Z. Yang, X.X. Feng, A note on the trace inequality for products of Hermitian matrix power. J. Inequal. Pure Appl. Math. **3**, 78 (2002), 12 pp
46. F. Zhang, Trace inequality for positive block matrices, IMAGE Solution 50-3.1. Bull. Int. Linear Algebra Soc. (2013)
47. F. Kittaneh, M. Lin, Trace inequalities for positive semidefinite block matrices. Linear Algebra Appl. **524**, 153–158 (2017)
48. F. Gaines, On the arithmetic mean-geometric mean inequality. Amer. Math. Monthly **74**, 305–306 (1967)
49. F. Kittaneh, Inequalities for the Schatten *p-norm. II.* Glasgow Math. J. **29**, 99–104 (1987)
50. R. Bhatia, X. Zhan, Norm inequalities for operators with positive real part. J. Operator Theory **50**, 67–76 (2003)

51. C. King, Inequalities for trace norms of 2×2 block matrices. Commun. Math. Phys. **242**, 531–545 (2003)
52. K.M.R. Audenaert, A norm compression inequality for block partitioned positive semidefinite matrices. Linear Algebra Appl. **413**, 155–176 (2006)
53. E.A. Carlen, E.H. Lieb, A Minkowski type trace inequality and strong subadditivity of quantum entropy. II. Convexity and concavity. Lett. Math. Phys. **83**, 107–126 (2008)
54. K.M.R. Audenaert, J. Calsamiglia, R. Muñoz-Tapia, E. Bagan, Ll. Masanes, A. Acin, F. Verstraete, Discriminating states: the quantum Chernoff bound. Phys. Rev. Lett. **98**, 160501 (2007)
55. M. Kian, M.S. Moslehian, H. Osaka, *Generalizations of the Powers–Størmer's inequality*. arxiv: 2302.07818 [quant-ph]
56. R. Bhatia, F. Kittaneh, Norm inequalities for positive operators. Lett. Math. Phys. **43**, 225–231 (1998)
57. D. Petz, J. Zemánek, Characterizations of the trace. Linear Algebra Appl. **111**, 43–52 (1988)

Chapter 3
Trace Inequalities for Positive Semidefinite Matrices

In this chapter, we address two questions posed by Jean-Christophe Bourin regarding matrix subadditivity inequalities. We provide a solution to a generalized version of the first question for the trace norm and the Hilbert–Schmidt norm. Moreover, we give partial answers to the Hilbert–Schmidt version of the second question by utilizing the Araki–Lieb–Thirring and Ando–Hiai–Okubo trace inequalities. We then generalize the Ando–Hiai–Okubo trace inequality to complex exponents and present a new proof of this famous inequality using complex interpolation in connection with the Hadamard three lines theorem. Finally, we consider various applications of trace inequalities.

Throughout this chapter, all matrices are assumed to be complex square matrices of the same size and $|||\cdot|||$ denotes any unitarily invariant norm.

3.1 Partial Solutions to the First Main Question

If f is a nonnegative concave function on $[0, \infty)$, then f is subadditive, that is,

$$f(a+b) \leq f(a) + f(b)$$

holds for all $a, b \geq 0$.

A noncommutative version of this inequality for positive semidefinite matrices is presented in [1]. This version says that if A and B are positive semidefinite matrices, then

$$|||f(A+B)||| \leq |||f(A) + f(B)|||$$

for every unitarily invariant norm. The trace norm case, which is a nontrivial result, is the Rotfel'd trace inequality [2]. The proof for the operator norm turns out to be

A. M. Bikchentaev et al., *Trace Inequalities*, Forum for Interdisciplinary Mathematics,
https://doi.org/10.1007/978-981-97-6520-1_3

the simplest one, which was established much later by Kosem [3]. This inequality can be generalized to normal matrices as follows [4]:

If A and B are normal matrices and $f : [0, \infty) \to [0, \infty)$ is a concave function, then

$$|||f(|A+B|)||| \le |||f(|A|) + f(|B|)|||$$

for every unitarily invariant norm.

An important special case of this inequality is the following

$$\left|\left|\left| |A+B|^p \right|\right|\right| \le \left|\left|\left| \, |A|^p + |B|^p \right|\right|\right|$$

for $0 < p \le 1$ (see, also [5, 6]). This inequality promoted Bourin [4] to ask the following related question.

Question 3.1.1 Let A and B be positive semidefinite matrices and $p, q > 0$. Is it true that

$$\left|\left|\left| A^{p+q} + B^{p+q} \right|\right|\right| \le \left|\left|\left| \left(A^p + B^p\right)\left(A^q + B^q\right) \right|\right|\right| ?$$

He also asked whether the stronger inequality

$$\left|\left|\left| A^{p+q} + B^{p+q} \right|\right|\right| \le \left|\left|\left| \left(A^p + B^p\right)^{\frac{1}{2}} \left(A^q + B^q\right) \left(A^p + B^p\right)^{\frac{1}{2}} \right|\right|\right| \tag{3.1.1}$$

holds. We call Question 3.1.1 the *first main question of Bourin.*

The authors of [7] settled this question affirmatively for the trace norm $\|\cdot\|_1$ and the Hilbert–Schmidt norm $\|\cdot\|_2$. They provided the solution for arbitrary commuting positive semidefinite matrices. They also conjectured that the first main question is true for all unitarily invariant norms and all commuting positive semidefinite matrices.

Conjecture 3.1.2 Let A_1, A_2, B_1, B_2 be positive semidefinite matrices with $A_1A_2 = A_2A_1$ and $B_1B_2 = B_2B_1$. Then, for every unitarily invariant norm,

$$|||A_1A_2 + B_1B_2||| \le |||(A_1 + B_1)(A_2 + B_2)||| \,. \tag{3.1.2}$$

They also asked whether the stronger inequality

$$|||A_1A_2 + B_1B_2||| \le \left|\left|\left|(A_1 + B_1)^{\frac{1}{2}}(A_2 + B_2)(A_1 + B_1)^{\frac{1}{2}}\right|\right|\right| \tag{3.1.3}$$

holds.

To prove our main results, we recall the following useful facts:

(i) From Proposition 2.1.11, if A and B are positive semidefinite matrices, then

$$\operatorname{tr}(AB) \ge 0. \tag{3.1.4}$$

(ii) From Theorem 1.4.2, if A and B are positive semidefinite matrices and $AB = BA$, then

$$AB \geq 0. \tag{3.1.5}$$

The following lemma follows immediately from Lemma 2.4.7 and the fact that $|||\mathrm{Re}\, X||| \leq |||X|||$ for any matrix X.

Lemma 3.1.3 *Let A and B be positive semidefinite matrices. Then*

$$\left|\left|\left| A^{\frac{1}{2}} B A^{\frac{1}{2}} \right|\right|\right| \leq |||AB|||$$

for every unitarily invariant norm.

Lemma 3.1.4 *Let X be a matrix and $m \in \mathbb{N}$. Then*

$$\left|\mathrm{tr}\ X^m\right| \leq \mathrm{tr}\ |X|^m. \tag{3.1.6}$$

Proof Let $\lambda_1, \lambda_2, \ldots, \lambda_n$ be the eigenvalues of X, and $s_1, s_2, \ldots, s_n$ be the singular values of X. Recall that the Weyl majorant theorem states that the absolute values of the eigenvalues are log-majorized by the singular value (see [8, Theorem II.3.6] or [9, Theorem 2.4]). Hence,

$$|\mathrm{tr}\ X^m| = \left|\sum_{i=1}^{n} \lambda_i^m\right| \leq \sum_{i=1}^{n} |\lambda_i|^m \leq \sum_{i=1}^{n} s_i^m = \mathrm{tr}\ |X|^m$$

for any positive integer m. □

The special case of $m = 1$ of inequality (3.1.6) is given in Corollary 2.1.24.

We use the previous facts to answer the above question for the trace norm in a more general setting in terms of commuting positive semidefinite matrices.

Theorem 3.1.5 *Let A_1, A_2, B_1, B_2 be positive semidefinite matrices with $A_1A_2 = A_2A_1$ and $B_1B_2 = B_2B_1$. Then*

$$\|A_1A_2 + B_1B_2\|_1 \leq \left\|(A_1 + B_1)^{\frac{1}{2}}(A_2 + B_2)(A_1 + B_1)^{\frac{1}{2}}\right\|_1.$$

Proof We have

$$\begin{aligned}
\left\|(A_1 + B_1)^{\frac{1}{2}}(A_2 + B_2)(A_1 + B_1)^{\frac{1}{2}}\right\|_1 &= \mathrm{tr}\left((A_1 + B_1)^{\frac{1}{2}}(A_2 + B_2)(A_1 + B_1)^{\frac{1}{2}}\right) \\
&= \mathrm{tr}\left((A_1 + B_1)(A_2 + B_2)\right) \\
&= \mathrm{tr}\left((A_1A_2 + B_1B_2) + (A_1B_2 + B_1A_2)\right) \\
&\geq \mathrm{tr}\left(A_1A_2 + B_1B_2\right) \ \text{(by (3.1.4))} \\
&= \|A_1A_2 + B_1B_2\|_1 \ \text{(by (3.1.5))}.
\end{aligned}$$

□

By Lemma 3.1.3, we have the following corollary.

Corollary 3.1.6 *Let A_1, A_2, B_1, B_2 be positive semidefinite matrices with $A_1A_2 = A_2A_1$ and $B_1B_2 = B_2B_1$. Then*

$$\|A_1A_2 + B_1B_2\|_1 \le \|(A_1 + B_1)(A_2 + B_2)\|_1 .$$

Corollary 3.1.7 *Let A and B be positive semidefinite matrices and $p, q > 0$. Then*

$$\left\|A^{p+q} + B^{p+q}\right\|_1 \le \left\|(A^p + B^p)^{\frac{1}{2}}(A^q + B^q)(A^p + B^p)^{\frac{1}{2}}\right\|_1 .$$

Corollary 3.1.8 *Let A and B be positive semidefinite matrices and $p, q > 0$. Then*

$$\left\|A^{p+q} + B^{p+q}\right\|_1 \le \left\|(A^p + B^p)(A^q + B^q)\right\|_1 .$$

The following theorem answers the same question for the case of the Hilbert–Schmidt norm in more general setting of commuting positive semidefinite matrices.

Theorem 3.1.9 *Let A_1, A_2, B_1, B_2 be positive semidefinite matrices with $A_1A_2 = A_2A_1$ and $B_1B_2 = B_2B_1$. Then*

$$\|A_1A_2 + B_1B_2\|_2 \le \left\|(A_1 + B_1)^{\frac{1}{2}}(A_2 + B_2)(A_1 + B_1)^{\frac{1}{2}}\right\|_2 .$$

Proof We have

$$\begin{aligned}
&\left\|(A_1 + B_1)^{\frac{1}{2}}(A_2 + B_2)(A_1 + B_1)^{\frac{1}{2}}\right\|_2^2 \\
=&\operatorname{tr}\left(((A_1 + B_1)^{\frac{1}{2}}(A_2 + B_2)(A_1 + B_1)^{\frac{1}{2}})^2\right) \\
=&\operatorname{tr}\left((A_1 + B_1)^{\frac{1}{2}}(A_2 + B_2)(A_1 + B_1)(A_2 + B_2)(A_1 + B_1)^{\frac{1}{2}}\right) \\
=&\operatorname{tr}\left(((A_2 + B_2)(A_1 + B_1))^2\right) \\
=&\operatorname{tr}\left(((A_1A_2 + B_1B_2) + (B_2A_1 + A_2B_1))^2\right) \\
=&\operatorname{tr}\left((A_1A_2 + B_1B_2)^2\right) + 2\operatorname{tr}\left((A_1A_2 + B_1B_2)(A_1B_2 + B_1A_2)\right) + \operatorname{tr}\left((A_1B_2 + B_1A_2)^2\right) \\
=&\|A_1A_2 + B_1B_2\|_2^2 + 2\operatorname{tr}(A_1A_2A_1B_2 + A_1A_2B_1A_2 + B_1B_2A_1B_2 + B_1B_2B_1A_2) \\
&+\operatorname{tr}(A_1B_2A_1B_2 + A_1B_2B_1A_2 + B_1A_2A_1B_2 + B_1A_2B_1A_2) \text{ (by (3.1.5))} \\
\ge&\|A_1A_2 + B_1B_2\|_2^2 \qquad \text{(by (3.1.4))} .
\end{aligned}$$

□

Applying Lemma 3.1.3, we have the following corollary.

Corollary 3.1.10 *Let A_1, A_2, B_1, B_2 be positive semidefinite matrices with $A_1A_2 = A_2A_1$ and $B_1B_2 = B_2B_1$. Then*

$$\|A_1A_2 + B_1B_2\|_2 \le \|(A_1 + B_1)(A_2 + B_2)\|_2 .$$

Corollary 3.1.11 *Let A and B be positive semidefinite matrices and* $p, q > 0$*. Then*

$$\left\| A^{p+q} + B^{p+q} \right\|_2 \leq \left\| (A^p + B^p)^{\frac{1}{2}} (A^q + B^q)(A^p + B^p)^{\frac{1}{2}} \right\|_2.$$

Corollary 3.1.12 *Let A and B be positive semidefinite matrices and* $p, q > 0$*. Then*

$$\left\| A^{p+q} + B^{p+q} \right\|_2 \leq \left\| (A^p + B^p)(A^q + B^q) \right\|_2.$$

In Chap. 4, we give a complete answer to the first main Question 3.1.1. To present some related trace inequalities, we start with two lemmas. The second lemma contains a generalization of (3.1.4).

Lemma 3.1.13 *Let A and B be positive semidefinite matrices and* $n \in \mathbb{N}$*. Then*

$$(AB)^n A \geq 0 \text{ and } B(AB)^n \geq 0.$$

Proof We prove the inequality $(AB)^n A \geq 0$ by induction. The proof of the second inequality is similar.

For $n = 1$, it is clear that $(AB)^1 A = ABA \geq 0$. Assume that the statement is true for all $k \leq n$, that is,

$$(AB)^k A \geq 0$$

for all $k \leq n$. So, in particular,

$$(AB)^{n-1} A \geq 0. \tag{3.1.7}$$

Now, let $k = n + 1$. Then by (3.1.7),

$$(AB)^{n+1} A = (AB) \left((AB)^{n-1} A \right) (AB)^* \geq 0. \tag{3.1.8}$$

This proves the inequality for $k = n + 1$, and completes the proof of the lemma. □

Proposition 3.1.14 *Let* A, B, C *be positive semidefinite matrices with* $BC = CB$ *and* $n \in \mathbb{N}$*. Then*

$$\operatorname{tr} \left((AB)^n C \right) \geq 0.$$

Proof By Lemma 3.1.13, we have $(AB)^{n-1} A \geq 0$. This, together with (3.1.4) and (3.1.5), implies that $\operatorname{tr}((AB)^n C) = \operatorname{tr}\left((AB)^{n-1} ABC \right) \geq 0$. □

Corollary 3.1.15 *Let A and B be positive semidefinite matrices and* $n \in \mathbb{N}$*. Then*

$$\operatorname{tr} \left((AB)^n \right) \geq 0.$$

This result can also be proved using the spectral mapping theorem and the fact that the trace of a matrix is the sum of its eigenvalues.

We conclude this section with the following trace inequalities, which are related to the first main Question 3.1.1.

Theorem 3.1.16 *Let A_1, A_2, B_1, B_2 be positive semidefinite matrices with $A_1A_2 = A_2A_1$ and $B_1B_2 = B_2B_1$. Then*

$$\operatorname{tr}\left((A_1B_2 + B_1A_2)^2\right) \leq \operatorname{tr}\left((A_1B_2 + B_1A_2)(A_2B_1 + B_2A_1)\right).$$

Proof We have

$$\begin{aligned}\operatorname{tr}\left((A_1B_2 + B_1A_2)^2\right) &= \operatorname{tr}\left((A_1B_2)^2\right) + \operatorname{tr}\left((B_1A_2)^2\right) + \operatorname{tr}(A_1B_2B_1A_2) + \operatorname{tr}\left(B_1A_2A_1B_2\right)\\ &= \operatorname{tr}\left((A_1B_2)^2\right) + \left((B_1A_2)^2\right) + 2\operatorname{tr}\left(A_1A_2B_1B_2\right).\end{aligned}$$

Applying (3.1.4) and Corollary 3.1.15, we see that $\operatorname{tr}\left((A_1B_2 + B_1A_2)^2\right) \geq 0$. Now, by Lemma 3.1.4 with the matrix $A_1B_2 + B_1A_2$ and $m = 2$, we obtain the result. □

Corollary 3.1.17 *Let A and B be positive semidefinite matrices and $p, q > 0$. Then*

$$\operatorname{tr}\left((A^pB^q + B^pA^q)^2\right) \leq \operatorname{tr}\left((A^pB^q + B^pA^q)(A^qB^p + B^qA^p)\right).$$

Applying an argument similar to that of the proof of Theorem 3.1.16, we can show the following trace inequalities.

Theorem 3.1.18 *Let A_1, A_2, B_1, B_2 be positive semidefinite matrices with $A_1A_2 = A_2A_1$ and $B_1B_2 = B_2B_1$. Then*

$$\operatorname{tr}\left((A_1B_1 + B_2A_2)^2\right) \leq \operatorname{tr}\left((A_1B_1 + B_2A_2)(B_1A_1 + A_2B_2)\right).$$

Corollary 3.1.19 *Let A and B be positive semidefinite matrices and $p, q > 0$. Then*

$$\operatorname{tr}\left((A^pB^p + B^qA^q)^2\right) \leq \operatorname{tr}\left((A^pB^p + B^qA^q)(B^pA^p + A^qB^q)\right).$$

3.2 Lieb–Thirring Trace Inequalities and the Second Main Question

In his investigation of matrix subadditivity inequalities, Bourin [10] posed the following question for positive semidefinite matrices A and B and unitarily invariant norms $|||\cdot|||$.

Question 3.2.1 Let A and B be positive semidefinite matrices and $p, q > 0$. Is it true that

$$|||A^pB^q + B^pA^q||| \leq |||A^{p+q} + B^{p+q}|||?$$

We refer to this inquiry as the *second main question of Bourin*.

In what follows, we apply the celebrated Heinz inequality. Let $0 \leq \nu \leq 1$, and let a and b be two nonnegative real numbers. The *Heinz means* denoted as $H_\nu(a, b)$ are defined as follows:

$$H_\nu(a,b) = \frac{a^\nu b^{1-\nu} + a^{1-\nu} b^\nu}{2} \qquad (\nu \in [0,1]).$$

It can be observed that $H_0(a,b) = H_1(a,b) = \frac{a+b}{2}$, and $H_{1/2}(a,b) = \sqrt{ab}$.

The arithmetic–geometric mean inequality for two nonnegative real numbers admits a refinement. This refinement states that the Heinz means interpolate between the geometric mean and arithmetic mean. In other words, $H_{1/2}(a,b) \le H_\nu(a,b) \le H_0(a,b)$ for $0 \le \nu \le 1$. More precisely, if a and b are nonnegative real numbers, then

$$\sqrt{ab} \le \frac{a^t b^{1-t} + a^{1-t} b^t}{2} \le \frac{a+b}{2} \tag{3.2.1}$$

for all $t \in [0,1]$.

The noncommutative version of the scalar inequalities in (3.2.1) is credited to Bhatia and Davis [5]. It asserts that for positive semidefinite matrices A and B, for any matrix X, and for every unitarily invariant norm $|||\cdot|||$, the double inequality

$$2\left|\left|\left|A^{\frac{1}{2}} X B^{\frac{1}{2}}\right|\right|\right| \le \left|\left|\left|A^\nu X B^{1-\nu} + A^{1-\nu} X B^\nu\right|\right|\right| \le |||AX + XB||| \tag{3.2.2}$$

holds for $\nu \in [0,1]$. The second part of the inequality is called the *Heinz inequality*; see [11, 12]. For the operator norm, McIntosh [13] deduced the Heinz inequality from the arithmetic–geometric mean inequality (2.4.11). Fujii et al. [14] proved that the Heinz inequality is equivalent to the *Corach–Porta–Recht inequality* $|||AXA^{-1} + A^{-1}XA||| \ge 2\,|||X|||$ in which A is a Hermitian invertible matrix and X is a Hermitian matrix; see also [15, 16]. Audenaert [17] showed that if $A, B \in \mathbb{M}_n$ are positive semidefinite and $0 \le \nu \le 1$, then

$$s_j(A^\nu B^{1-\nu} + A^{1-\nu} B^\nu) \le s_j(A+B)$$

for $j = 1, \ldots, n$. See [18, 19] for other refinements of the Heinz inequality.

It follows from the Heinz inequality in (3.2.2) that for $A, B \ge 0$ and $p, q > 0$, and for every unitarily invariant norm,

$$\left|\left|\left|A^p B^q + A^q B^p\right|\right|\right| \le \left|\left|\left|A^{p+q} + B^{p+q}\right|\right|\right|, \tag{3.2.3}$$

which is closely related to Question 3.2.1.

To attempt solving the second main Question 3.2.1 by applying inequality (3.2.3), the following conjecture can be proposed; see [20].

Conjecture 3.2.2 Let A and B be positive semidefinite matrices and $p, q > 0$. Then, for every unitarily invariant norm,

$$\left|\left|\left|A^p B^q + B^p A^q\right|\right|\right| \le \left|\left|\left|A^p B^q + A^q B^p\right|\right|\right|. \tag{3.2.4}$$

More generally, the following stronger inequality for commuting positive semidefinite matrices is formulated in [20].

Conjecture 3.2.3 Let $A_1, A_2, B_1, B_2 \geq 0$ with $A_1 A_2 = A_2 A_1$ and $B_1 B_2 = B_2 B_1$. Then

$$|||A_1 B_2 + B_1 A_2||| \leq |||A_1 B_2 + A_2 B_1||| . \tag{3.2.5}$$

Inequalities (3.2.4) and (3.2.5) are equivalent in the case of the Hilbert–Schmidt norm to the following Lieb–Thirring-type trace inequalities

$$\operatorname{tr} \operatorname{Re} \left(A^p B^q A^q B^p\right) \leq \operatorname{tr} \left(A^{p+q} B^{p+q}\right) \tag{3.2.6}$$

and

$$\operatorname{tr} \operatorname{Re} \left(A_1 B_2 A_2 B_1\right) \leq \operatorname{tr} \left(A_1 A_2 B_1 B_2\right) . \tag{3.2.7}$$

Recall that for all matrices X, Y, $\|X\|_2 = (\operatorname{tr} (X^* X))^{\frac{1}{2}}$ and $\operatorname{tr} (XY) = \operatorname{tr} (YX)$. Moreover, if $X, Y \geq 0$, then $\operatorname{tr}(XY) \geq 0$.

In this section, we prove inequality (3.2.6) for $p = 1, 2, 3$ and $q = 1$. This, together with inequality (3.2.3), settles the second main Question 3.2.1 affirmatively for $p = 1, 2, 3$ and $q = 1$ in the case of the Hilbert–Schmidt norm.

An important special case of inequality (3.2.7) is the following trace inequality:

$$\operatorname{tr} \operatorname{Re} \left(A^s B^n A^t B^m\right) \leq \operatorname{tr} \left(A^{s+t} B^{n+m}\right) , \tag{3.2.8}$$

where $A, B \geq 0$ and $n, m, s, t \in \mathbb{N}$.

To prove our main results, we apply the following useful trace inequality (see, for example, [8, p. 279]).

Lemma 3.2.4 (Lieb–Thirring inequality) *Let A and B be positive semidefinite matrices, and let $m \in \mathbb{N}$. Then*

$$\operatorname{tr} \left((AB)^m\right) \leq \operatorname{tr} \left(A^m B^m\right) .$$

The famous Lieb–Thirring inequality and its generalizations have numerous applications in mathematics and mathematical physics; see [21–24], and references therein. The case of Hermitian matrices A and B and $m = 2$ has been treated in Theorem 2.1.12.

We consider the following particular cases of inequality (3.2.8).

Case 1:

$$\operatorname{tr} \operatorname{Re} \left(A^n B^m A^n B^m\right) \leq \operatorname{tr} \left(A^{2n} B^{2m}\right) , \tag{3.2.9}$$

which holds by Lemma 3.2.4.

Case 2:

$$\operatorname{tr} \left(A B^n A B^m\right) \leq \operatorname{tr} \left(A^2 B^{n+m}\right) . \tag{3.2.10}$$

We prove (3.2.10) using number theory tools and an algorithmic approach.

Inequality (3.2.10) is equivalent to the following norm inequality:

$$\left\|AB^m + B^n A\right\|_2 \leq \left\|AB^m + AB^n\right\|_2, \tag{3.2.11}$$

which corresponds to the case of $A_1 = A_2 = A$, $B_1 = B^n$, and $B_2 = B^m$ in (3.2.5).

Case 3:

$$\operatorname{tr} \operatorname{Re}\left(A^n B^n A^m B^m\right) \leq \operatorname{tr}\left(A^{n+m} B^{n+m}\right). \tag{3.2.12}$$

The following trace inequality is a special case of inequality (3.2.12) with $m = 1$

$$\operatorname{tr} \operatorname{Re}\left(A^n B^n AB\right) \leq \operatorname{tr}\left(A^{n+1} B^{n+1}\right). \tag{3.2.13}$$

Note that for $n = 1$, inequality (3.2.13) is the following special case of Lemma 3.2.4.

$$\operatorname{tr}\left((AB)^2\right) \leq \operatorname{tr}\left(A^2 B^2\right). \tag{3.2.14}$$

In Theorems 3.2.5 and 3.2.8, inequality (3.2.13) is proved for the case of $n = 2, 3$, that is,

$$\operatorname{tr} \operatorname{Re}\left(A^2 B^2 AB\right) \leq \operatorname{tr}\left(A^3 B^3\right) \tag{3.2.15}$$

and

$$\operatorname{tr} \operatorname{Re}\left(A^3 B^3 AB\right) \leq \operatorname{tr}\left(A^4 B^4\right). \tag{3.2.16}$$

Now, we answer the second main Question 3.2.1 in the case of the Hilbert–Schmidt norm when $p = 1, 2, 3$ and $q = 1$.

Actually, we prove Conjecture 3.2.2 in the case of the Hilbert–Schmidt norm when $p = 1, 2, 3$ and $q = 1$. That is,

$$\|AB + BA\|_2 \leq \|AB + AB\|_2, \tag{3.2.17}$$

$$\left\|A^2 B + B^2 A\right\|_2 \leq \left\|A^2 B + AB^2\right\|_2, \tag{3.2.18}$$

and

$$\left\|A^3 B + B^3 A\right\|_2 \leq \left\|A^3 B + AB^3\right\|_2. \tag{3.2.19}$$

These inequalities, in virtue of inequality (3.2.3), give a partial answer to the second main Question 3.2.1 when $p = 1, 2, 3$ and $q = 1$:

$$\|AB + BA\|_2 \leq \left\|A^2 + B^2\right\|_2, \tag{3.2.20}$$

$$\left\|A^2 B + B^2 A\right\|_2 \leq \|A^3 + B^3\|_2, \tag{3.2.21}$$

and

$$\left\|A^3B+B^3A\right\|_2 \le \left\|A^4+B^4\right\|_2. \tag{3.2.22}$$

However, the second main Question 3.2.1 for all $p, q > 0$ remains open.

Inequality (3.2.20) is well known. It follows from arithmetic–geometric mean inequality (2.4.11).

In fact,

$$\begin{aligned}\left|\!\left|\!\left|AB+BA\right|\!\right|\!\right| &= 2\left|\!\left|\!\left|\mathrm{Re}\,(AB)\right|\!\right|\!\right|\\ &\le 2\left|\!\left|\!\left|AB\right|\!\right|\!\right|\\ &\le \left|\!\left|\!\left|A^2+B^2\right|\!\right|\!\right| \text{ (by inequality (2.4.11)).}\end{aligned}$$

Theorem 3.2.5 *Let A and B be positive semidefinite matrices. Then*

$$\left\|A^2B+B^2A\right\|_2 \le \left\|A^2B+AB^2\right\|_2.$$

Proof Let $C = B^{\frac{3}{2}}A - B^{\frac{1}{2}}AB$. Since $C^*C \ge 0$, it follows that $\mathrm{tr}\,(AC^*C) \ge 0$. Note that

$$\begin{aligned}&\mathrm{tr}\left(A\left(AB^{\frac{3}{2}} - BAB^{\frac{1}{2}}\right)\left(B^{\frac{3}{2}}A - B^{\frac{1}{2}}AB\right)\right) \ge 0\\ &\iff \mathrm{tr}\left(A^3B^3\right) + \mathrm{tr}\left((AB)^3\right) \ge \mathrm{tr}\left(BAB^2A^2 + A^2B^2AB\right)\\ &\iff \mathrm{tr}\left(A^3B^3\right) + \mathrm{tr}\left((AB)^3\right) \ge 2\,\mathrm{tr}\left(\mathrm{Re}\left(A^2B^2AB\right)\right).\end{aligned}$$

But, by Lemma 3.2.4, we have $\mathrm{tr}\left(A^3B^3\right) \ge \mathrm{tr}\left((AB)^3\right)$. Thus, $\mathrm{tr}\left(A^3B^3\right) \ge \mathrm{tr}\left(\mathrm{Re}\left(A^2B^2AB\right)\right)$, which is equivalent to inequality (3.2.18). □

Corollary 3.2.6 *Let A and B be positive semidefinite matrices. Then*

$$\left\|A^2B+B^2A\right\|_2 \le \left\|A^3+B^3\right\|_2.$$

Lemma 3.2.7 *Let A and B be positive semidefinite matrices. Then*

$$\mathrm{tr}\left(B^2(ABA)^2\right) \le \mathrm{tr}\left(A^4B^4\right).$$

Proof Let $C = AB^2A - BA^2B$. Then $C^* = C$, and so $C^2 \ge 0$. Note that

$$\mathrm{tr}\left(\left(AB^2A - BA^2B\right)^2\right) \ge 0 \iff \mathrm{tr}\left(\left(A^2B^2\right)^2\right) \ge \mathrm{tr}\left(B^2(ABA)^2\right).$$

Thus, by Lemma 3.2.4, we have $\mathrm{tr}\left(A^4B^4\right) \ge \mathrm{tr}\left(B^2(ABA)^2\right)$. □

Theorem 3.2.8 *Let A and B be positive semidefinite matrices. Then*

$$\left\|A^3B+B^3A\right\|_2 \le \left\|A^3B+AB^3\right\|_2.$$

Proof Let $C = B^2A - BAB$. Since $C^*C \geq 0$, it follows that $\operatorname{tr}\left(A^2C^*C\right) \geq 0$. Note that

$$\begin{aligned}
&\operatorname{tr}\left(A^2\left(AB^2 - BAB\right)\left(B^2A - BAB\right)\right) \geq 0 \\
&\iff \operatorname{tr}\left(A^4B^4\right) + \operatorname{tr}\left(BAB^2ABA^2\right) \geq \operatorname{tr}\left(A^3B^3AB + BAB^3A^3\right) \\
&\iff \operatorname{tr}\left(A^4B^4\right) + \operatorname{tr}\left(B^2\left(ABA\right)^2\right) \geq 2 \operatorname{tr} \operatorname{Re}\left(A^3B^3AB\right).
\end{aligned}$$

Moreover, by Lemma 3.2.7, we have $\operatorname{tr}\left(B^2\left(ABA\right)^2\right) \leq \operatorname{tr}\left(A^4B^4\right)$, and so

$$\operatorname{tr}\left(A^4B^4\right) \geq \operatorname{tr} \operatorname{Re}\left(A^3B^3AB\right),$$

which is equivalent to inequality (3.2.19). □

Corollary 3.2.9 *Let A and B be positive semidefinite matrices. Then*

$$\left\|A^3B + B^3A\right\|_2 \leq \left\|A^4 + B^4\right\|_2.$$

The following discussion is devoted to proving a special case of Theorem 3.2.16 by utilizing certain number theory tools. The proof proceeds algorithmically.

For any $a, b, c \in \mathbb{Z}$, recall that
(1) $a \equiv b \pmod{c}$ means that $a = b + kc$ for some $k \in \mathbb{Z}$;
(2) $a|b$ means that a divides b;
(3) $a|b \iff b \equiv 0 \pmod{a}$;
(4) $a|bc \iff \frac{a}{(a,c)}|b$, where (a, c) is the greatest common divisor of a and c;
(5) a is a unit in $\mathbb{Z}_n \iff a^m \equiv 1 \pmod{n}$ for some $m \in \mathbb{Z}$.

The following remark is needed to prove inequality (3.2.10). Let $p, a_0 \in \mathbb{Z}$ be such that $0 \leq a_0 < p$, and let a_k be the sequence in $\mathbb{Z}$ defined as

$$a_{k+1} = \begin{cases} 2a_k & \text{if } 2a_k \leq p, \\ 2a_k - p & \text{if } 2a_k > p. \end{cases} \tag{3.2.23}$$

Remark 3.2.10 For all k, we have
(1) $a_{k+1} \equiv 2a_k \pmod{p}$;
(2) $a_{k+1} \equiv 2^k a_0 \pmod{p}$;
(3) $0 \leq a_k \leq p$.

Part (3) can be seen through induction as follows. First note that for $k = 0$, $0 \leq a_0 \leq p$.

Now let us assume that it is true for some k. To prove that it is also true for $k + 1$, we have the following two cases:
Case 1: If $a_k \leq \frac{p}{2}$, then $0 \leq a_{k+1} = 2a_k \leq p$.
Case 2: If $a_k > \frac{p}{2}$, then by the induction assumption, we have

$$0 \leq a_{k+1} = 2a_k - p \leq 2p - p = p.$$

Lemma 3.2.11 *Let $p, a_0 \in \mathbb{Z}$ be such that $0 \leq a_0 < p$, and let a_k be the sequence in $\mathbb{Z}$ as in (3.2.23). Then there exists an $s \neq 0$ such that*

$$a_s = a_0 \Longleftrightarrow \frac{p}{(p, a_0)} \text{ is an odd number.}$$

Proof We have

$$\begin{aligned}
a_s = a_0 \text{ for some } s \neq 0 &\Longleftrightarrow a_s \equiv a_0 \,(\text{mod} p) \text{ for some } s \neq 0 \text{ because } a_s \leq p \\
&\Longleftrightarrow 2^s a_0 \equiv a_0 \,(\text{mod} p) \\
&\Longleftrightarrow \left(2^s - 1\right) a_0 \equiv 0 \,(\text{mod} p) \\
&\Longleftrightarrow p | \left(2^s - 1\right) a_0 \\
&\Longleftrightarrow \frac{p}{(p, a_0)} | \left(2^s - 1\right) \\
&\Longleftrightarrow 2^s - 1 \equiv 0 \left(\text{mo d}\left(\frac{p}{(p, a_0)}\right)\right)
\end{aligned}$$

$$\begin{aligned}
&\Longleftrightarrow 2^s \equiv 1 \left(\text{mod } \left(\frac{p}{(p, a_0)}\right)\right) \\
&\Longleftrightarrow 2 \text{is unit in } \mathbb{Z}_{\frac{p}{(p,a_0)}} \\
&\Longleftrightarrow \left(2, \frac{p}{(p, a_0)}\right) = 1 \\
&\Longleftrightarrow \frac{p}{(p, a_0)} \text{ is an odd number.}
\end{aligned}$$

□

Theorem 3.2.12 *Let A and B be positive semidefinite matrices. Then*

$$2 \text{ tr}\left(AB^n AB^m\right) \leq \text{tr}\left(A^2 B^{n+m}\right) + \text{tr}\left(AB^{2n} AB^{m-n}\right),$$

where $m, n \in \mathbb{N}$ and $m > n$.

Proof Let $C = AB^n - B^n A$. Then $C^* = -C$, and so $C^2 \leq 0$ and $\text{tr}\left(C^2 B^{m-n}\right) \leq 0$. Note that

$$\begin{aligned}
&\text{tr}((AB^n - B^n A)^2 B^{m-n}) \leq 0 \\
&\Longleftrightarrow \text{tr}\left(\left(AB^n AB^n + B^n AB^n A - AB^{2n} A - B^n A^2 B^n\right) B^{m-n}\right) \leq 0 \\
&\Longleftrightarrow \text{tr}\left(AB^n AB^m + B^n AB^n AB^{m-n} - AB^{2n} AB^{m-n} - B^n A^2 B^m\right) \leq 0 \\
&\Longleftrightarrow 2 \text{ tr}\left(AB^n AB^m\right) \leq \text{tr}\left(A^2 B^{n+m}\right) + \text{tr}\left(AB^{2n} AB^{m-n}\right).
\end{aligned}$$

□

Corollary 3.2.13 *Let A and B be positive semidefinite matrices. If $p, a_0 \in \mathbb{Z}$ are such that $0 \leq a_0 < p$ and a_k is the sequence as in (3.2.23), then*

$$2 \operatorname{tr}\left(AB^{a_k}AB^{p-a_k}\right) \leq \operatorname{tr}\left(A^2B^p\right) + \operatorname{tr}\left(AB^{a_{k+1}}AB^{p-a_{k+1}}\right).$$

Proof **Case 1:** If $2a_k \leq p$, then by replacing n with a_k and m with $p - a_k$ in Theorem 3.2.12, we obtain $n + m = p$, $2n = 2a_k = a_{k+1}$, and $m - n = p - a_k - a_k = p - 2a_k = p - a_{k+1}$, and the result follows.

Case 2: If $2a_k > p$, then by replacing n with $p - a_k$ and m with a_k in Theorem 3.2.12, we achieve $n + m = p$, $2n = 2p - 2a_k = p + (p - 2a_k) = p - a_{k+1}$, and $m - n = a_k - p + a_k = 2a_k - p = a_{k+1}$, and the result follows. □

Theorem 3.2.14 *Let A and B be positive semidefinite matrices. If $p, a_0 \in \mathbb{Z}$ are such that $0 \leq a_0 < p$ and a_k is defined as in (3.2.23) and $s \in \mathbb{Z}$, then*

$$2^s \operatorname{tr}\left(AB^{a_0}AB^{p-a_0}\right) \leq \left(\sum_{i=1}^{k} 2^{s-i}\right) \operatorname{tr}\left(A^2B^p\right) + 2^{s-k} \operatorname{tr}\left(AB^{a_k}AB^{p-a_k}\right)$$

for $k = 1, \ldots, s$.

Proof First, note that the result is true for $k = 1$ by Corollary 3.2.13. Now assume that it is true for some k, that is,

$$2^s \operatorname{tr}\left(AB^{a_0}AB^{p-a_0}\right) \leq \left(\sum_{i=1}^{k} 2^{s-i}\right) \operatorname{tr}\left(A^2B^p\right) + 2^{s-k} \operatorname{tr}\left(AB^{a_k}AB^{p-a_k}\right).$$

By induction assumption and Corollary 3.2.13, we have

$$\begin{aligned}
&2^s \operatorname{tr}\left(AB^{a_0}AB^{p-a_0}\right)\\
&\quad\leq \left(\sum_{i=1}^{k} 2^{s-i}\right) \operatorname{tr}\left(A^2B^p\right) + 2^{s-k-1} \operatorname{tr}\left(A^2B^p\right) + 2^{s-k-1} \operatorname{tr}\left(AB^{a_{k+1}}AB^{p-a_{k+1}}\right)\\
&\quad= \left(\sum_{i=1}^{k+1} 2^{s-i}\right) \operatorname{tr}\left(A^2B^p\right) + 2^{s-k-1} \operatorname{tr}\left(AB^{a_{k+1}}AB^{p-a_{k+1}}\right).
\end{aligned}$$

This completes the induction. □

Lemma 3.2.15 *Let A and B be positive semidefinite matrices. Then*

$$\operatorname{tr}\left(AB^nAB^m\right) \leq \operatorname{tr}\left(A^2B^{n+m}\right),$$

where $m, n \in \mathbb{N}$ and $n + m$ is an odd number.

Proof Let $a_0 = n$, $p = n + m$, and a_k be the sequence as in (3.2.23). Since $p = n + m$ is an odd number, it follows that $\frac{p}{(p,a_0)}$ is an odd number, and so there exists a number $s \neq 0$ such that $a_s = a_0$.

In Theorem 3.2.14, if we take $k = s$, then

$$2^s \operatorname{tr}\left(AB^{a_0}AB^{p-a_0}\right) \le \left(\sum_{i=1}^{s} 2^{s-i}\right) \operatorname{tr}\left(A^2B^p\right) + \operatorname{tr}\left(AB^{a_s}AB^{p-a_s}\right).$$

Since $a_s = a_0$, it follows that

$$\left(2^s - 1\right) \operatorname{tr}\left(AB^{a_0}AB^{p-a_0}\right) \le \left(\sum_{i=1}^{s} 2^{s-i}\right) \operatorname{tr}\left(A^2B^p\right).$$

Note that $2^s - 1 = \sum_{i=1}^{s} 2^{s-i}$. Thus,

$$\operatorname{tr}\left(AB^nAB^m\right) \le \operatorname{tr}\left(A^2B^{n+m}\right).$$

□

Theorem 3.2.16 *Let A and B be positive semidefinite matrices. Then*

$$\operatorname{tr}\left(AB^nAB^m\right) \le \operatorname{tr}\left(A^2B^{n+m}\right), \tag{3.2.24}$$

where $m, n \in \mathbb{N}$.

Proof We prove the theorem by induction on $n + m$. If $n + m = 2$, then $m = n = 1$, and in this case, inequality (3.2.24) is true by Lemma 3.2.4. Assume that inequality (3.2.24) holds for $n + m \le k$ for some $k \ge 2$. Then we have to show that inequality (3.2.24) is true for $n + m = k + 1$.

Let $\nu, \mu \in \mathbb{N}$ be such that $\nu + \mu = k + 1$, and let $C, D \ge 0$. Then we have to show that

$$\operatorname{tr}\left(CD^\nu CD^\mu\right) \le \operatorname{tr}\left(C^2D^{\nu+\mu}\right). \tag{3.2.25}$$

Case 1: If exactly one of ν, μ is even and the other is odd, then (3.2.25) follows by Lemma 3.2.15.

Case 2: If ν, μ are both odd or even, then we can assume that $\mu > \nu$ because

$$\operatorname{tr}\left(CD^\nu CD^\mu\right) = \operatorname{tr}\left(CD^\mu CD^\nu\right).$$

It follows from Theorem 3.2.12 that $2\operatorname{tr}\left(CD^\nu CD^\mu\right) \le \operatorname{tr}\left(CD^{2\nu}CD^{\mu-\nu}\right) + \operatorname{tr}\left(C^2D^{\nu+\mu}\right)$, and so it suffices to prove that $\operatorname{tr}\left(CD^{2\nu}CD^{\mu-\nu}\right) \le \operatorname{tr}\left(C^2D^{\nu+\mu}\right)$. In fact, this follows from the inductive assumption with $A = C$, $B = D^2$, $n = \nu$, $m = \frac{\mu-\nu}{2}$, where $n + m = \frac{\mu+\upsilon}{2} = \frac{k+1}{2} \le k$, and so the proof of inequality (3.2.24) is completed. □

Now, we give the following alternative proof of Theorem 3.2.16, which is valid for all $p, q > 0$. This proof is based on Hölder's inequality for the trace asserting that if A and B are positive semidefinite matrices and $p, q > 0$ with $\frac{1}{p} + \frac{1}{q} = 1$, then

$$\operatorname{tr}(AB) \leq \left(\operatorname{tr} A^p\right)^{1/p} \left(\operatorname{tr} B^q\right)^{1/p} \tag{3.2.26}$$

(see Problem 2.5.29), and on the Araki–Lieb–Thirring trace inequality, which says that if $A, B \geq 0$ and $p > 0, q \geq 1$, then

$$\operatorname{tr}\left(\left(B^{\frac{1}{2}} A B^{\frac{1}{2}}\right)^{pq}\right) \leq \operatorname{tr}\left(\left(B^{q/2} A^q B^{q/2}\right)^p\right) \tag{3.2.27}$$

(see [21]). This is a consequence of the Araki's log-majorization (Theorem 6.2.4) explained later in Chap. 6.

Now, if A and B are positive semidefinite matrices and $p, q > 0$, then by inequality (3.2.26), we have

$$\begin{aligned} \operatorname{tr}\left(AB^pAB^q\right) \leq & \left(\operatorname{tr}\left(A^{p/(p+q)} B^p A^{p/(p+q)}\right)^{(p+q)/p}\right)^{p/(p+q)} \\ & \times \left(\operatorname{tr}\left(A^{q/(p+q)} B^q A^{q/(p+q)}\right)^{(p+q)/q}\right)^{q/(p+q)}. \end{aligned} \tag{3.2.28}$$

Since $\frac{p+q}{p}, \frac{p+q}{q} > 1$, it follows by inequality (3.2.27) that

$$\left(\operatorname{tr}\left(A^{p/(p+q)} B^p A^{p/(p+q)}\right)\right)^{(p+q)/p} \leq \operatorname{tr}\left(A^2 B^{p+q}\right) \tag{3.2.29}$$

and

$$\left(\operatorname{tr}\left(A^{q/(p+q)} B^q A^{q/(p+q)}\right)\right)^{(p+q)/q} \leq \operatorname{tr}\left(A^2 B^{p+q}\right). \tag{3.2.30}$$

Now, inequalities (3.2.28), (3.2.29), and (3.2.30) yield the desired inequality

$$\operatorname{tr}\left(AB^pAB^q\right) \leq \operatorname{tr}\left(A^2 B^{p+q}\right). \tag{3.2.31}$$

Inequality (3.2.31) is equivalent to the norm inequality

$$\left\|AB^q + B^pA\right\|_2 \leq \left\|AB^q + AB^p\right\|_2. \tag{3.2.32}$$

It should be mentioned here that inequality (3.2.31) can also be inferred from the Chebyshev-type inequality

$$\operatorname{tr}\left(Z^* f(C) Z g(C)\right) \leq \operatorname{tr}\left(Z^* Z f(C) g(C)\right)$$

for every normal matrix Z, any positive semidefinite matrix C, and all pairs of nondecreasing functions $f, g : [0, \infty) \to \mathbb{R}$ (see [25]).

3.3 Ando–Hiai–Okubo Trace Inequalities

In [20], the authors investigate the Lieb–Thirring trace inequalities and make a conjecture. The conjecture states that if A_1, A_2, B_1, B_2 are positive semidefinite matrices with $A_1A_2 = A_2A_1$ and $B_1B_2 = B_2B_1$, then for every unitarily invariant norm,

$$|||A_1B_2 + B_1A_2||| \leq |||A_1B_2 + A_2B_1||| .$$

An important special case of this inequality is the inequality

$$\left|\left|\left|A^s B^p + B^q A^t\right|\right|\right| \leq \left|\left|\left|A^s B^p + A^t B^q\right|\right|\right|, \tag{3.3.1}$$

where A and B are positive semidefinite matrices and s, t, p, q are positive real numbers.

We now investigate the Hilbert–Schmidt norm version of (3.3.1), that is, the inequality

$$\left\|A^s B^p + B^q A^t\right\|_2 \leq \left\|A^s B^p + A^t B^q\right\|_2 . \tag{3.3.2}$$

In fact, we prove inequality (3.3.2) under the following condition:

$$\left|\frac{s}{s+t} - \frac{1}{2}\right| + \left|\frac{p}{p+q} - \frac{1}{2}\right| \leq \frac{1}{2}. \tag{3.3.3}$$

A special case of inequality (3.3.2) when $s = q, t = p$ is the inequality

$$\left\|A^q B^p + B^q A^p\right\|_2 \leq \left\|A^q B^p + A^p B^q\right\|_2 , \tag{3.3.4}$$

that was proved by Bhatia [26] under the condition that

$$\frac{1}{4} \leq \frac{p}{p+q} \leq \frac{3}{4}. \tag{3.3.5}$$

This is a significant improvement of the cases $q = 1, 2$ or 3 and $p = 1$ mentioned in the previous section.

In Theorem 3.3.3, we prove that inequality (3.3.2) holds under the condition given in (3.3.3). This is a generalization of the result given by Bhatia in [26], where the particular inequality (3.3.4) is proved under the condition given in (3.3.5).

Another special case of (3.3.2) when $s = t = 1$ is the inequality

$$\left\|AB^p + B^q A\right\|_2 \leq \left\|AB^p + AB^q\right\|_2 . \tag{3.3.6}$$

We clarify that inequality (3.3.6) can also be concluded from Theorem 2.2 in [25]. In [20], it is proved (3.3.6) by applying some number theory tools in an algorithmic way.

Remark 3.3.1 By taking $p = 1$, $q = 3$, $s = 3$, and $t = 2$ in inequality (3.3.2), we have the following special case in view of the condition given in (3.3.3)

$$\left\| A^3 B + B^3 A^2 \right\|_2 \le \left\| A^3 B + A^2 B^3 \right\|_2,$$

which does not follow from inequalities (3.3.4) and (3.3.6). This demonstrates the power of the four parameters' inequality (3.3.2) under the condition given in (3.3.3).

Now, we discuss the second main Question 3.2.1 for the Hilbert–Schmidt norm. To do this, we need the following trace inequality due to Ando–Hiai–Okubo [27]. This lemma played a crucial role in the proof of the main result in [26]. For more details on the results of this section, we refer to [20, 28, 29].

Lemma 3.3.2 *Let A and B be positive semidefinite matrices, and let $\mu, \nu \in [0, 1]$ be such that $\left|\mu - \frac{1}{2}\right| + \left|\nu - \frac{1}{2}\right| \le \frac{1}{2}$. Then*

$$\left|\operatorname{tr}\left(A^{\mu} B^{\nu} A^{1-\mu} B^{1-\nu}\right)\right| \le \operatorname{tr}(AB).$$

Our main result of this section can be stated as follows.

Theorem 3.3.3 *Let A and B be positive semidefinite matrices, and let s, t, p, q be positive real numbers such that $\left|\frac{s}{s+t} - \frac{1}{2}\right| + \left|\frac{p}{p+q} - \frac{1}{2}\right| \le \frac{1}{2}$. Then*

$$\left\| A^s B^p + B^q A^t \right\|_2 \le \left\| A^s B^p + A^t B^q \right\|_2. \tag{3.3.7}$$

Proof We have

$$\left\| A^s B^p + B^q A^t \right\|_2^2 = \operatorname{tr}\left(A^{2s} B^{2p} + A^s B^q A^t B^p + A^t B^q A^s B^p + A^{2t} B^{2q}\right)$$

and

$$\left\| A^s B^p + A^t B^q \right\|_2^2 = \operatorname{tr}\left(A^{2s} B^{2p} + 2A^{s+t} B^{p+q} + A^{2t} B^{2q}\right).$$

Here, we have used the fact that $\|X\|_2 = (\operatorname{tr}(X^* X))^{\frac{1}{2}}$ for any matrix X as well as the cyclicity of the trace.

Therefore, inequality (3.3.7) is equivalent to the statement

$$\operatorname{Re}\operatorname{tr}\left(A^s B^p A^t B^q\right) \le \operatorname{tr}\left(A^{s+t} B^{p+q}\right). \tag{3.3.8}$$

Replacing A and B with $A^{\frac{1}{s+t}}$ and $B^{\frac{1}{p+q}}$, respectively, we see that this is equivalent to

$$\operatorname{Re}\operatorname{tr}\left(A^{\mu} B^{\nu} A^{1-\mu} B^{1-\nu}\right) \le \operatorname{tr}(AB) \tag{3.3.9}$$

for $\mu, \nu \in [0, 1]$.

By Lemma 3.3.2, inequality (3.3.9) holds provided

$$\left|\mu - \frac{1}{2}\right| + \left|\nu - \frac{1}{2}\right| \leq \frac{1}{2}.$$

Replacing μ and ν with $\frac{s}{s+t}$ and $\frac{p}{p+q}$ in (3.3.9), we see that inequality (3.3.8) is true if

$$\left|\frac{s}{s+t} - \frac{1}{2}\right| + \left|\frac{p}{p+q} - \frac{1}{2}\right| \leq \frac{1}{2}.$$

□

As mentioned in Lemma 3.3.2, and in their investigation of trace inequalities for multiple products of powers of two matrices, Ando, Hiai, and Okubo [27] proved that

$$\left|\operatorname{tr}\left(A^{\mu} B^{\nu} A^{1-\mu} B^{1-\nu}\right)\right| \leq \operatorname{tr}(AB), \tag{3.3.10}$$

where A and B are positive semidefinite matrices and μ, ν are positive real numbers for which

$$\left|\mu - \frac{1}{2}\right| + \left|\nu - \frac{1}{2}\right| \leq \frac{1}{2}. \tag{3.3.11}$$

An important special case of inequality (3.2.5) is the inequality

$$\left|\left|\left|A^{s} B^{p} + B^{q} A^{t}\right|\right|\right| \leq \left|\left|\left|A^{s} B^{p} + A^{t} B^{q}\right|\right|\right|,$$

where A and B are positive semidefinite matrices and s, t, p, q are positive real numbers. Replacing A and B with $A^{\frac{1}{s+t}}$ and $B^{\frac{1}{p+q}}$, respectively, we see that this inequality is equivalent to

$$\left|\left|\left|A^{\mu} B^{\nu} + B^{1-\nu} A^{1-\mu}\right|\right|\right| \leq \left|\left|\left|A^{\mu} B^{\nu} + A^{1-\mu} B^{1-\nu}\right|\right|\right| \tag{3.3.12}$$

for $\mu, \nu \in [0, 1]$.

The Hilbert–Schmidt norm version of (3.3.12) is the inequality

$$\left\|A^{\mu} B^{\nu} + B^{1-\nu} A^{1-\mu}\right\|_2 \leq \left\|A^{\mu} B^{\nu} + A^{1-\mu} B^{1-\nu}\right\|_2. \tag{3.3.13}$$

Inequality (3.3.13) was proved in [30] under condition (3.3.11). A special case of inequality (3.3.13) when $\nu = 1 - \mu$ is the inequality

$$\left\|A^{\mu} B^{1-\mu} + B^{\mu} A^{1-\mu}\right\|_2 \leq \left\|A^{\mu} B^{1-\mu} + A^{1-\mu} B^{\mu}\right\|_2. \tag{3.3.14}$$

In [26], Bhatia proved inequality (3.3.14) under the condition

$$\frac{1}{4} \leq \mu \leq \frac{3}{4}, \tag{3.3.15}$$

which is a significant improvement of an earlier result of [20].

In this section, applying complex interpolation related to the Hadamard three lines theorem, we generalize inequality (3.3.10) by the inequality

$$\left|\operatorname{tr}\left(A^{w} B^{z} A^{1-w} B^{1-z}\right)\right| \leq \operatorname{tr}(A B) \tag{3.3.16}$$

for all complex numbers w, z for which

$$\left|\operatorname{Re} w-\frac{1}{2}\right|+\left|\operatorname{Re} z-\frac{1}{2}\right| \leq \frac{1}{2}, \tag{3.3.17}$$

where for a positive definite matrix X, $X^{z}=\exp(z \log X)$. This generalization provides a new method for proving the Ando–Hiai–Okubo trace inequality [27] without resorting to the log-majorization technique or the use of antisymmetric tensor products.

A special case of inequality (3.3.16) when $w=z$ is the inequality

$$\left|\operatorname{tr}\left(A^{z} B^{z} A^{1-z} B^{1-z}\right)\right| \leq \operatorname{tr}(A B). \tag{3.3.18}$$

Recently, Bottazzi et al. [31] proved inequality (3.3.18) under the condition that

$$\frac{1}{4} \leq \operatorname{Re} z \leq \frac{3}{4}. \tag{3.3.19}$$

It is important to note that inequality (3.3.16) under condition (3.3.17) is a generalization of inequality (3.3.18) under condition (3.3.19).

Now, as an application of inequality (3.3.16) under condition (3.3.17), we can also generalize inequality (3.3.13) to complex values under condition (3.3.17). In fact, we prove that the inequality

$$\left\|A^{w} B^{z}+B^{1-\bar{z}} A^{1-\bar{w}}\right\|_{2} \leq\left\|A^{w} B^{z}+A^{1-\bar{w}} B^{1-\bar{z}}\right\|_{2} \tag{3.3.20}$$

holds for the complex numbers w, z under condition (3.3.17).

Bottazzi et al. [31] gave a counterexample to the following special case of inequality (3.2.5):

$$\left|\left|\left|A^{t} B^{1-t}+B^{t} A^{1-t}\right|\right|\right| \leq\left|\left|\left|A^{t} B^{1-t}+A^{1-t} B^{t}\right|\right|\right|, \tag{3.3.21}$$

where $0 \leq t \leq 1$. They answered it in the negative for the operator norm by giving a pair of positive semidefinite matrices such that the claim does not hold. Although inequality (3.3.21) is not true for the operator norm, in view of inequality (3.3.14), which is valid for the Hilbert–Schmidt norm under condition (3.3.15), it would be interesting to discuss inequality (3.3.21) for other unitarily invariant norms like the Schatten p-norms.

To prove inequality (3.3.20), we introduce some notations regarding vertical strips in the complex plane. Let

$$\mathcal{S} = \{z \in \mathbb{C} : 0 \leq \mathrm{Re}\, z \leq 1\},$$

$$\mathcal{S}_1 = \left\{z \in \mathbb{C} : 0 \leq \mathrm{Re}\, z \leq \frac{1}{2}\right\},$$

and

$$\mathcal{S}_2 = \left\{z \in \mathbb{C} : \frac{1}{2} \leq \mathrm{Re}\, z \leq 1\right\}.$$

For $0 \leq s \leq 1$, we define the following strips:

$$\begin{aligned}\mathcal{A}_s &= \left\{z \in \mathbb{C} : \frac{1}{2} - s \leq \mathrm{Re}\, z \leq \frac{1}{2} + s\right\} \\ &= \left\{z \in \mathbb{C} : \left|\mathrm{Re}\, z - \frac{1}{2}\right| \leq s\right\}\end{aligned}$$

and

$$\begin{aligned}\mathcal{B}_s &= \left\{z \in \mathbb{C} : s - \frac{1}{2} \leq \mathrm{Re}\, z \leq \frac{3}{2} - s\right\} \\ &= \left\{z \in \mathbb{C} : \left|\mathrm{Re}\, z - \frac{1}{2}\right| \leq 1 - s\right\}.\end{aligned}$$

The generalized Hölder inequality for the Schatten p-norm is frequently used in establishing the main results of this section. This inequality states that for every matrices X, Y, Z and every real numbers $p, q, r \geq 1$ with $\frac{1}{p} + \frac{1}{q} + \frac{1}{r} = 1$, it holds that

$$|\mathrm{tr}\,(XYZ)| \leq \|XYZ\|_1 \leq \|X\|_p \|Y\|_q \|Z\|_r \tag{3.3.22}$$

(see [32, Theorem 2.8]).

As an immediate consequence of the famous Araki–Lieb–Thirring trace inequality [21], we have the following lemma [31].

Lemma 3.3.4 *Let A and B be positive semidefinite matrices, and let $r \geq 2$. Then*

$$\left\| A^{\frac{1}{r}} B^{\frac{1}{r}} \right\|_r \leq (\mathrm{tr}\,(AB))^{\frac{1}{r}}.$$

The following lemma was proved in [33].

Lemma 3.3.5 *Let A and B be positive semidefinite matrices, and let X be any matrix. Then, for $0 \leq \nu \leq 1$ and for every unitarily invariant norm,*

$$\left|\left|\left| A^{\nu} X B^{1-\nu} \right|\right|\right| \leq |||AX|||^{\nu}\, |||XB|||^{1-\nu}.$$

Another useful lemma that we need is the following.

Lemma 3.3.6 *Let A and B be positive semidefinite matrices, and let $w, z \in \mathbb{C}$ with $s = \operatorname{Re} w$ and $r = \operatorname{Re} z$. If $0 \le r, s \le 1$, then*

$$\left|\operatorname{tr}\left(A^{w} B^{z} A^{1-w} B^{1-z}\right)\right| \le \min\left\{\|A\| \, \|B\|_{1}, \|A\|_{1} \, \|B\|\right\}.$$

Proof Without loss of generality, we can assume that A and B are invertible. The general case follows by a continuity argument.

Let $w = s + \mathrm{i}x$ and $z = r + \mathrm{i}y$ with $0 \le r, s \le 1$ and $x, y \in \mathbb{R}$. Then

$$\begin{aligned}
\left|\operatorname{tr}\left(A^{w} B^{z} A^{1-w} B^{1-z}\right)\right| &\le \left\|A^{w} B^{z} A^{1-w} B^{1-z}\right\|_{1} \text{ (by inequality (3.3.22))}\\
&\le \left\|A^{w}\right\| \, \left\|B^{z} A^{1-w} B^{1-z}\right\|_{1} \text{ (by inequality (2.3.4))}\\
&= \left\|A^{s}\right\| \, \left\|B^{r} A^{1-w} B^{1-r}\right\|_{1} \text{ (since } A^{\mathrm{i}x} \text{ and } B^{\mathrm{i}y} \text{ are unitary)}\\
&\le \|A\|^{s} \left\|B A^{1-w}\right\|_{1}^{r} \left\|A^{1-w} B\right\|_{1}^{1-r} \text{ (by Lemma 3.3.5)}\\
&= \|A\|^{s} \left\|B A^{1-s}\right\|_{1}^{r} \left\|A^{1-s} B\right\|_{1}^{1-r}\\
&\le \|A\|^{s} \|B\|_{1}^{r} \left\|A^{1-s}\right\|^{r} \left\|A^{1-s}\right\|^{1-r} \|B\|_{1}^{1-r} \text{ (by inequality (2.3.4))}\\
&= \|A\|^{s} \|B\|_{1}^{r} \|A\|^{(1-s)r} \|A\|^{(1-s)(1-r)} \|B\|_{1}^{1-r}\\
&= \|A\| \, \|B\|_{1}.
\end{aligned}$$

Similarly, it can be shown that

$$\left|\operatorname{tr}\left(A^{w} B^{z} A^{1-w} B^{1-z}\right)\right| \le \|A\|_{1} \, \|B\|.$$

Thus, $\left|\operatorname{tr}\left(A^{w} B^{z} A^{1-w} B^{1-z}\right)\right| \le \min\left\{\|A\| \, \|B\|_{1}, \|A\|_{1} \, \|B\|\right\}$. □

We are in a position to present our second main result of this section, which is applied in the proof of inequality (3.3.20). In the proof of this result, we make use of the Hadamard three lines theorem (see, for example, [34, p. 33] or [35, p. 387]).

Theorem 3.3.7 *Let A and B be positive semidefinite matrices, and let $w, z \in \mathbb{C}$ with $s = \operatorname{Re} w$. Then the following hold:*

(a) If $w \in \mathcal{S}_1$ and $z \in \mathcal{A}_s$, then

$$\left|\operatorname{tr}\left(A^{w} B^{z} A^{1-w} B^{1-z}\right)\right| \le \operatorname{tr}(AB).$$

(b) If $w \in \mathcal{S}_2$ and $z \in \mathcal{B}_s$, then

$$\left|\operatorname{tr}\left(A^{w} B^{z} A^{1-w} B^{1-z}\right)\right| \le \operatorname{tr}(AB).$$

Proof Without loss of generality, we can assume that A and B are invertible. The general case follows by a continuity argument. To prove part (a), let $w = s + \mathrm{i}x \in \mathcal{S}_1, x \in \mathbb{R}$. Then $0 \le s \le \frac{1}{2}$.

For $z=\frac{1}{2}-s+\mathrm{i}y$, applying inequality (3.3.22) for the partition $\frac{1}{p}+\frac{1}{q}+\frac{1}{r}=s+\left(\frac{1}{2}-s\right)+\frac{1}{2}=1$ and noting that with $p=\frac{1}{s}, q=\frac{1}{\frac{1}{2}-s}, r=2\geq 2$, we have

$$\begin{aligned}
&|\operatorname{tr}\left(A^{w}B^{z}A^{1-w}B^{1-z}\right)|\\
&=\left|\operatorname{tr}\left(A^{s+\mathrm{i}x}B^{\frac{1}{2}-s+\mathrm{i}y}A^{1-s-\mathrm{i}x}B^{\frac{1}{2}+s-\mathrm{i}y}\right)\right|\\
&=\left|\operatorname{tr}\left(A^{s}A^{\mathrm{i}x}B^{\mathrm{i}y}B^{\frac{1}{2}-s}A^{\frac{1}{2}-s}A^{-\mathrm{i}x}A^{\frac{1}{2}}B^{\frac{1}{2}}B^{-\mathrm{i}y}B^{s}\right)\right|\\
&=\left|\operatorname{tr}\left(B^{s}A^{s}A^{\mathrm{i}x}B^{\mathrm{i}y}B^{\frac{1}{2}-s}A^{\frac{1}{2}-s}A^{-\mathrm{i}x}A^{\frac{1}{2}}B^{\frac{1}{2}}B^{-\mathrm{i}y}\right)\right|\\
&\leq\left\|B^{s}A^{s}A^{\mathrm{i}x}B^{\mathrm{i}y}B^{\frac{1}{2}-s}A^{\frac{1}{2}-s}A^{-\mathrm{i}x}A^{\frac{1}{2}}B^{\frac{1}{2}}B^{-\mathrm{i}y}\right\|_{1}\\
&\leq\left\|B^{s}A^{s}A^{\mathrm{i}x}\right\|_{p}\left\|B^{\mathrm{i}y}B^{\frac{1}{2}-s}A^{\frac{1}{2}-s}A^{-\mathrm{i}x}\right\|_{q}\left\|A^{\frac{1}{2}}B^{\frac{1}{2}}B^{-\mathrm{i}y}\right\|_{r}\\
&=\left\|B^{s}A^{s}\right\|_{p}\left\|B^{\frac{1}{2}-s}A^{\frac{1}{2}-s}\right\|_{q}\left\|A^{\frac{1}{2}}B^{\frac{1}{2}}\right\|_{r}\quad(\text{since }A^{\mathrm{i}x}\text{ and }B^{\mathrm{i}y}\text{ are unitary})\\
&\leq(\operatorname{tr}(AB))^{s+\left(\frac{1}{2}-s\right)+\frac{1}{2}}\quad(\text{by Lemma 3.3.4})\\
&=\operatorname{tr}(AB).
\end{aligned}$$

For $z=\frac{1}{2}+s+\mathrm{i}y$, applying inequality (3.3.22) for the partition $\frac{1}{p}+\frac{1}{q}+\frac{1}{r}=s+\frac{1}{2}+\left(\frac{1}{2}-s\right)=1$ and noting that with $p=\frac{1}{s}, q=2, r=\frac{1}{\frac{1}{2}-s}\geq 2$, we have

$$\begin{aligned}
\left|\operatorname{tr}\left(A^{w}B^{z}A^{1-w}B^{1-z}\right)\right|&=\left|\operatorname{tr}\left(A^{s+\mathrm{i}x}B^{\frac{1}{2}+s+\mathrm{i}y}A^{1-s-\mathrm{i}x}B^{\frac{1}{2}-s-\mathrm{i}y}\right)\right|\\
&=\left|\operatorname{tr}\left(A^{\mathrm{i}x}A^{s}B^{s}B^{\mathrm{i}y}B^{\frac{1}{2}}A^{\frac{1}{2}}A^{-\mathrm{i}x}A^{\frac{1}{2}-s}B^{\frac{1}{2}-s}B^{-\mathrm{i}y}\right)\right|\\
&\leq\left\|A^{\mathrm{i}x}A^{s}B^{s}B^{\mathrm{i}y}B^{\frac{1}{2}}A^{\frac{1}{2}}A^{-\mathrm{i}x}A^{\frac{1}{2}-s}B^{\frac{1}{2}-s}B^{-\mathrm{i}y}\right\|_{1}\\
&\leq\left\|A^{\mathrm{i}x}A^{s}B^{s}B^{\mathrm{i}y}\right\|_{p}\left\|B^{\frac{1}{2}}A^{\frac{1}{2}}A^{-\mathrm{i}x}\right\|_{q}\left\|A^{\frac{1}{2}-s}B^{\frac{1}{2}-s}B^{-\mathrm{i}y}\right\|_{r}\\
&=\left\|A^{s}B^{s}\right\|_{p}\left\|B^{\frac{1}{2}}A^{\frac{1}{2}}\right\|_{q}\left\|A^{\frac{1}{2}-s}B^{\frac{1}{2}-s}\right\|_{r}\quad(\text{since }A^{\mathrm{i}x}\text{ and }B^{\mathrm{i}y}\text{ are unitary})\\
&\leq(\operatorname{tr}(AB))^{s+\frac{1}{2}+\left(\frac{1}{2}-s\right)}\quad(\text{by Lemma 3.3.4})\\
&=\operatorname{tr}(AB).
\end{aligned}$$

Since for a positive definite matrix X, the matrix-valued function $\psi(z)=X^{z}=\exp(z\log X)=\sum_{k=0}^{\infty}\frac{z^{k}(\log X)^{k}}{k!}$ is entire, it follows that the complex-valued function

$$\varphi(z)=\operatorname{tr}\left(A^{w}B^{z}A^{1-w}B^{1-z}\right)$$

is entire. Moreover, Lemma 3.3.6 shows that the function is bounded on the vertical strip $0\leq\operatorname{Re}z\leq 1$. Since $[\frac{1}{2}-s,\frac{1}{2}+s]\subseteq[0,1]$, it follows that the function is bounded on the vertical strip $\frac{1}{2}-s\leq\operatorname{Re}z\leq\frac{1}{2}+s$. Since $|\varphi(z)|\leq\operatorname{tr}(AB)$ on the edges $\operatorname{Re} z = \frac{1}{2} - s$ and $\operatorname{Re} z = \frac{1}{2} + s$, it follows by the Hadamard three lines theorem that $|\varphi(z)| \le \operatorname{tr}(AB)$ for all z in the vertical strip $\frac{1}{2} - s \le \operatorname{Re} z \le \frac{1}{2} + s$. This proves part (a).

To prove part (b), let $w = s + \mathrm{i}x \in \mathcal{S}_2$, $x \in \mathbb{R}$. Then $\frac{1}{2} \le s \le 1$. For $z = s - \frac{1}{2} + \mathrm{i}y$, applying inequality (3.3.22) for the partition $\frac{1}{p} + \frac{1}{q} + \frac{1}{r} = \left(s - \frac{1}{2}\right) + (1-s) + \frac{1}{2} = 1$ and noting that with $p = \frac{1}{s-\frac{1}{2}}, q = \frac{1}{1-s}, r = 2 \ge 2$, we have

$$\begin{aligned}
\left|\operatorname{tr}\left(A^w B^z A^{1-w} B^{1-z}\right)\right| &= \left|\operatorname{tr}\left(A^{s+\mathrm{i}x} B^{s-\frac{1}{2}+\mathrm{i}y} A^{1-s-\mathrm{i}x} B^{\frac{3}{2}-s-\mathrm{i}y}\right)\right| \\
&= \left|\operatorname{tr}\left(A^s A^{\mathrm{i}x} B^{s-\frac{1}{2}} B^{\mathrm{i}y} A^{1-s} A^{-\mathrm{i}x} B^{\frac{3}{2}-s} B^{-\mathrm{i}y}\right)\right| \\
&= \left|\operatorname{tr}\left(A^{\mathrm{i}x} A^{s-\frac{1}{2}} B^{s-\frac{1}{2}} B^{\mathrm{i}y} A^{-\mathrm{i}x} A^{1-s} B^{1-s} B^{-\mathrm{i}y} B^{\frac{1}{2}} A^{\frac{1}{2}}\right)\right| \\
&\le \left\|A^{\mathrm{i}x} A^{s-\frac{1}{2}} B^{s-\frac{1}{2}} B^{\mathrm{i}y} A^{-\mathrm{i}x} A^{1-s} B^{1-s} B^{-\mathrm{i}y} B^{\frac{1}{2}} A^{\frac{1}{2}}\right\|_1 \\
&\le \left\|A^{\mathrm{i}x} A^{s-\frac{1}{2}} B^{s-\frac{1}{2}} B^{\mathrm{i}y}\right\|_p \left\|A^{-\mathrm{i}x} A^{1-s} B^{1-s} B^{-\mathrm{i}y}\right\|_q \left\|B^{\frac{1}{2}} A^{\frac{1}{2}}\right\|_r \\
&= \left\|A^{s-\frac{1}{2}} B^{s-\frac{1}{2}}\right\|_p \left\|A^{1-s} B^{1-s}\right\|_q \left\|B^{\frac{1}{2}} A^{\frac{1}{2}}\right\|_r \\
&\le (\operatorname{tr}(AB))^{\left(s-\frac{1}{2}\right)+(1-s)+\frac{1}{2}} \quad \text{(by Lemma 3.3.4)} \\
&= \operatorname{tr}(AB).
\end{aligned}$$

For $z = \frac{3}{2} - s + \mathrm{i}y$, by inequality (3.3.22) for the partition $\frac{1}{p} + \frac{1}{q} + \frac{1}{r} = \frac{1}{2} + (1-s) + \left(s - \frac{1}{2}\right) = 1$ and by noting that with $p = 2, q = \frac{1}{1-s}, r = \frac{1}{s-\frac{1}{2}} \ge 2$, we have

$$\begin{aligned}
&\left|\operatorname{tr}\left(A^w B^z A^{1-w} B^{1-z}\right)\right| \\
&= \left|\operatorname{tr}\left(A^{s+\mathrm{i}x} B^{\frac{3}{2}-s+\mathrm{i}y} A^{1-s-\mathrm{i}x} B^{-\frac{1}{2}+s-\mathrm{i}y}\right)\right| \\
&= \left|\operatorname{tr}\left(A^s A^{\mathrm{i}x} B^{\frac{3}{2}-s} B^{\mathrm{i}y} A^{1-s} A^{-\mathrm{i}x} B^{s-\frac{1}{2}} B^{-\mathrm{i}y}\right)\right| \\
&= \left|\operatorname{tr}\left(A^{\mathrm{i}x} A^{\frac{1}{2}} B^{\frac{1}{2}} B^{\mathrm{i}y} B^{1-s} A^{1-s} A^{-\mathrm{i}x} B^{-\mathrm{i}y} B^{s-\frac{1}{2}} A^{s-\frac{1}{2}}\right)\right| \\
&\le \left\|A^{\mathrm{i}x} A^{\frac{1}{2}} B^{\frac{1}{2}} B^{\mathrm{i}y} B^{1-s} A^{1-s} A^{-\mathrm{i}x} B^{-\mathrm{i}y} B^{s-\frac{1}{2}} A^{s-\frac{1}{2}}\right\|_1 \\
&\le \left\|A^{\mathrm{i}x} A^{\frac{1}{2}} B^{\frac{1}{2}} B^{\mathrm{i}y}\right\|_p \left\|B^{1-s} A^{1-s} A^{-\mathrm{i}x}\right\|_q \left\|B^{-\mathrm{i}y} B^{s-\frac{1}{2}} A^{s-\frac{1}{2}}\right\|_r \\
&= \left\|A^{\frac{1}{2}} B^{\frac{1}{2}}\right\|_p \left\|B^{1-s} A^{1-s}\right\|_q \left\|B^{s-\frac{1}{2}} A^{s-\frac{1}{2}}\right\|_r \quad \text{(since } A^{\mathrm{i}x} \text{ and } B^{\mathrm{i}y} \text{ are unitary)} \\
&\le (\operatorname{tr}(AB))^{\frac{1}{2}+(1-s)+\left(s-\frac{1}{2}\right)} \quad \text{(by Lemma 3.3.4)} \\
&= \operatorname{tr}(AB).
\end{aligned}$$

Since the function

$$\varphi(z) = \operatorname{tr}\left(A^w B^z A^{1-w} B^{1-z}\right)$$

is bounded on the vertical strip $0 \le \operatorname{Re} z \le 1$, and since $[s - \frac{1}{2}, \frac{3}{2} - s] \subseteq [0, 1]$, it follows that the function is bounded on the vertical strip $s - \frac{1}{2} \le \operatorname{Re} z \le \frac{3}{2} - s$. Since $|\varphi(z)| \le \operatorname{tr}(AB)$ on the edges $\operatorname{Re} z = s - \frac{1}{2}$ and $\operatorname{Re} z = \frac{3}{2} - s$, it follows by the Hadamard three lines theorem that $|\varphi(z)| \le \operatorname{tr}(AB)$ for all z in the vertical strip $s - \frac{1}{2} \le \operatorname{Re} z \le \frac{3}{2} - s$. This proves part (b). □

To understand the conditions related to the strips given in Theorem 3.3.7 more clearly, we need the following lemma.

Lemma 3.3.8 *Let w, z be complex numbers with $s = \operatorname{Re} w$. Then*

$$\left|\operatorname{Re} w - \frac{1}{2}\right| + \left|\operatorname{Re} z - \frac{1}{2}\right| \le \frac{1}{2} \Longleftrightarrow (w \in \mathcal{S}_1 \text{ and } z \in \mathcal{A}_s) \text{ or } (w \in \mathcal{S}_2 \text{ and } z \in \mathcal{B}_s).$$

Proof ($\Longrightarrow$)

Assume that $\left|\operatorname{Re} w - \frac{1}{2}\right| + \left|\operatorname{Re} z - \frac{1}{2}\right| \le \frac{1}{2}$. Then $\left|\operatorname{Re} w - \frac{1}{2}\right| \le \frac{1}{2}$. Therefore, $0 \le \operatorname{Re} w \le 1$, and so we have the following two cases.

Case 1: If $w \in S_1$, then

$$\begin{aligned}\left|\operatorname{Re} w - \frac{1}{2}\right| + \left|\operatorname{Re} z - \frac{1}{2}\right| \le \frac{1}{2} &\Longleftrightarrow \frac{1}{2} - \operatorname{Re} w + \left|\operatorname{Re} z - \frac{1}{2}\right| \le \frac{1}{2} \\ &\Longleftrightarrow \left|\operatorname{Re} z - \frac{1}{2}\right| \le \operatorname{Re} w \\ &\Longleftrightarrow z \in \mathcal{A}_s.\end{aligned}$$

Case 2: If $w \in \mathcal{S}_2$, then

$$\begin{aligned}\left|\operatorname{Re} w - \frac{1}{2}\right| + \left|\operatorname{Re} z - \frac{1}{2}\right| \le \frac{1}{2} &\Longleftrightarrow \operatorname{Re} w - \frac{1}{2} + \left|\operatorname{Re} z - \frac{1}{2}\right| \le \frac{1}{2} \\ &\Longleftrightarrow \left|\operatorname{Re} z - \frac{1}{2}\right| \le 1 - \operatorname{Re} w \\ &\Longleftrightarrow z \in \mathcal{B}_s.\end{aligned}$$

($\Longleftarrow$)

Case 1: If $w \in \mathcal{S}_1$ and $z \in \mathcal{A}_s$, then

$$\begin{aligned}\left|\operatorname{Re} w - \frac{1}{2}\right| + \left|\operatorname{Re} z - \frac{1}{2}\right| &= \frac{1}{2} - \operatorname{Re} w + \left|\operatorname{Re} z - \frac{1}{2}\right| \\ &\le \frac{1}{2} - \operatorname{Re} w + \operatorname{Re} w \\ &= \frac{1}{2}.\end{aligned}$$

Case 2: If $w \in \mathcal{S}_2$ and $z \in \mathcal{B}_s$, then

$$\begin{aligned}\left|\text{Re } w - \frac{1}{2}\right| + \left|\text{Re } z - \frac{1}{2}\right| &= \text{Re } w - \frac{1}{2} + \left|\text{Re } z - \frac{1}{2}\right| \\ &\leq \text{Re } w - \frac{1}{2} + 1 - \text{Re } w \\ &= \frac{1}{2}.\end{aligned}$$

□

Now, by Lemma 3.3.8 and Theorem 3.3.7, we have the following corollary, which is a generalization of the Ando–Hiai–Okubo trace inequality [27].

Corollary 3.3.9 *Let A and B be positive semidefinite matrices, and let w, z be complex numbers such that*

$$\left|\text{Re } w - \frac{1}{2}\right| + \left|\text{Re } z - \frac{1}{2}\right| \leq \frac{1}{2}.$$

Then

$$\left|\text{tr}\left(A^w B^z A^{1-w} B^{1-z}\right)\right| \leq \text{tr}\,(AB).$$

As an application of Corollary 3.3.9, we obtain our third main result in this section, which is a generalization of norm inequality (3.3.13), under condition (3.3.11), to complex values. That is, we prove inequality (3.3.20) under condition (3.3.17).

Theorem 3.3.10 *Let A and B be positive semidefinite matrices, and let w, z be complex numbers such that*

$$\left|\text{Re } w - \frac{1}{2}\right| + \left|\text{Re } z - \frac{1}{2}\right| \leq \frac{1}{2}.$$

Then

$$\left\| A^w B^z + B^{1-\overline{z}} A^{1-\overline{w}} \right\|_2 \leq \left\| A^w B^z + A^{1-\overline{w}} B^{1-\overline{z}} \right\|_2.$$

Proof We can see that the square of the left-hand side of the desired norm inequality is equal to

$$\text{tr}\left(A^{w+\overline{w}} B^{z+\overline{z}} + B^{\overline{z}} A^{\overline{w}} B^{1-\overline{z}} A^{1-\overline{w}} + A^{1-w} B^{1-z} A^w B^z + A^{2-(w+\overline{w})} B^{2-(z+\overline{z})}\right)$$

and the square of the right-hand side is equal to

$$\text{tr}\left(A^{w+\overline{w}} B^{z+\overline{z}} + 2AB + A^{2-(w+\overline{w})} B^{2-(z+\overline{z})}\right).$$

Therefore, the norm inequality is equivalent to the statement

$$\text{Re tr}\left(A^w B^z A^{1-w} B^{1-z}\right) \le \text{tr}\,(AB). \tag{3.3.23}$$

By Corollary 3.3.9 and the fact that for every matrix X, Re (tr X) $\le |\text{tr}\,X|$, inequality (3.3.23) holds provided

$$\left|\text{Re}\,w - \frac{1}{2}\right| + \left|\text{Re}\,z - \frac{1}{2}\right| \le \frac{1}{2}.$$

□

As another application of (3.3.16) and (3.3.17), we obtain the following reverse-type inequality.

Theorem 3.3.11 *Let A and B be positive semidefinite matrices, and let w, z be complex numbers such that*

$$\left|\text{Re}\,w - \frac{1}{2}\right| + \left|\text{Re}\,z - \frac{1}{2}\right| \le \frac{1}{2}.$$

Then

$$\left\|A^w B^z - B^{1-\overline{z}} A^{1-\overline{w}}\right\|_2 \ge \left\|A^w B^z - A^{1-\overline{w}} B^{1-\overline{z}}\right\|_2.$$

Proof We can see that the square of the left-hand side of the desired norm inequality is equal to

$$\text{tr}\left(A^{w+\overline{w}} B^{z+\overline{z}} - 2\text{Re}\left(A^{1-w} B^{1-z} A^w B^z\right) + A^{2-(w+\overline{w})} B^{2-(z+\overline{z})}\right)$$

and the square of the right-hand side is equal to

$$\text{tr}\left(A^{w+\overline{w}} B^{z+\overline{z}} - 2AB + A^{2-(w+\overline{w})} B^{2-(z+\overline{z})}\right).$$

Here, we have used $\|X\|_2 = (\text{tr}\,(X^*X))^{\frac{1}{2}}$ and as the cyclicity of the trace.

Therefore, the desired norm inequality is equivalent to the statement

$$\text{Re tr}\left(A^w B^z A^{1-w} B^{1-z}\right) \le \text{tr}\,(AB). \tag{3.3.24}$$

By Corollary 3.3.9 and the fact that Re (tr X) $\le |\text{tr}\,X|$ for any matrix X, the reverse inequality (3.3.24) holds provided

$$\left|\text{Re}\,w - \frac{1}{2}\right| + \left|\text{Re}\,z - \frac{1}{2}\right| \le \frac{1}{2}.$$

□

We present trace inequalities related to the norm inequalities presented in this section. To achieve it, we need the classical Young's inequality saying that

$$a^v b^{1-v} \leq va + (1-v)b \tag{3.3.25}$$

for every $a, b \geq 0$ and $v \in [0, 1]$; see [36, p. 266] for an integral form.

Theorem 3.3.12 *Let A and B be positive semidefinite matrices, and let*

$$\mathcal{S} = \{z \in \mathbb{C} : 0 \leq \operatorname{Re} z \leq 1\}.$$

Then

$$\left|\operatorname{tr}\left(\left(A^z B^{1-z}\right)^2 + \left(B^z A^{1-z}\right)^2\right)\right| \leq \operatorname{tr}\left(A^2 + B^2\right) \tag{3.3.26}$$

for all $z \in \mathcal{S}$.

Proof Let $z = x + iy$, $0 \leq x \leq 1$. Then

$$\begin{aligned}
&\left|\operatorname{tr}\left(\left(A^z B^{1-z}\right)^2 + \left(B^z A^{1-z}\right)^2\right)\right| = \left|\operatorname{tr}\left(A^z B^{1-z}\right)^2 + \operatorname{tr}\left(B^z A^{1-z}\right)^2\right| \\
&\leq \left|\operatorname{tr}\left(A^z B^{1-z}\right)^2\right| + \left|\operatorname{tr}\left(B^z A^{1-z}\right)^2\right| \\
&\leq \left\|\left(A^z B^{1-z}\right)^2\right\|_1 + \left\|\left(B^z A^{1-z}\right)^2\right\|_1 \quad \text{(by inequality (3.3.22))} \\
&= \left\|A^z B^{1-z} A^z B^{1-z}\right\|_1 + \left\|B^z A^{1-z} B^z A^{1-z}\right\|_1 \\
&= \left\|A^x B^{1-x} B^{-\mathrm{i}y} A^x A^{\mathrm{i}y} B^{1-x}\right\|_1 \\
&+ \left\|B^x A^{1-x} A^{-\mathrm{i}y} B^x B^{\mathrm{i}y} A^{1-x}\right\|_1 \\
&\leq \left\|A^x B^{1-x} B^{-\mathrm{i}y}\right\|_2 \left\|A^{\mathrm{i}y} A^x B^{1-x}\right\|_2 + \left\|B^x A^{1-x} A^{-\mathrm{i}y}\right\|_2 \left\|B^{\mathrm{i}y} B^x A^{1-x}\right\|_2 \\
&\quad \text{(by inequality (3.3.25))} \\
&= \left\|A^x B^{1-x}\right\|_2 \left\|A^x B^{1-x}\right\|_2 + \left\|B^x A^{1-x}\right\|_2 \left\|B^x A^{1-x}\right\|_2 \\
&\quad \text{(since } A^{\mathrm{i}y} \text{ and } B^{\mathrm{i}y} \text{ are unitary)} \\
&= \left\|A^x B^{1-x}\right\|_2^2 + \left\|B^x A^{1-x}\right\|_2^2 \\
&= \operatorname{tr}\left(B^{1-x} A^{2x} B^{1-x}\right) + \operatorname{tr}\left(A^{1-x} B^{2x} A^{1-x}\right) \\
&= \operatorname{tr}\left(A^{2x} B^{2(1-x)}\right) + \operatorname{tr}\left(A^{2(1-x)} B^{2x}\right) \\
&\leq \operatorname{tr}\left(A^2\right)^x \operatorname{tr}\left(B^2\right)^{1-x} + \operatorname{tr}\left(A^2\right)^{1-x} \operatorname{tr}\left(B^2\right)^x \\
&\leq \operatorname{tr}\left(A^2\right) + \operatorname{tr}\left(B^2\right) \\
&\quad \text{(by the Young inequality (3.3.25))} \\
&= \operatorname{tr}\left(A^2 + B^2\right).
\end{aligned}$$

□

Based on Theorem 3.3.12, we have the following related trace inequality.

Corollary 3.3.13 *Let A and B be positive semidefinite matrices, and let*

$$\mathcal{S} = \{z \in \mathbb{C} : 0 \leq \operatorname{Re} z \leq 1\}.$$

Then

$$\operatorname{Re}\left(\operatorname{tr}\left(A^z B^{1-z} + B^z A^{1-z}\right)^2\right) \leq \operatorname{tr}\left((A+B)^2\right) \tag{3.3.27}$$

for all $z \in \mathcal{S}$.

Proof We can see that inequality (3.3.27) is equivalent to the statement

$$\operatorname{Re}\left(\operatorname{tr}\left(\left(A^z B^{1-z}\right)^2 + \left(B^z A^{1-z}\right)^2\right)\right) \leq \operatorname{tr}\left(A^2 + B^2\right). \tag{3.3.28}$$

Thus, our goal is to show that inequality (3.3.28) holds provided $z \in \mathcal{S}$. For $z \in \mathcal{S}$, we have

$$\begin{aligned}\operatorname{Re}\left(\operatorname{tr}\left(\left(A^z B^{1-z}\right)^2 + \left(B^z A^{1-z}\right)^2\right)\right) &\leq \left|\operatorname{tr}\left(\left(A^z B^{1-z}\right)^2 + \left(B^z A^{1-z}\right)^2\right)\right| \\ &\leq \operatorname{tr}\left(A^2 + B^2\right) \text{ (by Theorem 3.3.12).}\end{aligned}$$

Hence, inequality (3.3.27) is valid provided $z \in \mathcal{S}$. □

3.4 Some Classical Trace Inequalities

In their investigation of trace inequalities for multiple products of powers of two positive semidefinite matrices, Ando, Hiai, and Okubo [27] proved that if A and B are positive semidefinite matrices, then

$$\operatorname{tr}\left(\left(A^{\frac{1}{2}} B\right)^2\right) \leq \operatorname{tr}\left(A^t B A^{1-t} B\right) \leq \operatorname{tr}\left(A B^2\right) \tag{3.4.1}$$

for $0 \leq t \leq 1$; see Corollary 2.2 in [27].

We generalize inequality (3.4.1) by proving that

$$\operatorname{tr}\left(A^s B A^{1-s} B\right) \leq \operatorname{tr}\left(A^t B A^{1-t} B\right) \tag{3.4.2}$$

holds for $\frac{1}{2} \leq s \leq t \leq 1$, where A is a positive semidefinite matrix and B is a Hermitian matrix.

To achieve this, we consider the function $f(t) = \operatorname{tr}\left(A^t B A^{1-t} B\right)$ for $0 \leq t \leq 1$. Note that $f(t) = f(1-t)$, and so $f(t)$ is symmetric about $t = \frac{1}{2}$. The Cauchy–Schwarz inequality for the inner product $\langle X, Y\rangle = \operatorname{tr}(Y^* X)$ ensures that for any matrices X and Y, we have

$$|\mathrm{tr}\,(XY)| \le \left(\mathrm{tr}\left(X^*X\right)\right)^{\frac{1}{2}}\left(\mathrm{tr}\left(Y^*Y\right)\right)^{\frac{1}{2}}.$$

By this inequality, we can prove that $f(t)$ is logarithmically convex (and hence it is convex) for $0 \le t \le 1$. In fact, if $0 \le s \le t \le 1$, then

$$\begin{aligned}
f\left(\frac{s+t}{2}\right) &= \mathrm{tr}\left(A^{\frac{s+t}{2}}BA^{1-\left(\frac{s+t}{2}\right)}B\right)\\
&= \mathrm{tr}\left(\left(A^{\frac{t}{2}}BA^{\frac{1-t}{2}}\right)\left(A^{\frac{1-s}{2}}BA^{\frac{s}{2}}\right)\right)\\
&\le \left(\mathrm{tr}\left(A^{\frac{1-t}{2}}BA^{t}BA^{\frac{1-t}{2}}\right)\right)^{1/2}\left(\mathrm{tr}\left(A^{\frac{s}{2}}BA^{1-s}BA^{\frac{s}{2}}\right)\right)^{1/2}\\
&= \left(\mathrm{tr}\left(A^{t}BA^{1-t}B\right)\right)^{1/2}\left(\mathrm{tr}\left(A^{s}BA^{1-s}B\right)\right)^{1/2}\\
&= (f(t))^{1/2}(f(s))^{1/2}\\
&\le \frac{1}{2}(f(s)+f(t)).
\end{aligned}$$

Thus, $f(t)$ is decreasing for $0 \le t \le \frac{1}{2}$, increasing for $\frac{1}{2} \le t \le 1$, attains its minimum at $t = \frac{1}{2}$, and attains its maximum at $t = 0$ and $t = 1$.

Another proof of inequality (3.4.2) can be concluded from Lemma 2 in [37]. We remark here that inequality (3.4.2) is equivalent to saying that

$$\mathrm{tr}\left(A^{\alpha}BA^{\beta}B\right) \le \mathrm{tr}\left(A^{\gamma}BA^{\delta}B\right)$$

for $\alpha, \beta, \gamma, \delta \ge 0$ with $\alpha + \beta = \gamma + \delta$ and

$$\max\{\alpha, \beta\} \le \max\{\gamma, \delta\}.$$

Related classical trace inequalities, based on log convexity results, can be found in [38–40].

The second inequality in (3.4.1) is a particular case of the inequality

$$\left|\mathrm{tr}\left(A^{s}B^{t}A^{1-s}B^{1-t}\right)\right| \le \mathrm{tr}\,(AB), \tag{3.4.3}$$

where A and B are positive semidefinite matrices and $0 \le s, t \le 1$.

Ando, Hiai, and Okubo [27] proved that inequality (3.4.3) holds for all nonnegative real numbers s and t for which

$$\left|s - \frac{1}{2}\right| + \left|t - \frac{1}{2}\right| \le \frac{1}{2}.$$

It seems natural to ask what is the complete range of validity of inequality (3.4.3). Plevnik [39] gave a counterexample to inequality (3.4.3). He showed that it fails for $s = \frac{4}{5}, t = \frac{1}{5}$.

This section is devoted to proving the following Hilbert–Schmidt norm inequality as the first application of inequality (3.4.2)

$$\left\|A^s B + BA^{1-s}\right\|_2 \leq \left\|A^t B + BA^{1-t}\right\|_2$$

for $\frac{1}{2} \leq s \leq t \leq 1$, where A is a positive semidefinite matrix and B is a Hermitian matrix.

Moreover, we prove the following trace inequality as the second application of inequality (3.4.2)

$$\operatorname{tr}\left(A^t BA^{1-t} (\log A)\, B\right) \leq \operatorname{tr}\left(A^t (\log A)\, BA^{1-t} B\right),$$

where A is a positive definite matrix, B is a Hermitian matrix, and $\frac{1}{2} \leq t \leq 1$. As a consequence of this trace inequality, we prove that the inequality

$$\left\|A^t B + BA^{1-t} \log A\right\|_2 \leq \left\|A^t (\log A)\, B + BA^{1-t}\right\|_2$$

holds for $\frac{1}{2} \leq t \leq 1$, where A is a positive definite matrix with $\operatorname{sp}(A) \subseteq \left[e^{-1}, 1\right] \cup [e, \infty)$ and B is a Hermitian matrix.

It is tempting to investigate the following conjectures concerning the generalizations of these Hilbert–Schmidt norm inequalities to the wider class of unitarily invariant norms (see [41]).

Conjecture 3.4.1 Let A be a positive semidefinite matrix and B be a Hermitian matrix. Then, for $\frac{1}{2} \leq s \leq t \leq 1$ and for every unitarily invariant norm, we have

$$\left|\left|\left|A^s B + BA^{1-s}\right|\right|\right| \leq \left|\left|\left|A^t B + BA^{1-t}\right|\right|\right|.$$

Conjecture 3.4.2 Let A be a positive definite matrix such that $\operatorname{sp}(A) \subseteq \left[e^{-1}, 1\right] \cup [e, \infty)$ and B be a Hermitian matrix. Then, for $\frac{1}{2} \leq t \leq 1$ and for every unitarily invariant norm, we have

$$\left|\left|\left|A^t B + BA^{1-t} \log A\right|\right|\right| \leq \left|\left|\left|A^t (\log A)\, B + BA^{1-t}\right|\right|\right|.$$

We affirmatively settle Conjecture 3.4.1 for the Hilbert–Schmidt norm as an application of inequality (3.4.2) and the following lemma. This application is a Hilbert–Schmidt norm inequality, which asserts that

$$\left\|A^s B + BA^{1-s}\right\|_2 \leq \left\|A^t B + BA^{1-t}\right\|_2,$$

where A is a positive semidefinite matrix, B is a Hermitian matrix, and $\frac{1}{2} \leq s \leq t \leq 1$.

A useful lemma reads as follows.

Lemma 3.4.3 *Let A and C be positive semidefinite matrices. Then the function $g(t) = \operatorname{tr}\left(\left(A^t + A^{1-t}\right) C\right)$ is increasing for $\frac{1}{2} \leq t \leq 1$.*

Proof Without loss of generality, we can assume that A is a positive definite matrix. The general case follows by a continuity argument.

By the spectral mapping theorem, it is clear that the matrix $\left(A^t - A^{1-t}\right)\log A$ is a positive semidefinite matrix for $\frac{1}{2} \le t \le 1$. Since C is a positive semidefinite matrix, it follows that

$$\frac{d}{dt} g(t) = \operatorname{tr}\left(\left(A^t \log A - A^{1-t}\log A\right) C\right) \ge 0.$$

Thus, $g(t)$ is increasing for $\frac{1}{2} \le t \le 1$. □

Theorem 3.4.4 *Let A be a positive semidefinite matrix and B be a Hermitian matrix. Then*

$$\left\|A^s B + BA^{1-s}\right\|_2 \le \left\|A^t B + BA^{1-t}\right\|_2 \ \textit{for}\ \frac{1}{2} \le s \le t \le 1.$$

That is, the function $h(t) = \left\|A^t B + BA^{1-t}\right\|_2$ is increasing for $\frac{1}{2} \le t \le 1$.

Proof Since for any matrix X, $\|X\|_2^2 = \operatorname{tr}(X^*X)$, we have

$$\begin{aligned}(h(t))^2 &= \left\|A^t B + BA^{1-t}\right\|_2^2 \\ &= \operatorname{tr}\left(\left(BA^t + A^{1-t}B\right)\left(A^t B + BA^{1-t}\right)\right) \\ &= \operatorname{tr}\left(A^{2t}B^2 + A^{2(1-t)}B^2\right) + 2\operatorname{tr}\left(A^t BA^{1-t}B\right) \\ &= \operatorname{tr}\left(\left(A^{2t} + A^{2(1-t)}\right)B^2\right) + 2\operatorname{tr}\left(A^t BA^{1-t}B\right).\end{aligned}$$

Replacing A with A^2 and taking $C = B^2$ in Lemma 3.4.3, we observe that the function $\operatorname{tr}\left(\left(A^{2t} + A^{2(1-t)}\right)B^2\right)$ is increasing for $\frac{1}{2} \le t \le 1$. Since $\operatorname{tr}\left(A^t BA^{1-t}B\right)$ is increasing for $\frac{1}{2} \le t \le 1$ by (3.4.2), it follows that $h(t)$ is increasing for $\frac{1}{2} \le t \le 1$. □

It follows from the arithmetic–geometric mean inequality for unitarily invariant norms (see Theorem 2.4.9) that if S and T are Hermitian matrices and X is any matrix, then, for every unitarily invariant norm, we have

$$2\,|||SXT||| \le \left|\left|\left|S^2X + XT^2\right|\right|\right|.$$

By the triangle inequality, the self-adjointness of unitarily invariant norms, and the arithmetic–geometric mean inequality for unitarily invariant norms, we have

$$\left|\left|\left|A^{\frac{1}{2}}B + BA^{\frac{1}{2}}\right|\right|\right| \le \left|\left|\left|A^{\frac{1}{2}}B\right|\right|\right| + \left|\left|\left|BA^{\frac{1}{2}}\right|\right|\right| = 2\left|\left|\left|A^{\frac{1}{2}}B\right|\right|\right| \le |||AB + B|||,$$

where A is a positive semidefinite matrix and B is a Hermitian matrix. This affirmatively settles Conjecture 3.4.1 in the case of $s = \frac{1}{2}$ and $t = 1$.

The following trace inequality is the second application of inequality (3.4.2).

Theorem 3.4.5 *Let A be a positive definite matrix and B be a Hermitian matrix. Then, for $\frac{1}{2} \le t \le 1$,*

$$\operatorname{tr}\left(A^t B A^{1-t} (\log A) B\right) \le \operatorname{tr}\left(A^t (\log A) B A^{1-t} B\right).$$

Proof Consider $f(t) = \operatorname{tr}\left(A^t B A^{1-t} B\right)$. Then we have

$$\begin{aligned}\frac{d}{dt} f(t) &= \operatorname{tr}\left(\frac{d}{dt}\left(A^t B\right) A^{1-t} B + A^t B \frac{d}{dt}\left(A^{1-t} B\right)\right)\\ &= \operatorname{tr}\left(-A^t B A^{1-t} (\log A) B + A^t (\log A) B A^{1-t} B\right).\end{aligned}$$

Since the function $f(t) = \operatorname{tr}\left(A^t B A^{1-t} B\right)$ is increasing for $\frac{1}{2} \le t \le 1$, it follows that $\frac{d}{dt} f(t) \ge 0$. Thus,

$$\operatorname{tr}\left(A^t B A^{1-t} (\log A) B\right) \le \operatorname{tr}\left(A^t (\log A) B A^{1-t} B\right).$$

□

Letting $t = 1$ in Theorem 3.4.5, we have the following corollary.

Corollary 3.4.6 *Let A be a positive definite matrix and B be a Hermitian matrix. Then*

$$\operatorname{tr}(AB(\log A) B) \le \operatorname{tr}\left(A (\log A) B^2\right). \tag{3.4.4}$$

We remark here that inequality (3.4.4) can also be concluded from Theorem 1.2 in [25].

The following norm inequality is another consequence of Theorem 3.4.5. This affirmatively settles Conjecture 3.4.2 for the Hilbert–Schmidt norm.

Theorem 3.4.7 *Let A be a positive definite matrix such that $\operatorname{sp}(A) \subseteq \left[e^{-1}, 1\right] \cup [e, \infty)$ and B be a Hermitian matrix. Then, for $\frac{1}{2} \le t \le 1$,*

$$\left\|A^t B + B A^{1-t} (\log A)\right\|_2 \le \left\|A^t (\log A) B + B A^{1-t}\right\|_2.$$

Proof The square of the right-hand side of the desired norm inequality is equal to

$$\operatorname{tr}\left(A^{2t} (\log A)^2 B^2 + A^{2(1-t)} B^2\right) + 2\operatorname{tr}\left(A^t (\log A) B A^{1-t} B\right)$$

and the square of the left-hand side is equal to

$$\operatorname{tr}\left(A^{2t} B^2 + A^{2(1-t)} (\log A)^2 B^2\right) + 2\operatorname{tr}\left(A^t B A^{1-t} (\log A) B\right).$$

Note that $\operatorname{tr}\left(A^t B A^{1-t} (\log A) B\right) \le \operatorname{tr}\left(A^t (\log A) B A^{1-t} B\right)$ by Theorem 3.4.5. Thus, it suffices to show that

$$\operatorname{tr}\left(A^{2t} B^2 + A^{2(1-t)} (\log A)^2 B^2\right) \le \operatorname{tr}\left(A^{2t} (\log A)^2 B^2 + A^{2(1-t)} B^2\right). \tag{3.4.5}$$

By the spectral mapping theorem, it is clear that $\operatorname{sp}(A) \subseteq \left[e^{-1}, 1\right] \cup [e, \infty)$ implies that the matrix $\left(A^{2t} - A^{2(1-t)}\right)\left((\log A)^2 - I\right)$ is a positive semidefinite matrix. Since B^2 is also positive semidefinite, it follows that

$$\operatorname{tr}\left(\left(A^{2t} - A^{2(1-t)}\right)\left((\log A)^2 - I\right) B^2\right) \geq 0.$$

This leads us to inequality (3.4.5). Thus,

$$\begin{aligned} &\operatorname{tr}\left(A^{2t} (\log A)^2 B^2 + A^{2(1-t)} B^2\right) + 2\operatorname{tr}\left(A^t (\log A) B A^{1-t} B\right) \\ \geq\; &\operatorname{tr}\left(A^{2t} B^2 + A^{2(1-t)} (\log A)^2 B^2\right) \quad + 2\operatorname{tr}\left(A^t B A^{1-t} ((\log A) B\right). \end{aligned}$$

Hence, the desired norm inequality is valid for $\frac{1}{2} \leq t \leq 1$. □

Note that if we set $t = \frac{1}{2}$ in Theorem 3.4.7, inequality turns into equality, but if we set $t = 1$, we obtain the following inequality.

Corollary 3.4.8 *Let A be a positive definite matrix such that* $\operatorname{sp}(A) \subseteq \left[e^{-1}, 1\right] \cup [e, \infty)$*, and let B be a Hermitian matrix. Then*

$$\|AB + B(\log A)\|_2 \leq \|A(\log A) B + B\|_2 .$$

3.5 Further Applications

In this section, we discuss more applications of inequality (3.4.2). These applications focus on trace inequalities that deal with the means of two nonnegative real numbers. These inequalities also include a generalization of the Ando–Hiai–Okubo trace inequalities (3.4.1). In this context, we assume that A is a positive semidefinite matrix, B is a Hermitian matrix, $a, b \geq 0$, and $\frac{1}{2} \leq r \leq 1$ (see [41]).

Remark 3.5.1 Let $f(a, b)$ and $g(a, b)$ be means of a and b. Then

$$\operatorname{tr}\left(A^{g\left(r,\frac{1}{2}\right)} B A^{1-g\left(r,\frac{1}{2}\right)} B\right) \leq \operatorname{tr}\left(A^r B A^{1-r} B\right) \leq \operatorname{tr}\left(A^{f(r,1)} B A^{1-f(r,1)} B\right). \quad (3.5.1)$$

In fact, since $f(a, b)$ and $g(a, b)$ are means of a and b and $\frac{1}{2} \leq r \leq 1$, it follows by the internality property that

$$\frac{1}{2} \leq g\left(r, \frac{1}{2}\right) \leq r \leq f(r, 1) \leq 1.$$

Therefore, by inequality (3.4.2), we have

$$\operatorname{tr}\left(A^{g\left(r,\frac{1}{2}\right)} B A^{1-g\left(r,\frac{1}{2}\right)} B\right) \leq \operatorname{tr}\left(A^r B A^{1-r} B\right) \leq \operatorname{tr}\left(A^{f(r,1)} B A^{1-f(r,1)} B\right).$$

The following example is stemmed from inequality (3.4.1).

Example 3.5.2 Let $f(a,b) = \max\{a,b\}$ and $g(a,b) = \min\{a,b\}$ in inequality (3.4.1). Then

$$\operatorname{tr}\left(\left(A^{\frac{1}{2}}B\right)^2\right) \le \operatorname{tr}\left(A^r BA^{1-r}B\right) \le \operatorname{tr}\left(AB^2\right). \tag{3.5.2}$$

Inequalities (3.5.2) yield inequalities (3.4.1) when B is a positive semidefinite matrix.

Another related trace inequality is

$$\operatorname{tr}\left(A^{\alpha}BA^{\beta}B\right) \le \frac{1}{2}\operatorname{tr}\left(A^{\alpha+\eta}BA^{\beta-\eta}B + A^{\alpha-\eta}BA^{\beta+\eta}B\right), \tag{3.5.3}$$

where A is a positive semidefinite matrix, B is a Hermitian matrix, and $\alpha, \beta \ge \eta \ge 0$.

To prove inequality (3.5.3), let $C = BA^{\frac{\beta+\eta}{2}} - A^{\eta}BA^{\frac{\beta-\eta}{2}}$ and $R = A^{\alpha-\eta}$. Since $\operatorname{tr}(RCC^*) \ge 0$, it follows that

$$\operatorname{tr}\left(A^{\alpha-\eta}\left(BA^{\frac{\beta+\eta}{2}} - A^{\eta}BA^{\frac{\beta-\eta}{2}}\right)\left(A^{\frac{\beta+\eta}{2}}B - A^{\frac{\beta-\eta}{2}}BA^{\eta}\right)\right) \ge 0,$$

which is equivalent to inequality (3.5.3).

We mention that inequality (3.5.3) gives another proof of the convexity of the function $f(t)$. To see this, replace A with $A^{\frac{1}{\alpha+\beta}}$ in inequality (3.5.3) and set $s = \frac{\alpha+\eta}{\alpha+\beta}$, $t = \frac{\alpha-\eta}{\alpha+\beta}$ to get $f\left(\frac{s+t}{2}\right) \le \frac{1}{2}\left(f(s) + f(t)\right)$.

Remark 3.5.3 Since the function $f(t) = \operatorname{tr}\left(A^t BA^{1-t}B\right)$ is logarithmically convex (and hence it is convex) for $0 \le t \le 1$, it follows that $\frac{d^2}{dt^2} f(t) \ge 0$. Thus, for a positive definite matrix A and a Hermitian matrix B, we have the trace inequality

$$\operatorname{tr}\left(A^t(\log A)BA^{1-t}(\log A)B\right) \le \frac{1}{2}\operatorname{tr}\left(A^t BA^{1-t}(\log A)^2 B + A^{1-t}BA^t(\log A)^2 B\right). \tag{3.5.4}$$

Letting $t = \frac{1}{2}$ in inequality (3.5.4), we obtain the inequality

$$\operatorname{tr}\left(\left(A^{\frac{1}{2}}(\log A)B\right)^2\right) \le \operatorname{tr}\left(A^{\frac{1}{2}}BA^{\frac{1}{2}}(\log A)^2 B\right).$$

Letting $t = 0$ or $t = 1$ in inequality (3.5.4), we obtain the inequality

$$\operatorname{tr}\left((\log A)BA(\log A)B\right) \le \frac{1}{2}\operatorname{tr}\left(AB(\log A)^2 B + BA(\log A)^2 B\right).$$

Remark 3.5.4 We remark here that the functions $g(t)$ given in Lemma 3.4.3 and $h(t)$ given in Theorem 3.4.4 are also logarithmically convex (and hence they are convex) for $0 \le t \le 1$, symmetric about $t = \frac{1}{2}$, decreasing for $0 \le t \le \frac{1}{2}$, increasing for $\frac{1}{2} \leq t \leq 1$, attain their minima at $t = \frac{1}{2}$, and attain their maxima at $t = 0$ and $t = 1$.

We conclude this section with a general trace inequality, from which we obtain a trace inequality related to earlier inequalities.

Theorem 3.5.5 *Let T be a positive definite matrix, X, Y be positive semidefinite matrices, and B be a Hermitian matrix. Then*

$$\begin{aligned}&\operatorname{tr}\left(T^{\frac{1}{2}}YT^{-\frac{1}{2}}BXB + T^{-\frac{1}{2}}YT^{\frac{1}{2}}BXB\right)\\ &\qquad\leq \operatorname{tr}\left(T^{-\frac{1}{2}}YT^{-\frac{1}{2}}BX^{\frac{1}{2}}TX^{\frac{1}{2}}B + T^{\frac{1}{2}}YT^{\frac{1}{2}}BX^{\frac{1}{2}}T^{-1}X^{\frac{1}{2}}B\right).\end{aligned}$$

If, in addition, T commutes with X and Y, then

$$\operatorname{tr}(YBXB) \leq \frac{1}{2}\operatorname{tr}\left(YT^{-1}BXTB + YTBXT^{-1}B\right).$$

Proof Let $C = BX^{\frac{1}{2}}T^{\frac{1}{2}} - TBX^{\frac{1}{2}}T^{-\frac{1}{2}}$ and $R = T^{-\frac{1}{2}}YT^{-\frac{1}{2}}$. Since $\operatorname{tr}(RCC^*) \geq 0$, it follows that

$$\operatorname{tr}\left(T^{-\frac{1}{2}}YT^{-\frac{1}{2}}\left(BX^{\frac{1}{2}}T^{\frac{1}{2}} - TBX^{\frac{1}{2}}T^{-\frac{1}{2}}\right)\left(T^{\frac{1}{2}}X^{\frac{1}{2}}B - T^{-\frac{1}{2}}X^{\frac{1}{2}}BT\right)\right) \geq 0,$$

which is equivalent to

$$\begin{aligned}&\operatorname{tr}\left(T^{\frac{1}{2}}YT^{-\frac{1}{2}}BXB + T^{-\frac{1}{2}}YT^{\frac{1}{2}}BXB\right)\\ &\qquad\leq \operatorname{tr}\left(T^{-\frac{1}{2}}YT^{-\frac{1}{2}}BX^{\frac{1}{2}}TX^{\frac{1}{2}}B + T^{\frac{1}{2}}YT^{\frac{1}{2}}BX^{\frac{1}{2}}T^{-1}X^{\frac{1}{2}}B\right).\end{aligned}$$

□

Based on Theorem 3.5.5, we have the following trace inequality, which is closely related to the inequality given in inequality (3.5.3). In this inequality, the positivity of the matrix A is strengthened, while the positivity of the exponents is released.

Corollary 3.5.6 *Let A be a positive definite matrix and B be a Hermitian matrix. Then, for the real numbers α, β, η,*

$$\operatorname{tr}\left(A^{\alpha}BA^{\beta}B\right) \leq \frac{1}{2}\operatorname{tr}\left(A^{\alpha+\eta}BA^{\beta-\eta}B + A^{\alpha-\eta}BA^{\beta+\eta}B\right).$$

Proof This follows immediately by replacing X, Y, T with $A^{\beta}, A^{\alpha}, A^{\eta}$, respectively, in Theorem 3.5.5. □

If we restrict the values of α, β, η in Corollary 3.5.6 such that $\alpha, \beta \geq \eta \geq 0$ and if we use a continuity argument, then we retain inequality (3.5.3).

3.6 Exercises and Problems

Exercise 3.6.1 Prove that if $a, b \geq 0$, then

$$\left(a^r + b^r\right) \leq (a+b)^r \leq 2^{r-1}\left(a^r + b^r\right) \text{ for } r \geq 1$$

and

$$2^{r-1}\left(a^r + b^r\right) \leq (a+b)^r \leq \left(a^r + b^r\right) \text{ for } 0 < r \leq 1.$$

Exercise 3.6.2 Let A and B be normal matrices such that AB is normal. Prove that BA is also normal.

Exercise 3.6.3 Let A and B be matrices. Prove that

$$|\operatorname{tr}(AB)| \leq \frac{\operatorname{tr}(A^*A) + \operatorname{tr}(B^*B)}{2}.$$

Exercise 3.6.4 Let A and B be matrices such that AB is Hermitian, and let $1 < p < \infty$. Prove that

$$\|\operatorname{Re}(BA)\|_p = \|AB\|_p$$

if and only if BA is Hermitian.

Exercise 3.6.5 Let A, B, and X be matrices such that A and B are positive semidefinite, and let $1 < p < \infty$. Prove that

$$2\left\|A^{\frac{1}{2}}XB^{\frac{1}{2}}\right\|_p = \|AX + XB\|_p$$

if and only if $AX = XB$.

Exercise 3.6.6 Let A, B, and X be matrices such that A and B are positive semidefinite, and let $1 < p < \infty$. Prove that

$$\left\|A^{v}XB^{1-v} + A^{1-v}XB^{v}\right\|_p = \|AX + XB\|_p$$

for some v, with $0 < v < 1$, if and only if $AX = XB$.

Exercise 3.6.7 Let A and B be matrices such that A is invertible and Hermitian. Prove that

$$2\,\|\|B\|\| \leq \left\|\left\|ABA^{-1} + A^{-1}BA\right\|\right\|$$

for every unitarily invariant norm.

Exercise 3.6.8 Let A, B, and C be matrices such that A is invertible. Prove that

$$2\,\|\|C\|\| \leq \left\|\left\|A^*CB^{-1} + A^{-1}CB^*\right\|\right\|$$

for every unitarily invariant norm.

Exercise 3.6.9 Let A and B be positive semidefinite matrices, and let X be any matrix. Then, for $\frac{1}{2} \leq s \leq t \leq 1$, prove that

$$\left\|A^s X + X B^{1-s}\right\|_2^2 + \left\|A^{1-s} X + X B^s\right\|_2^2 \leq \left\|A^t X + X B^{1-t}\right\|_2^2 + \left\|A^{1-t} X + X B^t\right\|_2^2.$$

Exercise 3.6.10 Let A and B be positive definite matrices such that $\mathrm{sp}(A), \mathrm{sp}(B) \subseteq \left[e^{-1}, 1\right] \cup [e, \infty)$, and let X be any matrix. Prove that, for $\frac{1}{2} \leq t \leq 1$, we have

$$\left\|A^t X + X B^{1-t} \log B)\right\|_2^2 + \left\|X B^t + A^{1-t} (\log A) X\right\|_2^2$$

$$\leq \left\|A^t (\log A) X + X B^{1-t}\right\|_2^2 + \left\|X B^t \log B + A^{1-t} X\right\|_2^2.$$

Exercise 3.6.11 Let A and B be positive semidefinite matrices, and let X be any matrix. Prove that, for $r \geq 1$, we have

$$2 \left\|\left| A^{\frac{1}{2}} X + X B^{\frac{1}{2}} \right\|\right|^r \leq \||AX + X\||^r + \||X + XB\||^r$$

for every unitarily invariant norm.

Problem 3.6.12 Let A and B be matrices. Prove the following classical Clarkson inequalities for the Schatten p-norms:

(i) $2\left(\|A\|_p^p + \|B\|_p^p\right) \leq \|A + B\|_p^p + \|A - B\|_p^p \leq 2^{p-1}\left(\|A\|_p^p + \|B\|_p^p\right)$
for $2 \leq p < \infty$.

(ii) $2^{p-1}\left(\|A\|_p^p + \|B\|_p^p\right) \leq \|A + B\|_p^p + \|A - B\|_p^p \leq 2\left(\|A\|_p^p + \|B\|_p^p\right)$
for $0 < p \leq 2$.

(iii) $2\left(\|A\|_p^p + \|B\|_p^p\right)^{\frac{q}{p}} \leq \|A + B\|_p^q + \|A - B\|_p^q$
for $2 \leq p < \infty$ and $\frac{1}{p} + \frac{1}{q} = 1$.

(iv) $\|A + B\|_p^q + \|A - B\|_p^q \leq 2\left(\|A\|_p^p + \|B\|_p^p\right)^{\frac{q}{p}}$
for $1 < p \leq 2$ and $\frac{1}{p} + \frac{1}{q} = 1$.

3.7 Notes, Hints, and References

Let us present some hints and references for the exercises and problems in this section.

Exercise 3.6.1 follows from the convexity (concavity) of the function $f(t) = t^r$ on $[0, \infty)$ for $r \geq 1$ $(0 < r \leq 1)$. Exercise 3.6.2 can be solved by utilizing relation (1.1.7) and Theorem 2.1.18. Exercise 3.6.3 follows from Theorem 2.1.25 and the arithmetic–geometric mean inequality. A solution to Exercise 3.6.4 can be found in [42].

Exercises 3.6.5 and 3.6.6 are related to the convexity of the Heinz means discussed in [43]. Exercises 3.6.7 and 3.6.8 are related to the Corach–Porta–Recht inequalities given in [16]; see also [15]. Exercises 3.6.9, 3.6.10, and 3.6.11 involve recent norm inequalities in [44].

Problem 3.6.12 is the celebrated Clarkson inequalities for the Schatten p-norms, which can be found in [32]. These inequalities have been generalized to n-tuples of operators in [45, 46]. For other related inequalities; see [47].

References

1. J.-C. Bourin, M. Uchiyama, A matrix subadditivity inequality for $f(A + B)$ and $f(A) + f(B)$. Linear Algebra Appl. **423**, 512–518 (2007)
2. S.J. Rotfel'd, Remarks on the singular values of a sum of completely continuous operators (Russian). Funkcional Anal. i Priložen. **1**, 95–96 (1967)
3. T. Kosem, Inequalities between $||f(A + B)||$ and $||f(A) + f(B)||$. Linear Algebra Appl. **418**, 153–160 (2006)
4. J.-C. Bourin, Matrix subadditivity inequalities and block-matrices. Internat. J. Math. **20**, 979–691 (2009)
5. T. Ando, X. Zhan, Norm inequalities related to operator monotone functions. Math. Ann. **31**, 771–780 (1999)
6. R. Bhatia, F. Kittaneh, Norm inequalities for positive operators. Lett. Math. Phys. **43**, 225–231 (1998)
7. S. Hayajneh, F. Kittaneh, Trace inequalities and a question of Bourin. Bull. Austral. Math. Soc. **88**, 384–389 (2013)
8. R. Bhatia, *Matrix Analysis* (Springer, New York, 1997)
9. X. Zhan, *Matrix Inequalities*, Lecture Notes in Mathematics, vol. 1790. (Springer, Berlin, 2002)
10. J.-C. Bourin, A matrix subadditivity inequality for symmetric norms. Proc. Amer. Math. Soc. **138**, 495–504 (2010)
11. E. Heinz, Beiträge zur Störungstheorie der Spektralzerlegung (German). Math. Ann. **123**, 415–438 (1951)
12. F. Hiai, H. Kosaki, *Means of Hilbert Space Operators*, Lecture Notes in Mathematics, vol. 1820 (Springer, 2003)
13. A. McIntosh, *Heinz Inequalities and Perturbation of Spectral Families*, Macquarie Mathematical Reports, Macquarie University (1979)
14. J.I. Fujii, M. Fujii, T. Furuta, M. Nakamoto, Norm inequalities equivalent to Heinz inequality. Proc. Amer. Math. Soc. **118**, 827–830 (1993)
15. C. Conde, A. Seddik, M.S. Moslehian, Operator inequalities related to the Corach-Porta-Recht inequality. Linear Algebra Appl. **436**, 3008–3017 (2012)
16. F. Kittaneh, On some operator inequalities. Linear Algebra Appl. **208**(209), 19–28 (1994)
17. K.M.R. Audenaert, A singular value inequality for Heinz means. Linear Algebra Appl. **422**, 279–283 (2007)
18. R. Kaur, M.S. Moslehian, M. Singh, C. Conde, Further refinements of the Heinz inequality. Linear Algebra Appl. **447**, 26–37 (2014)
19. M. Sababheh, M.S. Moslehian, Advanced refinements of Young and Heinz inequalities. J. Number Theory **172**, 178–199 (2017)
20. S. Hayajneh, F. Kittaneh, Lieb–Thirring trace inequalities and a question of Bourin. J. Math. Phys. **54**, 033504 (2013), 8 pp
21. H. Araki, On an inequality of Lieb and Thirring. Lett. Math. Phys. **19**, 167–170 (1990)
22. K.M.R. Audenaert, On the Araki-Lieb-Thirring inequality. Int. J. Inform. Sys. Sci. **4**, 78–83 (2008)
23. K.M.R. Audenaert, A Lieb-Thirring inequality for singular values. Linear Algebra Appl. **430**, 3053–3057 (2009)

24. E. Lieb, W. Thirring, *Studies in Mathematical Physics*, ed. by E. Lieb, B. Simon, A. Wightman (Princeton University, 1976), pp. 301–302
25. J.-C. Bourin, Matrix versions of some classical inequalities. Linear Algebra Appl. **416**, 890–907 (2006)
26. R. Bhatia, *Trace inequalities for products of positive definite matrices*. J. Math. Phys. **55**, 013509 (2014), 3 pp
27. T. Ando, F. Hiai, K. Okubo, Trace inequalities for multiple products of two matrices. Math. Inequal. Appl. **3**, 307–318 (2000)
28. M. Hayajneh, S. Hayajneh, F. Kittaneh, On the Ando-Hiai-Okubo trace inequality. J. Operator Theory **77**, 77–86 (2017)
29. M. Hayajneh, S. Hayajneh, F. Kittaneh, Norm inequalities related to the arithmetic-geometric mean inequalities for positive semidefinite matrices. Positivity **22**, 1311–1324 (2018)
30. M. Hayajneh, S. Hayajneh, F. Kittaneh, Remarks on some norm inequalities for positive semidefinite matrices and questions of Bourin. Math. Inequal. Appl. **20**, 225–232 (2017)
31. T. Bottazzi, R. Elencwajg, G. Larotonda, A. Varela, Inequalities related to Bourin and Heinz means with complex parameter. J. Math. Anal. Appl. **426**, 765–773 (2015)
32. B. Simon, *Trace Ideals and Their Applications*, 2nd edn. Mathematical Surveys and Monographs, vol. 120 (American Mathematical Society, Providence, RI, 2005)
33. F. Kittaneh, Norm inequalities for fractional powers of positive operators. Lett. Math. Phys. **27**, 279–285 (1993)
34. M. Reed, B. Simon, *Methods of Modern Mathematical Physics II. Fourier Analysis, Self-adjointness* (Academic, New York-London, 1975)
35. D.C. Ullrich, *Complex Made Simple, Graduate Studies in Mathematics*, vol. 97 (American Mathematical Society, Providence, RI, 2008)
36. P.S. Bullen, *A Dictionary of Inequalities*, Pitman Monographs and Surveys in Pure and Applied Mathematics, 97 (Longman, Harlow, 1998)
37. J.-C. Bourin, Some inequalities for norms on matrices and operators. Linear Algebra Appl. **292**, 139–154 (1999)
38. J.-C. Bourin, E.Y. Lee, *Matrix inequalities from a two variables functional*. Int. J. Math. **27**, 1650071 (2016) 19 pp
39. L. Plevnik, On a matrix trace inequality due to Ando, Hiai and Okubo. Indian J. Pure Appl. Math. **47**, 491–500 (2016)
40. M. Sababheh, Log and harmonically log convex functions related to matrix norms. Oper. Matrices **10**, 453–465 (2016)
41. M. Hayajneh, S. Hayajneh, F. Kittaneh, On some classical trace inequalities and a new Hilbert-Schmidt norm inequality. Math. Inequal. Appl. **21**, 1175–1183 (2018)
42. F. Kittaneh, A note on the arithmetic-geometric-mean inequality for matrices. Linear Algebra Appl. **171**, 1–8 (1992)
43. F. Kittaneh, On the convexity of the Heinz means. Integral Equ. Oper. Theory **68**, 519–527 (2010)
44. M. Hayajneh, S. Hayajneh, F. Kittaneh, New Hilbert–Schmidt norm inequalities for positive semidefinite matrices. Adv. Oper. Theory **8**, 23 (2023), 12 pp
45. R. Bhatia, F. Kittaneh, Clarkson inequalities with several operators. Bull. London Math. Soc. **36**, 820–832 (2004)
46. O. Hirzallah, F. Kittaneh, Non-commutative Clarkson inequalities for n-tuples of operators. Integral Equ. Oper. Theory **60**, 369–379 (2008)
47. O. Hirzallah, F. Kittaneh, M.S. Moslehian, Schatten p-norm inequalities related to a characterization of inner product spaces. Math. Inequal. Appl. **13**, 235–241 (2010)

Chapter 4
Norm Inequalities for Positive Semidefinite Matrices

In this chapter, we first give a comprehensive answer to a generalized version of the first main Question 3.1.1 for all unitarily invariant norms. We then present several norm inequalities related to the second main Question 3.2.1, with a particular focus on the trace norm and the Hilbert–Schmidt norm. We provide a complete answer to the second main Question 3.2.1 for the trace norm, as well as a complete answer to a slightly weaker version of this question for the Hilbert–Schmidt norm. Moreover, we discuss majorization relations and unitarily invariant norm inequalities that involve the spectral radius.

Throughout this chapter, we assume that all matrices are complex square matrices of the same size. The symbol $|||\cdot|||$ represents any unitarily invariant norm.

4.1 A Complete Solution to the First Main Question

Recall that if X and Y are matrices such that XY is Hermitian, then

$$|||XY||| \leq |||\mathrm{Re}\,(YX)||| \leq |||YX||| \tag{4.1.1}$$

for every unitarily invariant norm; see Lemma 2.4.7 and [1].

In [2], Audenaert proved the following more general version of inequality (3.1.2) for all unitarily invariant norms:

$$\left|\left|\left|\sum_{i=1}^{k} A_i B_i\right|\right|\right| \leq \left|\left|\left|\left(\sum_{i=1}^{k} A_i\right)\left(\sum_{i=1}^{k} B_i\right)\right|\right|\right|, \tag{4.1.2}$$

where A_i, B_i are positive semidefinite matrices such that, for each i, A_i commutes with B_i; see [3, 4] for different proofs of inequality (4.1.2).

A. M. Bikchentaev et al., *Trace Inequalities*, Forum for Interdisciplinary Mathematics,
https://doi.org/10.1007/978-981-97-6520-1_4

In what follows, we prove the stronger inequality

$$\left|\left|\left|\sum_{i=1}^{k} A_i B_i\right|\right|\right| \leq \left|\left|\left|\left(\sum_{i=1}^{k} A_i\right)^{\frac{1}{2}} \left(\sum_{i=1}^{k} B_i\right) \left(\sum_{i=1}^{k} A_i\right)^{\frac{1}{2}}\right|\right|\right|. \tag{4.1.3}$$

As a consequence of our main result (Theorem 4.1.3), we obtain further refinements of inequality (4.1.2), which answers the first main Question 3.1.1 affirmatively.

To prove inequality (4.1.3), we need the following two lemmas from [2].

Lemma 4.1.1 *For $i = 1, \ldots, k$, let A_i, B_i be positive semidefinite matrices such that, for each i, A_i commutes with B_i. Then, for every unitarily invariant norm,*

$$\left|\left|\left|\sum_{i=1}^{k} A_i B_i\right|\right|\right| \leq \left|\left|\left|\left(\sum_{i=1}^{k} A_i^{\frac{1}{2}} B_i^{\frac{1}{2}}\right)^2\right|\right|\right|.$$

Lemma 4.1.2 *Let S be an $n \times m$ complex matrix, and let L and M be two diagonal, positive semidefinite $m \times m$ matrices. Then*

$$s\left(\left(S\,(LM)^{\frac{1}{2}}\,S^*\right)^2\right) \prec_w \lambda\left(SLS^*SMS^*\right).$$

Our main result of this section can be stated as follows (see [5]).

Theorem 4.1.3 *For $i = 1, \ldots, k$, let A_i, B_i be positive semidefinite matrices such that, for each i, A_i commutes with B_i. Then, for every unitarily invariant norm,*

$$\left|\left|\left|\left(\sum_{i=1}^{k} A_i^{\frac{1}{2}} B_i^{\frac{1}{2}}\right)^2\right|\right|\right| \leq \left|\left|\left|\left(\sum_{i=1}^{k} A_i\right)^{\frac{1}{2}} \left(\sum_{i=1}^{k} B_i\right) \left(\sum_{i=1}^{k} A_i\right)^{\frac{1}{2}}\right|\right|\right|.$$

Proof Let A_i, B_i have the spectral decompositions

$$A_i = U_i D_i U_i^* \quad \text{and} \quad B_i = U_i E_i U_i^*,$$

where U_i are unitary matrices, and D_i, E_i are positive semidefinite diagonal matrices.

Let

$$L = \begin{bmatrix} D_1 & 0 & \cdots & 0 & 0 \\ 0 & D_2 & \cdots & 0 & 0 \\ \vdots & \vdots & \ddots & \vdots & \vdots \\ 0 & 0 & \cdots & D_{k-1} & 0 \\ 0 & 0 & \cdots & 0 & D_k \end{bmatrix}, \quad M = \begin{bmatrix} E_1 & 0 & \cdots & 0 & 0 \\ 0 & E_2 & \cdots & 0 & 0 \\ \vdots & \vdots & \ddots & \vdots & \vdots \\ 0 & 0 & \cdots & E_{k-1} & 0 \\ 0 & 0 & \cdots & 0 & E_k \end{bmatrix},$$

and

$$S = \left[U_1 \, U_2 \cdots U_{k-1} \, U_k \right].$$

Then

$$\sum_{i=1}^{k} A_i = SLS^*, \quad \sum_{i=1}^{k} B_i = SMS^*, \text{ and } \sum_{i=1}^{k} A_i^{\frac{1}{2}} B_i^{\frac{1}{2}} = S \left(LM\right)^{\frac{1}{2}} S^*.$$

We need to show that

$$s \left(\left(S \left(LM\right)^{\frac{1}{2}} S^* \right)^2 \right) \prec_w s \left(\left(SLS^*\right)^{\frac{1}{2}} SMS^* \left(SLS^*\right)^{\frac{1}{2}} \right).$$

Now,

$$\begin{aligned} s \left(\left(S \left(LM\right)^{\frac{1}{2}} S^* \right)^2 \right) &\prec_w \lambda \left(SLS^* SMS^* \right) \text{ (by Lemma 4.1.2)} \\ &= \lambda \left(\left(SLS^*\right)^{\frac{1}{2}} SMS^* \left(SLS^*\right)^{\frac{1}{2}} \right) \\ &= s \left(\left(SLS^*\right)^{\frac{1}{2}} SMS^* \left(SLS^*\right)^{\frac{1}{2}} \right). \end{aligned}$$

Thus,

$$s \left(\left(\sum_{i=1}^{k} A_i^{\frac{1}{2}} B_i^{\frac{1}{2}} \right)^2 \right) \prec_w s \left(\left(\sum_{i=1}^{k} A_i \right)^{\frac{1}{2}} \left(\sum_{i=1}^{k} B_i \right) \left(\sum_{i=1}^{k} A_i \right)^{\frac{1}{2}} \right),$$

which is equivalent to

$$\left|\left|\left| \left(\sum_{i=1}^{k} A_i^{\frac{1}{2}} B_i^{\frac{1}{2}} \right)^2 \right|\right|\right| \le \left|\left|\left| \left(\sum_{i=1}^{k} A_i \right)^{\frac{1}{2}} \left(\sum_{i=1}^{k} B_i \right) \left(\sum_{i=1}^{k} A_i \right)^{\frac{1}{2}} \right|\right|\right|.$$

□

We are ready to prove inequality (4.1.3). Using Lemma 4.1.1, Theorem 4.1.3, and inequality (4.1.1), we have the following refinements of inequality (4.1.2), including inequality (4.1.3).

Corollary 4.1.4 *For $i = 1, \ldots, k$, let A_i, B_i be positive semidefinite matrices such that, for each i, A_i commutes with B_i. Then, for every unitarily invariant norm,*

$$\begin{aligned}
\left|\left|\left|\sum_{i=1}^{k} A_i B_i\right|\right|\right| &\le \left|\left|\left|\left(\sum_{i=1}^{k} A_i^{\frac{1}{2}} B_i^{\frac{1}{2}}\right)^2\right|\right|\right| \\
&\le \left|\left|\left|\left(\sum_{i=1}^{k} A_i\right)^{\frac{1}{2}} \left(\sum_{i=1}^{k} B_i\right) \left(\sum_{i=1}^{k} A_i\right)^{\frac{1}{2}}\right|\right|\right| \\
&\le \frac{1}{2}\left|\left|\left|\left(\sum_{i=1}^{k} A_i\right)\left(\sum_{i=1}^{k} B_i\right) + \left(\sum_{i=1}^{k} B_i\right)\left(\sum_{i=1}^{k} A_i\right)\right|\right|\right| \quad \text{(by Theorem 2.4.9)} \\
&\le \left|\left|\left|\left(\sum_{i=1}^{k} A_i\right)\left(\sum_{i=1}^{k} B_i\right)\right|\right|\right|.
\end{aligned}$$

Letting $k = 2$, $A_1 = A^p$, $A_2 = B^p$, $B_1 = A^q$, and $B_2 = B^q$ in Corollary 4.1.4, we have the following chain of norm inequalities for positive semidefinite matrices, which lead to a positive answer to the first main Question 3.1.1:

$$\begin{aligned}
|||A^{p+q} + B^{p+q}||| &\le \left|\left|\left|\left(A^{\frac{p+q}{2}} + B^{\frac{p+q}{2}}\right)^2\right|\right|\right| \quad (4.1.4) \\
&\le \left|\left|\left|\left(A^p + B^p\right)^{\frac{1}{2}}\left(A^q + B^q\right)\left(A^p + B^p\right)^{\frac{1}{2}}\right|\right|\right| \\
&\le \frac{1}{2}\left|\left|\left|\left(A^p + B^p\right)\left(A^q + B^q\right) + \left(A^q + B^q\right)\left(A^p + B^p\right)\right|\right|\right| \\
&\le \left|\left|\left|\left(A^p + B^p\right)\left(A^q + B^q\right)\right|\right|\right|.
\end{aligned}$$

We remark here that inequality (4.1.4) also follows from [6, Theorem 2].

4.2 Partial Solutions to the Second Main Question

In [7], the second main Question 3.2.1 has been affirmatively settled for $p = 1, 2, 3$ and $q = 1$ in the case of the Hilbert–Schmidt norm.

Replacing A and B with $A^{\frac{1}{p+q}}$ and $B^{\frac{1}{p+q}}$, respectively, we can see that the inequality in the second main Question 3.2.1 and inequality (3.2.3), respectively, are equivalent to the inequality

$$\left|\left|\left|A^t B^{1-t} + B^t A^{1-t}\right|\right|\right| \le |||A + B||| \quad (4.2.1)$$

and the Heinz inequality

$$\left|\left|\left|A^t B^{1-t} + A^{1-t} B^t\right|\right|\right| \le |||A + B||| \quad (4.2.2)$$

for $t \in [0, 1]$.

In [8], Bhatia proved inequality (4.2.1) for $t \in \left[\frac{1}{4}, \frac{3}{4}\right]$ in the case of the Hilbert–Schmidt norm.

In the case of the Hilbert–Schmidt norm, we see that inequality (3.3.21) is equivalent to

$$\operatorname{Re}\ \operatorname{tr}\left(A^t B^{1-t} A^{1-t} B^t\right) \leq \operatorname{tr}(AB) \tag{4.2.3}$$

for $t \in [0, 1]$.

In [8], Bhatia proved inequality (4.2.3) for $t \in \left[\frac{1}{4}, \frac{3}{4}\right]$. Recently, Bottazzi et al. [9] proved that this inequality holds for t in a vertical strip in the complex plane that contains the interval $\left[\frac{1}{4}, \frac{3}{4}\right]$. Moreover, they provide a counterexample showing that inequality (3.3.21) does not hold for all $t \in [0, 1]$ if $|||\cdot|||$ is the operator norm.

Related to inequality (4.2.1), we prove the following norm inequality for all unitarily invariant norms

$$\left|\left|\left|A^t X B^{1-t} + B^t X^* A^{1-t}\right|\right|\right| \leq |||AX||| + |||XB|||, \tag{4.2.4}$$

where A and B are positive semidefinite matrices, X is any matrix, and $t \in [0, 1]$.

As an application of inequality (4.2.4), we prove that for $t \in [0, 1]$ the inequality

$$\left\|A^t B^{1-t} + B^t A^{1-t}\right\|_1 \leq \|A + B\|_1$$

holds. This answers affirmatively the second main Question 3.2.1 in the case of the trace norm. However, the second main Question 3.2.1 remains open for other unitarily invariant norms such as the Schatten p-norms.

Then, we prove the following two norm inequalities, which are weaker than inequality (4.2.1):

$$\left|\left|\left|\operatorname{Re}\left(A^t B^{1-t} + B^t A^{1-t}\right)\right|\right|\right| \leq |||A + B|||$$

and

$$\left|\left|\left|\operatorname{Im}\left(A^t B^{1-t} + B^t A^{1-t}\right)\right|\right|\right| \leq |||A + B|||,$$

where $A, B \in \mathbb{M}_n$ are positive semidefinite and $t \in [0, 1]$. These inequalities give a partial answer to the second main Question 3.2.1 by inserting the real and imaginary parts in the left-hand side of inequality (4.2.1).

We can state and prove the main result of this section (see [10]), which can be regarded as a Heinz-type inequality; see (3.2.2).

Theorem 4.2.1 *Let A and B be positive semidefinite matrices, and let X be any matrix. Then, for $t \in [0, 1]$ and for every unitarily invariant norm,*

$$\left|\left|\left|A^t X B^{1-t} + B^t X^* A^{1-t}\right|\right|\right| \leq |||AX||| + |||XB|||.$$

Proof We have

$$\begin{aligned}
&|||A^t X B^{1-t} + B^t X^* A^{1-t}||| \\
&\quad \le |||A^t X B^{1-t}||| + |||B^t X^* A^{1-t}||| \text{ (by the triangle inequality)} \\
&\quad \le |||AX|||^t \, |||XB|||^{1-t} + |||BX^*|||^t \, |||X^* A|||^{1-t} \text{ (by Lemma 3.3.5)} \\
&\quad = |||AX|||^t \, |||XB|||^{1-t} + |||XB|||^t \, |||AX|||^{1-t} \\
&\qquad\qquad \left(\text{since } |||Z||| = |||Z^*||| \text{ for every } Z\right) \\
&\quad \le |||AX||| + |||XB||| \qquad \text{(by 3.2.1).}
\end{aligned}$$

□

Utilizing Theorem 4.2.1, we have the following corollary.

Corollary 4.2.2 *Let A and B be positive semidefinite matrices, and let X be any matrix. Then, for* $t \in [0, 1]$,

$$\left\| A^t X B^{1-t} + B^t X^* A^{1-t} \right\|_1 \le \|A + B\|_1 \, \|X\| .$$

Proof We have

$$\begin{aligned}
&\left\| A^t X B^{1-t} + B^t X^* A^{1-t} \right\|_1 \\
&\quad \le \|AX\|_1 + \|XB\|_1 \text{ (by Theorem 4.2.1)} \\
&\quad \le \|A\|_1 \, \|X\| + \|X\| \, \|B\|_1 \text{ (by inequality (3.3.22))} \\
&\quad = (\|A\|_1 + \|B\|_1) \, \|X\| \\
&\quad = \operatorname{tr}(A + B) \, \|X\| \\
&\quad = \|A + B\|_1 \, \|X\| .
\end{aligned}$$

□

Letting $X = I$ in Corollary 4.2.2, we have the following result answering the second main Question 3.2.1 affirmatively for the trace norm.

Corollary 4.2.3 *Let A and B be positive semidefinite matrices. Then, for* $t \in [0, 1]$,

$$\left\| A^t B^{1-t} + B^t A^{1-t} \right\|_1 \le \|A + B\|_1 .$$

Using Theorem 4.2.1 again, we have the following remark.

Remark 4.2.4 It follows from Theorem 4.2.1 that for every unitarily invariant norm,

$$|||A^t B^{1-t} + B^t A^{1-t}||| \le |||A||| + |||B||| ,$$

where A and B are positive semidefinite matrices and $t \in [0, 1]$.

To pursue our investigation of the second main Question 3.2.1, we need the following lemma, which was proved in [11]. It concerns commutators of positive semidefinite matrices.

Lemma 4.2.5 *Let X and Y be positive semidefinite matrices. Then*

$$\|XY - YX\| \leq \frac{1}{2} \|X\| \, \|Y\| .$$

A different version of the Heinz inequality (4.2.2) can be stated as follows (see for example, [12]).

Lemma 4.2.6 *Let A and B be positive semidefinite matrices. Then, for $t \in [0, 1]$ and for every unitarily invariant norm,*

$$\left|\left|\left| A^t B^{1-t} - A^{1-t} B^t \right|\right|\right| \leq |2t - 1| \; |||A - B||| .$$

We introduce some notations regarding the results that are related to the second main Question 3.2.1: Let A and B be positive semidefinite matrices, and set $b_t = A^t B^{1-t} + B^t A^{1-t}$ and $h_t = A^t B^{1-t} + A^{1-t} B^t$ for $t \in [0, 1]$.

Theorem 4.2.7 *Let A and B be positive semidefinite matrices. Then, for $t \in [0, 1]$ and for every unitarily invariant norm,*

$$|||\mathrm{Re}\, b_t||| \leq |||A + B||| .$$

Proof We have

$$\begin{aligned}
|||\mathrm{Re}\, b_t||| &= \frac{1}{2} \left|\left|\left| A^t B^{1-t} + B^t A^{1-t} + B^{1-t} A^t + A^{1-t} B^t \right|\right|\right| \\
&= \frac{1}{2} \left|\left|\left| \left(A^t B^{1-t} + A^{1-t} B^t\right) + \left(B^{1-t} A^t + B^t A^{1-t}\right) \right|\right|\right| \\
&= \frac{1}{2} \left|\left|\left| h_t + h_t^* \right|\right|\right| \\
&= |||\mathrm{Re}\, h_t||| \\
&\leq |||h_t||| \\
&\leq |||A + B||| \text{ (by inequality (4.2.2)).}
\end{aligned}$$

□

Theorem 4.2.8 *Let A and B be positive semidefinite matrices. Then, for $t \in [0, 1]$ and for every unitarily invariant norm,*

$$|||\mathrm{Im}\; b_t||| \leq |||A + B||| .$$

Proof We have

$$\begin{aligned}
\|\|\mathrm{Im}\ b_t\|\| &= \frac{1}{2}\left\|\left\|A^t B^{1-t} + B^t A^{1-t} - B^{1-t}A^t - A^{1-t}B^t\right\|\right\| \\
&= \frac{1}{2}\left\|\left\|A^t B^{1-t} - A^{1-t}B^t - \left(A^t B^{1-t} - A^{1-t}B^t\right)^*\right\|\right\| \\
&\le \left\|\left\|A^t B^{1-t} - A^{1-t}B^t\right\|\right\| \\
&\le |2t-1|\ \|\|A - B\|\| \text{ (by Lemma 4.2.6).}
\end{aligned} \tag{4.2.5}$$

Since $\|\|A - B\|\| \le \|\|A + B\|\|$ (see inequality (2.5) in [13]) and since $|2t - 1| \le 1$ for $t \in [0, 1]$, it follows from inequality (4.2.5) that $\|\|\mathrm{Im}\ b_t\|\| \le \|\|A + B\|\|$. □

In the case of the operator norm, we have the following estimates for $\|\mathrm{Im} b_t\|$.

Theorem 4.2.9 *Let A and B be positive semidefinite matrices. Then, for $t \in [0, 1]$,*

$$\|\mathrm{Im}\ b_t\| \le \frac{1}{4}\left(\|A\| + \|B\|\right).$$

Proof We have

$$\begin{aligned}
\|\mathrm{Im}\ b_t\| &= \frac{1}{2}\left\|A^t B^{1-t} + B^t A^{1-t} - B^{1-t}A^t - A^{1-t}B^t\right\| \\
&\le \frac{1}{2}\left(\left\|A^t B^{1-t} - B^{1-t}A^t\right\| + \left\|B^t A^{1-t} - A^{1-t}B^t\right\|\right) \\
&\le \frac{1}{2}\left(\frac{1}{2}\left\|A^t\right\|\left\|B^{1-t}\right\| + \frac{1}{2}\left\|B^t\right\|\left\|A^{1-t}\right\|\right) \text{ (by Lemma 4.2.5)} \\
&= \frac{1}{4}\left(\|A\|^t\|B\|^{1-t} + \|B\|^t\|A\|^{1-t}\right) \\
&\le \frac{1}{4}\left(\|A\| + \|B\|\right) \text{ (by 3.2.1).}
\end{aligned}$$

□

If A and B are positive semidefinite matrices, then $\max\{\|A\|, \|B\|\} \le \|A + B\|$. This, together with Theorem 4.2.9, yields the following result, which is an improvement over Theorem 4.2.8 for the operator norm.

Corollary 4.2.10 *Let A and B be positive semidefinite matrices. Then*

$$\|\mathrm{Im}\ b_t\| \le \frac{1}{2}\|A + B\|$$

for all $t \in [0, 1]$.

4.3 Related Trace Inequalities

Let A and B be positive semidefinite matrices. For $t \in [0, 1]$, recall that

$$b_t = A^t B^{1-t} + B^t A^{1-t},$$

and for $t \in \left[\frac{1}{2}, 1\right]$, let

$$f_t = B^{1-t} A^{2t-1} B^{1-t} + A^{1-t} B^{2t-1} A^{1-t}.$$

We introduce the following new notation. Let A and B be positive semidefinite matrices, and let X be any matrix. For $t \in [0, 1]$, let $B_t = A^t X B^{1-t} + B^t X^* A^{1-t}$, and for $t \in \left[\frac{1}{2}, 1\right]$, let $F_t = B^{1-t} X^* A^{2t-1} X B^{1-t} + A^{1-t} X B^{2t-1} X^* A^{1-t}$.

We show that for all unitarily invariant norms,

$$|||B_t||| \leq |||A + B|||^{\frac{1}{2}} \, |||F_t|||^{\frac{1}{2}}, \tag{4.3.1}$$

where A and B are positive semidefinite matrices and $t \in \left[\frac{1}{2}, 1\right]$.

To address the second main Question 3.2.1, by using inequality (4.3.1), we ask the following question. For more details on the main results of this section, we refer to [14, 15].

Question 4.3.1 Let A and B be positive semidefinite matrices, and let $t \in \left[\frac{1}{2}, 1\right]$. Is it true that

$$|||f_t||| = \left|\left|\left|B^{1-t} A^{2t-1} B^{1-t} + A^{1-t} B^{2t-1} A^{1-t}\right|\right|\right| \leq |||A + B|||?$$

We affirmatively answer this question for $t = \frac{1}{4}$ and $\frac{3}{4}$ as well as for $t \in \left[\frac{1}{4}, \frac{3}{4}\right]$ in the Hilbert–Schmidt norm. This leads to an affirmative answer to the second main Question 3.2.1 in the case of $t = \frac{1}{4}$ and $t = \frac{3}{4}$. It provides a different approach to answering the second main Question 3.2.1 for $t \in \left[\frac{1}{4}, \frac{3}{4}\right]$ in the Hilbert–Schmidt norm.

Moreover, we prove the following related norm inequality for all unitarily invariant norms

$$|||B_t||| \leq \|X\| \sqrt{|||A + B||| \left(|||A||| + |||B|||\right)}, \tag{4.3.2}$$

where A and B are positive semidefinite matrices and $t \in [0, 1]$.

The following lemma concerning the positivity of a block matrix involving B_t and F_t is helpful in the sequel.

Lemma 4.3.2 *Let A and B be positive semidefinite matrices, and let $t \in \left[\frac{1}{2}, 1\right]$. Then the matrix $\begin{bmatrix} A + B & B_t \\ B_t^* & F_t \end{bmatrix}$ is positive semidefinite.*

Proof We have

$$\begin{bmatrix} A+B & B_t \\ B_t^* & F_t \end{bmatrix} = \begin{bmatrix} A^{\frac{1}{2}} & B^{\frac{1}{2}} \\ B^{1-t}X^*A^{\frac{2t-1}{2}} & A^{1-t}XB^{\frac{2t-1}{2}} \end{bmatrix} \begin{bmatrix} A^{\frac{1}{2}} & A^{\frac{2t-1}{2}}XB^{1-t} \\ B^{\frac{1}{2}} & B^{\frac{2t-1}{2}}X^*A^{1-t} \end{bmatrix}$$
$$= \begin{bmatrix} A^{\frac{1}{2}} & B^{\frac{1}{2}} \\ B^{1-t}X^*A^{\frac{2t-1}{2}} & A^{1-t}XB^{\frac{2t-1}{2}} \end{bmatrix} \begin{bmatrix} A^{\frac{1}{2}} & B^{\frac{1}{2}} \\ B^{1-t}X^*A^{\frac{2t-1}{2}} & A^{1-t}XB^{\frac{2t-1}{2}} \end{bmatrix}^*$$
$$\geq 0.$$

□

The following lemma can be found in [16, p. 253].

Lemma 4.3.3 *Let A and B be matrices such that the product AB is normal. Then, for every unitarily invariant norm,*

$$|||AB||| \leq |||BA|||\,.$$

Employing Lemma 4.3.2 and Theorem 1.6.1, we have the following lemma.

Lemma 4.3.4 *Let A and B be positive semidefinite matrices, and let $t \in \left[\frac{1}{2}, 1\right]$. Then there exists a contraction C such that $B_t = (A+B)^{\frac{1}{2}}\, C\, (F_t)^{\frac{1}{2}}$.*

Theorem 4.3.5 *Let A and B be positive semidefinite matrices, and let X be any matrix. Then*

$$|||B_t||| \leq |||A+B|||^{\frac{1}{2}}\, |||F_t|||^{\frac{1}{2}}$$

for all $t \in \left[\frac{1}{2}, 1\right]$ and for every unitarily invariant norm.

Proof We have

$$\begin{aligned} |||B_t||| &= \left|\left|\left|(A+B)^{\frac{1}{2}}\, CF_t^{\frac{1}{2}}\right|\right|\right| \quad \text{(by Lemma 4.3.4)} \\ &\leq |||(A+B)\,C|||^{\frac{1}{2}}\, |||CF_t|||^{\frac{1}{2}} \quad \text{(by Lemma 3.3.5)} \\ &\leq (|||A+B|||\ \|C\|)^{\frac{1}{2}}\, (\|C\|\ |||F_t|||)^{\frac{1}{2}} \quad \text{(by inequality (2.3.4))} \\ &\leq |||A+B|||^{\frac{1}{2}}\, |||F_t|||^{\frac{1}{2}} \quad \text{(since } C \text{ is contraction).} \end{aligned}$$

□

Theorem 4.3.6 *Let A and B be positive semidefinite matrices, and let X be any matrix. Then*

$$|||F_t||| \leq \|X\|\,(|||AX||| + |||XB|||)$$

for every $t \in \left[\frac{1}{2}, 1\right]$ and every unitarily invariant norm.

Proof Let $s = 2t - 1$. Then we have $s \in [0, 1]$. Now,

$$
\begin{aligned}
|||F_t||| &= \left|\left|\left| B^{1-t} X^* A^{2t-1} X B^{1-t} + A^{1-t} X B^{2t-1} X^* A^{1-t} \right|\right|\right| \\
&\le \left|\left|\left| B^{1-t} X^* A^{2t-1} X B^{1-t} \right|\right|\right| + \left|\left|\left| A^{1-t} X B^{2t-1} X^* A^{1-t} \right|\right|\right| \\
&\le \left|\left|\left| X^* A^{2t-1} X B^{2-2t} \right|\right|\right| + \left|\left|\left| A^{2-2t} X B^{2t-1} X^* \right|\right|\right| \qquad \text{(by Lemma 4.3.3)} \\
&\le \left\| X^* \right\| \left|\left|\left| A^{2t-1} X B^{2-2t} \right|\right|\right| + \left|\left|\left| A^{2-2t} X B^{2t-1} \right|\right|\right| \left\| X^* \right\| \\
&= \|X\| \left(\left|\left|\left| A^s X B^{1-s} \right|\right|\right| + \left|\left|\left| A^{1-s} X B^s \right|\right|\right| \right) \\
&\le \|X\| \left(|||AX|||^s \, |||XB|||^{1-s} + |||AX|||^{1-s} \, |||XB|||^s \right) \\
&\qquad\qquad \text{(by Lemma 3.3.5)} \\
&\le \|X\| \left(|||AX||| + |||XB||| \right) \\
&\qquad\qquad \text{(by the Heinz inequality 3.2.1).}
\end{aligned}
$$

□

Letting $X = I$ in Theorem 4.3.6, we have the following corollary.

Corollary 4.3.7 *Let A and B be positive semidefinite matrices, and let $t \in \left[\frac{1}{2}, 1\right]$. Then*

$$|||f_t||| \le |||A||| + |||B|||$$

for every unitarily invariant norm.

Using Corollary 4.3.7, we have the following corollary for the trace norm.

Corollary 4.3.8 *Let A and B be positive semidefinite matrices, and let $t \in \left[\frac{1}{2}, 1\right]$. Then*

$$\|f_t\|_1 \le \|A + B\|_1 .$$

We prove our second main result of this section, which refines a generalized version of the inequality given in Remark 4.2.4.

Theorem 4.3.9 *Let A and B be positive semidefinite matrices, and let X be any matrix. Then, for $t \in [0, 1]$ and for every unitarily invariant norm,*

$$|||B_t||| \le \|X\| \sqrt{|||A + B||| \left(|||A||| + |||B||| \right)}.$$

Proof **Case 1:** $t \in \left[\frac{1}{2}, 1\right]$. In this case,

$$
\begin{aligned}
|||B_t||| &\le |||A + B|||^{\frac{1}{2}} \, |||F_t|||^{\frac{1}{2}} \quad \text{(by Theorem 4.3.5)} \\
&\le |||A + B|||^{\frac{1}{2}} \, \|X\|^{\frac{1}{2}} \left(|||AX||| + |||XB||| \right)^{\frac{1}{2}} \quad \text{(by Theorem 4.3.6)} \\
&\le \|X\| \sqrt{|||A + B||| \left(|||A||| + |||B||| \right)} \qquad \text{(by inequality (2.3.4)).}
\end{aligned}
$$

Case 2: $t \in \left[0, \frac{1}{2}\right]$. In this case,

$$\begin{aligned}
|||B_t||| &= \left|\left|\left|A^t X B^{1-t} + B^t X^* A^{1-t}\right|\right|\right| \\
&= \left|\left|\left|\left(A^t X B^{1-t} + B^t X^* A^{1-t}\right)^*\right|\right|\right| \\
&= \left|\left|\left|B^{1-t} X^* A^t + A^{1-t} X B^t\right|\right|\right| \\
&= |||B_{1-t}||| \\
&\leq \|X\| \sqrt{|||A+B||| \left(|||A||| + |||B|||\right)} \\
&\quad \left(\text{since } 1-t \in \left[\frac{1}{2}, 1\right] \text{ and using Case 1}\right).
\end{aligned}$$

□

Letting $X = I$ in Theorem 4.3.9, we have the following corollary.

Corollary 4.3.10 *Let A and B be positive semidefinite matrices, and let $t \in [0, 1]$. Then, for every unitarily invariant norm,*

$$|||b_t||| \leq \sqrt{|||A+B||| \left(|||A||| + |||B|||\right)}.$$

The following theorem gives an affirmative answer to Question 4.3.1 in the case of $t = \frac{3}{4}$.

Theorem 4.3.11 *Let A and B be positive semidefinite matrices. Then, for every unitarily invariant norm,*

$$\left|\left|\left|f_{\frac{3}{4}}\right|\right|\right| \leq |||A+B||| \,.$$

Proof We have

$$\begin{aligned}
\left|\left|\left|f_{\frac{3}{4}}\right|\right|\right| &= \left|\left|\left|B^{\frac{1}{4}} A^{\frac{1}{2}} B^{\frac{1}{4}} + A^{\frac{1}{4}} B^{\frac{1}{2}} A^{\frac{1}{4}}\right|\right|\right| \\
&\leq \left|\left|\left|B^{\frac{1}{4}} A^{\frac{1}{2}} B^{\frac{1}{4}}\right|\right|\right| + \left|\left|\left|A^{\frac{1}{4}} B^{\frac{1}{2}} A^{\frac{1}{4}}\right|\right|\right| \\
&\leq \left|\left|\left|A^{\frac{1}{2}} B^{\frac{1}{2}}\right|\right|\right| + \left|\left|\left|A^{\frac{1}{2}} B^{\frac{1}{2}}\right|\right|\right| \qquad \text{(by Lemma 4.3.3)} \\
&= 2\left|\left|\left|A^{\frac{1}{2}} B^{\frac{1}{2}}\right|\right|\right| \\
&\leq |||A+B||| \qquad \text{(by Inequality (2.4.11))}.
\end{aligned}$$

□

Based on Theorems 4.3.5 and 4.3.11, we have the following norm inequality, which gives an affirmative answer to the second main Question 3.2.1 in the case of $t = \frac{3}{4}$.

Corollary 4.3.12 *Let A and B be positive semidefinite matrices. Then, for every unitarily invariant norm,*

$$\left|\left|\left| b_{\frac{3}{4}} \right|\right|\right| \leq |||A + B||| .$$

In virtue of Corollary 4.3.12 and the fact that $b_t = b^*_{1-t}$ and $|||T||| = |||T^*|||$ for any matrix T, we have the following norm inequality, which gives an affirmative answer to the second main Question 3.2.1 in the case of $t = \frac{1}{4}$.

Corollary 4.3.13 *Let A and B be positive semidefinite matrices. Then, for every unitarily invariant norm,*

$$\left|\left|\left| b_{\frac{1}{4}} \right|\right|\right| \leq |||A + B||| .$$

The following proposition gives an affirmative answer to the second main Question 3.2.1 in the case of $t = \frac{1}{2}$.

Proposition 4.3.14 *Let A and B be positive semidefinite matrices. Then, for every unitarily invariant norm,*

$$\left|\left|\left| b_{\frac{1}{2}} \right|\right|\right| \leq |||A + B||| .$$

Proof We have

$$\begin{aligned}
\left|\left|\left| b_{\frac{1}{2}} \right|\right|\right| &= \left|\left|\left| A^{\frac{1}{2}} B^{\frac{1}{2}} + B^{\frac{1}{2}} A^{\frac{1}{2}} \right|\right|\right| \\
&\leq \left|\left|\left| A^{\frac{1}{2}} B^{\frac{1}{2}} \right|\right|\right| + \left|\left|\left| B^{\frac{1}{2}} A^{\frac{1}{2}} \right|\right|\right| \\
&= \left|\left|\left| A^{\frac{1}{2}} B^{\frac{1}{2}} \right|\right|\right| + \left|\left|\left| \left(A^{\frac{1}{2}} B^{\frac{1}{2}}\right)^* \right|\right|\right| \\
&= 2 \left|\left|\left| A^{\frac{1}{2}} B^{\frac{1}{2}} \right|\right|\right| \\
&\leq |||A + B||| \qquad \text{(by inequality (2.4.11))}.
\end{aligned}$$

□

Remark 4.3.15 In this section, we have answered the second main Question 3.2.1 affirmatively for $t = \frac{1}{4}$, $\frac{1}{2}$, and $\frac{3}{4}$. This is the case for $t = 0$ and 1. The question remains open for other values of $t \in [0, 1]$.

We conclude this section by considering Question 4.3.1 for the Hilbert–Schmidt norm. To achieve our goal, we start with the following lemma. This lemma was proved by Ando, Hiai, and Okubo [17].

Lemma 4.3.16 *Let A and B be positive semidefinite matrices, and let $p_i, q_i \geq 0$ for $i = 1, \ldots, k$ be such that $\sum_{i=1}^k p_i = \sum_{i=1}^k q_i = 1$. If p_i, q_i satisfy the following assumptions:*

(i) $0 \leq \sum_{i=1}^j q_i - \sum_{i=1}^j p_i \leq \frac{1}{2}$ *for* $j = 1, \ldots, k-1$;

(ii) $0 \leq \sum_{i=1}^j p_i - \sum_{i=1}^{j-1} q_i \leq \frac{1}{2}$ *for* $j = 1, \ldots, k$,

then

$$\left|\operatorname{tr}\left(A^{p_1}B^{q_1}A^{p_2}B^{q_2}\ldots A^{p_k}B^{q_k}\right)\right| \le \operatorname{tr}(AB).$$

Here, we use the convention $\sum_{i=1}^{j-1} q_i = 0$ for $j = 1$.

The following lemma is a special case of the previous lemma.

Lemma 4.3.17 *Let A and B be positive semidefinite matrices, and let $p_1, p_2, p_3 \ge 0$, $q_1, q_2, q_3 \ge 0$ be such that $p_1 + p_2 + p_3 = 1$ and $q_1 + q_2 + q_3 = 1$. If p_1, p_2, p_3 and q_1, q_2, q_3 satisfy the following assumptions:*

(i) $0 \le p_1 \le \frac{1}{2}$.

(ii) $0 \le q_1 - p_1 \le \frac{1}{2}$.

(iii) $0 \le p_1 + p_2 - q_1 \le \frac{1}{2}$.

(iv) $0 \le q_1 + q_2 - (p_1 + p_2) \le \frac{1}{2}$.

(v) $0 \le p_1 + p_2 + p_3 - (q_1 + q_2) \le \frac{1}{2}$,

then

$$\left|\operatorname{tr}\left(A^{p_1}B^{q_1}A^{p_2}B^{q_2}A^{p_3}B^{q_3}\right)\right| \le \operatorname{tr}(AB).$$

To prove our main result, we state and prove the following two lemmas, which contain two interesting trace inequalities.

Lemma 4.3.18 *Let A and B be positive semidefinite matrices, and let $\nu \in \left[\frac{1}{2}, 1\right]$. Then*

$$\operatorname{tr}\left(B^{\frac{\nu}{2}}A^{1-\nu}B^{\frac{\nu}{2}}A^{\frac{\nu}{2}}B^{1-\nu}A^{\frac{\nu}{2}}\right) \le \operatorname{tr}(AB). \tag{4.3.3}$$

Proof Let $p_1 = \frac{\nu}{2}$, $p_2 = \frac{\nu}{2}$, $p_3 = 1 - \nu$ and $q_1 = 1 - \nu$, $q_2 = \frac{\nu}{2}$, $q_3 = \frac{\nu}{2}$. Then we have

$$0 \le p_1 = \frac{\nu}{2} \le \frac{1}{2} \Leftrightarrow 0 \le \nu \le 1,$$

$$0 \le q_1 - p_1 = 1 - \frac{3}{2}\nu \le \frac{1}{2} \Leftrightarrow \frac{1}{3} \le \nu \le \frac{2}{3},$$

$$0 \le p_1 + p_2 - q_1 = 2\nu - 1 \le \frac{1}{2} \Leftrightarrow \frac{1}{2} \le \nu \le \frac{3}{4}.$$

Also, we can easily see that $q_1 + q_2 - (p_1 + p_2) = 1 - \frac{3}{2}\nu$ and $p_1 + p_2 + p_3 - (q_1 + q_2) = \frac{\nu}{2}$.

Thus, by Lemma 4.3.17, inequality (4.3.3) holds provided $\nu \in \left[\frac{1}{2}, \frac{2}{3}\right]$.

Let $p_1 = \frac{\nu}{2}$, $p_2 = 1 - \nu$, $p_3 = \frac{\nu}{2}$ and $q_1 = \frac{\nu}{2}$, $q_2 = \frac{\nu}{2}$, $q_3 = 1 - \nu$. Then we have

$$0 \le p_1 = \frac{\nu}{2} \le \frac{1}{2} \Leftrightarrow 0 \le \nu \le 1,$$

$$0 \le p_1 + p_2 - q_1 = 1 - \nu \le \frac{1}{2} \Leftrightarrow \frac{1}{2} \le \nu \le 1,$$

$$0 \le q_1 + q_2 - (p_1 + p_2) = \frac{3}{2}\nu - 1 \le \frac{1}{2} \Leftrightarrow \frac{2}{3} \le \nu \le 1.$$

Also, we can easily see that $q_1 - p_1 = 0$ and $p_1 + p_2 + p_3 - (q_1 + q_2) = 1 - \nu$.

Thus, by Lemma 4.3.17, inequality (4.3.3) holds provided $\nu \in \left[\frac{2}{3}, 1\right]$. Therefore, inequality (4.3.3) holds provided $\nu \in \left[\frac{1}{2}, 1\right]$. □

The following lemma is based on the Araki–Lieb–Thirring trace inequality (3.2.27), the Weyl majorant theorem, and the Heinz inequality (3.2.2) for the trace norm.

Lemma 4.3.19 *Let A and B be positive semidefinite matrices, and let $\nu \in [0, 1]$. Then*

$$\operatorname{tr}\left(\left(B^{\frac{\nu}{2}} A^{1-\nu} B^{\frac{\nu}{2}}\right)^2 + \left(A^{\frac{\nu}{2}} B^{1-\nu} A^{\frac{\nu}{2}}\right)^2\right) \le \operatorname{tr}\left(A^2 + B^2\right).$$

Proof We have

$$\begin{aligned}
&\operatorname{tr}\left(\left(B^{\frac{\nu}{2}} A^{1-\nu} B^{\frac{\nu}{2}}\right)^2 + \left(A^{\frac{\nu}{2}} B^{1-\nu} A^{\frac{\nu}{2}}\right)^2\right) \\
&\quad = \operatorname{tr}\left(B^{\frac{\nu}{2}} A^{1-\nu} B^{\nu} A^{1-\nu} B^{\frac{\nu}{2}} + A^{\frac{\nu}{2}} B^{1-\nu} A^{\nu} B^{1-\nu} A^{\frac{\nu}{2}}\right) \\
&\quad = \operatorname{tr}\left(A^{1-\nu} B^{\nu} A^{1-\nu} B^{\nu} + A^{\nu} B^{1-\nu} A^{\nu} B^{1-\nu}\right) \\
&\quad = \operatorname{tr}\left(\left(A^{1-\nu} B^{\nu}\right)^2 + \left(A^{\nu} B^{1-\nu}\right)^2\right) \\
&\quad \le \operatorname{tr}\left(A^{2(1-\nu)} B^{2\nu} + A^{2\nu} B^{2(1-\nu)}\right) && \text{(by Lemma 3.2.4)} \\
&\quad = \left|\operatorname{tr}\left(A^{2(1-\nu)} B^{2\nu} + A^{2\nu} B^{2(1-\nu)}\right)\right| \\
&\quad \le \operatorname{tr}\left|A^{2(1-\nu)} B^{2\nu} + A^{2\nu} B^{2(1-\nu)}\right| && \text{(by Lemma 3.1.4)} \\
&\quad = \left\|A^{2(1-\nu)} B^{2\nu} + A^{2\nu} B^{2(1-\nu)}\right\|_1 \\
&\quad \le \left\|A^2 + B^2\right\|_1 && \text{(by inequality (4.2.2))} \\
&\quad = \operatorname{tr}\left(A^2 + B^2\right).
\end{aligned}$$

□

We are in a position to state and prove our last main result of this section.

Theorem 4.3.20 *Let A and B be positive semidefinite matrices, and let $t \in \left[\frac{1}{2}, \frac{3}{4}\right]$. Then*

$$\|f_t\|_2 \le \|A + B\|_2 . \tag{4.3.4}$$

Proof The square of the left-hand side of inequality (4.3.4) is equal to

$$\operatorname{tr}\left(\left(B^{1-t} A^{2t-1} B^{1-t}\right)^2 + \left(A^{1-t} B^{2t-1} A^{1-t}\right)^2 + 2B^{1-t} A^{2t-1} B^{1-t} A^{1-t} B^{2t-1} A^{1-t}\right)$$

and the square of the right-hand side is equal to

$$\operatorname{tr}\left(A^2 + B^2 + 2AB\right).$$

Letting $\nu = 2(1-t)$, we can see that the square of the left-hand side of inequality (4.3.4) is equal to

$$\operatorname{tr}\left(\left(B^{\frac{\nu}{2}} A^{1-\nu} B^{\frac{\nu}{2}}\right)^2 + \left(A^{\frac{\nu}{2}} B^{1-\nu} A^{\frac{\nu}{2}}\right)^2 + 2B^{\frac{\nu}{2}} A^{1-\nu} B^{\frac{\nu}{2}} A^{\frac{\nu}{2}} B^{1-\nu} A^{\frac{\nu}{2}}\right).$$

Again, we have used the definition $\|X\|_2 = (\operatorname{tr}(X^*X))^{\frac{1}{2}}$ and the cyclicity of the trace.

Therefore, inequality (4.3.4) is equivalent to the trace inequality

$$\operatorname{tr}\left(\left(B^{\frac{\nu}{2}} A^{1-\nu} B^{\frac{\nu}{2}}\right)^2 + \left(A^{\frac{\nu}{2}} B^{1-\nu} A^{\frac{\nu}{2}}\right)^2 + 2B^{\frac{\nu}{2}} A^{1-\nu} B^{\frac{\nu}{2}} A^{\frac{\nu}{2}} B^{1-\nu} A^{\frac{\nu}{2}}\right)$$
$$\leq \operatorname{tr}\left(A^2 + B^2 + 2AB\right). \tag{4.3.5}$$

Using Lemmas 4.3.18 and 4.3.19, inequality (4.3.5) holds provided $\nu \in \left[\frac{1}{2}, 1\right]$. Hence, inequality (4.3.4) is valid provided $t \in \left[\frac{1}{2}, \frac{3}{4}\right]$. □

Based on Theorem 4.3.20, and using inequality (4.3.1), we have the following corollary, which gives another answer to the second main Question 3.2.1 for the Hilbert–Schmidt norm under the condition $t \in \left[\frac{1}{4}, \frac{3}{4}\right]$.

Corollary 4.3.21 *Let A and B be positive semidefinite matrices, and let $t \in \left[\frac{1}{4}, \frac{3}{4}\right]$. Then*

$$\|b_t\|_2 \leq \|A + B\|_2.$$

Proof **Case 1:** $t \in \left[\frac{1}{2}, \frac{3}{4}\right]$. In this case,

$$\begin{aligned} \|b_t\|_2 &\leq \|A+B\|_2^{\frac{1}{2}} \|f_t\|_2^{\frac{1}{2}} && \text{(by inequality (4.3.1))} \\ &\leq \|A+B\|_2 && \text{(by Theorem 4.3.20)}. \end{aligned}$$

Case 2: $t \in \left[\frac{1}{4}, \frac{1}{2}\right]$. In this case,

$$\begin{aligned} \|b_t\|_2 &= \left\|b_t^*\right\|_2 \\ &= \|b_{1-t}\|_2 \\ &\leq \|A+B\|_2 \left(\text{since } 1-t \in \left[\frac{1}{2}, \frac{3}{4}\right] \text{ and using Case 1}\right). \end{aligned}$$

□

4.4 Related Majorization Relations

Let A and B be positive semidefinite matrices. Then, for $t \in [0, 1]$, recall that

$$b_t = A^t B^{1-t} + B^t A^{1-t},$$

and for $t \in [\frac{1}{2}, 1]$,

$$f_t = B^{1-t} A^{2t-1} B^{1-t} + A^{1-t} B^{2t-1} A^{1-t}.$$

In [18], Alakhrass proved the following inequality:

$$\||b_t\|| \leq 2^{\left|\frac{1}{2}-t\right|} \||A + B\|| \tag{4.4.1}$$

for $t \in [0, 1]$.

In an attempt to answer Question 4.3.1, we give the following norm inequality for any two positive semidefinite matrices A, B, and $t \in [\frac{1}{2}, 1]$

$$\||f_t\|| \leq 2^{2t-1} \||A + B\|| .$$

The following facts are used frequently in the sequel. For any two matrices A and B, we have

$$s(A + B) \prec_w s(A) + s(B) \tag{4.4.2}$$

and

$$s(AB) \prec_w s(A)s(B). \tag{4.4.3}$$

We need the next lemma (see examples given in Chap. 5 for its interpretation).

Lemma 4.4.1 *Let A and B be positive semidefinite matrices, and let $r \in [0, 1]$. Then*

$$A^r + B^r \leq 2^{1-r} (A + B)^r .$$

In view of Remark 2.3.2 and the pinching inequality for unitarily invariant norms (2.3.12), we have the following lemma.

Lemma 4.4.2 *Let A and B be positive semidefinite matrices. Then, for every unitarily invariant norm,*

$$\||A \oplus B\|| \leq \||A + B\|| .$$

The following lemma for commutators can be found in [11].

Lemma 4.4.3 *Let A and X be positive semidefinite matrices. Then, for every unitarily invariant norm,*

$$|||AX - XA||| \leq \frac{1}{2}\|A\| \; |||X \oplus X||| .$$

In particular,

$$\|AX - XA\| \leq \frac{1}{2}\|A\| \; \|X\|.$$

The following lemma can be derived from [19, p. 14].

Lemma 4.4.4 *Let A and B be positive semidefinite matrices, and let $p \geq 1$. Then*

$$\operatorname{tr}(A^p + B^p) \leq \operatorname{tr}(A + B)^p.$$

Our first main result of this section can be stated as follows [20].

Theorem 4.4.5 *Let A and B be positive semidefinite matrices, and let $t \in [\frac{1}{2}, 1]$. Then*

$$\|f_t\|_{(k)} \leq 2^{2t-1}\|A + B\|_{(k)} \text{ for } k = 1, \ldots, n.$$

Proof Let $Y = \begin{bmatrix} B^{1-t} & A^{1-t} \\ 0 & 0 \end{bmatrix}$, $D = \begin{bmatrix} A^{2t-1} & 0 \\ 0 & B^{2t-1} \end{bmatrix}$, and $Z = \begin{bmatrix} A^{2t-1}B^{1-t} & 0 \\ B^{2t-1}A^{1-t} & 0 \end{bmatrix}$. Then, for $k = 1, 2, \ldots, n$, we have

$$\begin{aligned}
\|f_t\|_{(k)} &= \|f_t \oplus 0\|_{(k)} \\
&= \|YZ\|_{(k)} \\
&\leq \|ZY\|_{(k)} \text{ (by Lemma 4.3.3)} \\
&= \|DY^*Y\|_{(k)} \\
&= \sum_{j=1}^{k} s_j\left(DY^*Y\right) \\
&\leq \sum_{j=1}^{k} s_j(D)s_j(Y^*Y) \text{ (by inequality (4.4.3))} \\
&= \sum_{j=1}^{k} s_j(D)s_j(YY^*) \\
&= \sum_{j=1}^{k} s_j\left((A^{2t-1} \oplus B^{2t-1})\right) s_j\left(A^{2(1-t)} + B^{2(1-t)}\right),
\end{aligned}$$

since $s_j(X \oplus 0) = s_j(X)$ for any matrix X. Hence,

$$\begin{aligned}
\|f_t\|_{(k)} &\le 2^{1-2(1-t)} \sum_{j=1}^{k} s_j\left((A+B)^{2(1-t)}\right) s_j\left((A \oplus B)^{2t-1}\right) \text{ (by Lemma 4.4.1)} \\
&= 2^{2t-1} \sum_{j=1}^{k} s_j^{2(1-t)}(A+B)\, s_j^{2t-1}(A \oplus B) \\
&\le 2^{2t-1} \sum_{j=1}^{k} 2(1-t)s_j(A+B) + (2t-1)s_j(A \oplus B) \\
&\qquad \text{(by Young's inequality (3.3.25))} \\
&= 2^{2t-1} \left(2(1-t) \sum_{j=1}^{k} s_j(A+B) + (2t-1) \sum_{j=1}^{k} s_j(A \oplus B) \right) \\
&= 2^{2t-1} \left(2(1-t)\|A+B\|_{(k)} + (2t-1)\|A \oplus B\|_{(k)} \right) \\
&\le 2^{2t-1} \|A+B\|_{(k)} \text{ (by Lemma 4.4.2).}
\end{aligned}$$

□

By the Fan dominance theorem, we have the following corollary.

Corollary 4.4.6 *Let A and B be positive semidefinite matrices, and let $t \in [\frac{1}{2}, 1]$. Then, for every unitarily invariant norm,*

$$|||f_t||| \le 2^{2t-1}\, |||A+B||| .$$

Taking $X = I$ in Theorem 4.3.5, and using the previous corollary, we have the following corollary.

Corollary 4.4.7 *Let A and B be positive semidefinite matrices, and let $t \in [\frac{1}{2}, 1]$. Then, for every unitarily invariant norm,*

$$|||b_t||| \le 2^{t-\frac{1}{2}}\, |||A+B||| .$$

Based on Corollary 4.4.7, we have the following corollary.

Corollary 4.4.8 *Let A and B be positive semidefinite matrices, and let $t \in [0, \frac{1}{2}]$. Then, for every unitarily invariant norm,*

$$|||b_t||| \le 2^{\frac{1}{2}-t}\, |||A+B||| .$$

Proof For $t \in [0, \frac{1}{2}]$, we have

$$\begin{aligned}
\left|\!\left|\!\left| b_t \right|\!\right|\!\right| &= \left|\!\left|\!\left| A^t B^{1-t} + B^t A^{1-t} \right|\!\right|\!\right| \\
&= \left|\!\left|\!\left| \left(A^t B^{1-t} + B^t A^{1-t}\right)^* \right|\!\right|\!\right| \\
&= \left|\!\left|\!\left| A^{1-t} B^t + B^{1-t} A^t \right|\!\right|\!\right| \\
&= \left|\!\left|\!\left| b_{1-t} \right|\!\right|\!\right| \\
&\le 2^{\frac{1}{2}-t} \left|\!\left|\!\left| A + B \right|\!\right|\!\right| \text{ (by Corollary 4.4.7 and since } 1-t \in [\frac{1}{2}, 1]).
\end{aligned}$$

□

Remark 4.4.9 Corollaries 4.4.7 and 4.4.8 represent inequality (4.4.1) due to Alakhrass [18].

The following lemma is applied in the proof of our second main result of this section.

Lemma 4.4.10 *Let A and B be positive semidefinite matrices. Then, for $p \ge 1$,*

$$\|A\|_p + \|B\|_p \le 2^{1-\frac{1}{p}} \|A + B\|_p.$$

Proof For $p \ge 1$, we have

$$\begin{aligned}
\|A\|_p + \|B\|_p &= \left(\operatorname{tr}|A|^p\right)^{\frac{1}{p}} + \left(\operatorname{tr}|B|^p\right)^{\frac{1}{p}} \\
&\le 2^{1-\frac{1}{p}} \left(\operatorname{tr}|A|^p + \operatorname{tr}|B|^p\right)^{\frac{1}{p}} \text{ (by the concavity of } f(t) = t^{\frac{1}{p}}) \\
&= 2^{1-\frac{1}{p}} \left(\operatorname{tr} A^p + \operatorname{tr} B^p\right)^{\frac{1}{p}} \\
&= 2^{1-\frac{1}{p}} \left(\operatorname{tr}(A^p + B^p)\right)^{\frac{1}{p}} \\
&\le 2^{1-\frac{1}{p}} \left(\operatorname{tr}(A + B)^p\right)^{\frac{1}{p}} \text{ (by Lemma 4.4.4)} \\
&= 2^{1-\frac{1}{p}} \|A + B\|_p.
\end{aligned}$$

□

We now state our second main result of this section.

Theorem 4.4.11 *Let A and B be positive semidefinite matrices, let X be any matrix, and let $t \in [0, 1]$. Then, for $p \ge 1$,*

$$\|B_t\|_p \le 2^{\frac{1}{2}-\frac{1}{2p}} \|X\| \, \|A + B\|_p.$$

Proof For $t \in [0, 1]$, we have

$$\begin{aligned}
\|B_t\|_p &\le \|X\|\sqrt{\|A+B\|_p\left(\|A\|_p+\|B\|_p\right)} \text{ (by Theorem 4.3.9} \\
&\le \|X\|\sqrt{\|A+B\|_p\left(2^{1-\frac{1}{p}}\|A+B\|_p\right)} \text{ (by Lemma 4.4.10)} \\
&= \|X\|\sqrt{2^{1-\frac{1}{p}}\|A+B\|_p^2} \\
&= 2^{\frac{1}{2}-\frac{1}{2p}}\|X\|\,\|A+B\|_p.
\end{aligned}$$

□

In Theorem 4.4.11, taking $X = I$, we have the following corollary.

Corollary 4.4.12 *Let A and B be positive semidefinite matrices, and let $t \in [0, 1]$. Then, for $p \ge 1$,*

$$\|b_t\|_p \le 2^{\frac{1}{2}-\frac{1}{2p}}\|A+B\|_p. \tag{4.4.4}$$

For $p = 1$, we obtain the trace inequality

$$\|b_t\|_1 \le \|A+B\|_1,$$

which has been obtained in Corollary 4.2.3.

For $p = 2$, we have the following corollary.

Corollary 4.4.13 *Let A and B be positive semidefinite matrices, and let $t \in [0, 1]$. Then*

$$\|b_t\|_2 \le 2^{\frac{1}{4}}\|A+B\|_2.$$

For the operator norm, we have the following corollary.

Corollary 4.4.14 *Let A and B be positive semidefinite matrices, and let $t \in [0, 1]$. Then*

$$\|b_t\| \le \sqrt{2}\|A+B\|.$$

As mentioned before, Bottazzi et al. [9] gave a counterexample to show that inequality (3.3.21) does not hold for the operator norm. The following theorem gives an inequality related to (3.3.21).

Theorem 4.4.15 *Let A and B be positive semidefinite matrices, and let $t \in [0, 1]$. Then*

$$\|b_t\| \le \|A^tB^{1-t}+A^{1-t}B^t\| + \frac{1}{2}\|A\|^{1-t}\|B\|^t.$$

Proof We have

$$\begin{aligned}
\|b_t\| &= \|A^t B^{1-t} + B^t A^{1-t}\| \\
&= \|A^t B^{1-t} + B^t A^{1-t} + A^{1-t} B^t - A^{1-t} B^t\| \\
&\le \|A^t B^{1-t} + A^{1-t} B^t\| + \|B^t A^{1-t} - A^{1-t} B^t\| \\
&\le \|A^t B^{1-t} + A^{1-t} B^t\| + \frac{1}{2}\|A^{1-t}\|\ \|B^t\| \text{ (by Lemma 4.4.3)} \\
&= \|A^t B^{1-t} + A^{1-t} B^t\| + \frac{1}{2}\|A\|^{1-t}\|B\|^t.
\end{aligned}$$

□

Using the Heinz inequality (4.2.2), we have the following corollary.

Corollary 4.4.16 *Let A and B be positive semidefinite matrices, and let $t \in [0, 1]$. Then*

$$\|b_t\| \le \|A + B\| + \frac{1}{2}\|A\|^{1-t}\|B\|^t.$$

We conclude this section with some related weak majorization relations.

Theorem 4.4.17 *Let A and B be positive semidefinite matrices. Then*

$$s(b_t) \prec_\omega s^{2(1-t)}(A + B)\,(s(A) + s(B))^{2t-1} \text{ for } t \in \left[\frac{1}{2}, 1\right]$$

and

$$s(b_t) \prec_\omega s^{2t}(A + B)\,(s(A) + s(B))^{1-2t} \text{ for } t \in \left[0, \frac{1}{2}\right].$$

Proof The result is obvious when $t = 0$ and $t = 1$ by inequality (4.4.2). Now, for $t \in [\frac{1}{2}, 1)$ and $k = 1, 2, \ldots, n$, we have

$$\begin{aligned}
\sum_{j=1}^{k} s_j(b_t) &= \sum_{j=1}^{k} s_j(A^t B^{1-t} + B^t A^{1-t}) \\
&\le \sum_{j=1}^{k} s_j(A^t B^{1-t}) + \sum_{j=1}^{k} s_j(B^t A^{1-t}) \text{ (by inequality (4.4.2))} \\
&= \sum_{j=1}^{k} s_j(A^{2t-1} A^{1-t} B^{1-t}) + \sum_{j=1}^{k} s_j(B^{2t-1} B^{1-t} A^{1-t}) \\
&\le \sum_{j=1}^{k} \left(s_j(A^{2t-1}) s_j(A^{1-t} B^{1-t})\right) + \sum_{j=1}^{k} \left(s_j(B^{2t-1}) s_j(B^{1-t} A^{1-t})\right) \\
&\qquad \text{(by inequality (4.4.3))}
\end{aligned}$$

$$= \sum_{j=1}^{k} \left(s_j \left(A^{1-t} B^{1-t}\right) \left(s_j(A^{2t-1}) + s_j(B^{2t-1})\right)\right)$$

$$\leq \frac{1}{2} \sum_{j=1}^{k} s_j \left(\left(A^{2(1-t)} + B^{2(1-t)}\right) \left(s_j^{2t-1}(A) + s_j^{2t-1}(B)\right)\right)$$

(by inequality 2.4.30)

$$\leq \frac{1}{2} \sum_{j=1}^{k} \left(s_j \left(A^{2(1-t)} + B^{2(1-t)}\right) \left(2^{2-2t} \left(s_j(A) + s_j(B)\right)^{2t-1}\right)\right)$$

(by Exercise 3.6.1)

$$\leq \frac{1}{2} \sum_{j=1}^{k} \left(2^{1-2(1-t)} s_j^{2(1-t)}(A + B) \left(2^{2-2t} (s_j(A) + s_j(B))^{2t-1}\right)\right)$$

(by Lemma 4.4.1)

$$= \frac{2^{1-2(1-t)} 2^{2-2t}}{2} \left(\sum_{j=1}^{k} \left(s_j^{2(1-t)}(A + B) \left(s_j(A) + s_j(B)\right)^{2t-1}\right)\right)$$

$$= \sum_{j=1}^{k} \left(s_j^{2(1-t)}(A + B) \left(s_j(A) + s_j(B)\right)^{2t-1}\right).$$

Hence, for $t \in \left[\frac{1}{2}, 1\right)$, we have

$$s(b_t) \prec_\omega s^{2(1-t)}(A + B) \left(s(A) + s(B)\right)^{2t-1}. \tag{4.4.5}$$

Now, if $t \in \left(0, \frac{1}{2}\right]$, then $1 - t \in \left[\frac{1}{2}, 1\right)$. Thus,

$$\begin{aligned} s(b_t) &= s(A^t B^{1-t} + B^t A^{1-t}) \\ &= s((A^t B^{1-t} + B^t A^{1-t})^*) \\ &= s(A^{1-t} B^t + B^{1-t} A^t) \\ &= s(b_{1-t}) \\ &\prec_\omega s^{2t}(A + B) \left(s(A) + s(B)\right)^{1-2t}. \\ &\quad \text{(replacing } t \text{ with } 1 - t \text{ in inequality (4.4.5))} \end{aligned}$$

□

Remark 4.4.18

(1) Equality in Theorem 4.4.5 holds for $A = B$.
(2) For $t = \frac{1}{2}$, we have

$$s(b_{\frac{1}{2}}) \prec_\omega s(A + B).$$

For $t = \frac{1}{4}$ and $t = \frac{3}{4}$, we have the following inequalities:

$$s(b_{\frac{1}{4}}) \prec_\omega \sqrt{s(A+B)}\sqrt{s(A)+s(B)} \tag{4.4.6}$$

and

$$s(b_{\frac{3}{4}}) \prec_\omega \sqrt{s(A+B)}\sqrt{s(A)+s(B)}. \tag{4.4.7}$$

(3) The inequalities (4.4.6) and (4.4.7) represent majorization versions of inequality (4.3.2) for $t = \frac{1}{4}$ and $t = \frac{3}{4}$.

Theorem 4.4.19 *Let A and B be positive semidefinite matrices, and let $t \in [0, 1]$. Then*

$$s(b_t) \prec_\omega s(A) + s(B).$$

Proof For $k = 1, 2, \ldots, n$, we have

$$\begin{aligned}
\sum_{j=1}^{k} s_j(b_t) &= \sum_{j=1}^{k} s_j(A^t B^{1-t} + B^t A^{1-t}) \\
&\le \sum_{j=1}^{k} s_j(A^t B^{1-t}) + \sum_{j=1}^{k} s_j(B^t A^{1-t}) \text{ (by inequality (4.4.2))} \\
&\le \sum_{j=1}^{k} s_j^t(A) s_j^{1-t}(B) + \sum_{j=1}^{k} s_j^t(B) s_j^{1-t}(A) \text{ (by inequality (4.4.3))} \\
&\le \sum_{j=1}^{k} t s_j(A) + (1-t) s_j(B) + \sum_{j=1}^{k} t s_j(B) + (1-t) s_j(A) \\
&\quad \text{(by Young's inequality (3.3.25))} \\
&= \sum_{j=1}^{k} \left(s_j(A) + s_j(B)\right).
\end{aligned}$$

□

Remark 4.4.20

(1) In the proof of Theorem 4.4.19, taking $k = n$, we have

$$\sum_{j=1}^{n} s_j(b_t) \le \sum_{j=1}^{n} s_j(A+B),$$

which is equivalent to the trace norm inequality mentioned in Corollary 4.2.3.

(2) Theorem 4.4.19 is a majorization version of the following inequality:

$$|||b_t||| \le |||A||| + |||B|||,$$

which has been mentioned in Remark 4.2.4.

4.5 Inequalities Involving the Spectral Radius

Let A and B be positive definite matrices. For $t \in [0, 1]$, recall that

$$b_t = A^t B^{1-t} + B^t A^{1-t},$$

and for $t \in \left[\frac{1}{2}, 1\right]$,

$$f_t = B^{1-t} A^{2t-1} B^{1-t} + A^{1-t} B^{2t-1} A^{1-t}. \tag{4.5.1}$$

In this section, and in order to try to answer the second main Question 3.2.1, we go further and introduce a sequence g_n of positive definite matrices such that $g_0 = f_t$ and

$$\begin{bmatrix} A+B & g_{n-1} \\ g_{n-1} & g_n \end{bmatrix} \geq 0 \text{ for } n = 1, 2, \ldots$$

Therefore, we have the following norm inequality for all unitarily invariant norms:

$$\|\|g_{n-1}\|\| \leq \|\|A+B\|\|^{\frac{1}{2}} \|\|g_n\|\|^{\frac{1}{2}},$$

where A and B are positive definite matrices.

Using the induction process, we get the following norm inequality for all unitarily invariant norms

$$\|\|b_t\|\| \leq \|\|A+B\|\|^{\sum_{k=1}^{n+1} \frac{1}{2^k}} \|\|g_n\|\|^{\frac{1}{2^{n+1}}} \text{ for } n = 1, 2, \ldots$$

where A and B are positive definite matrices and $t \in \left[\frac{1}{2}, 1\right]$.

The rest of this section is devoted to calculating the limit of the sequence $\left\{\|\|g_n\|\|^{\frac{1}{2^{n+1}}}\right\}_{n \geq 0}$. In a rather mysterious way, it turns out that the limit is written in terms of the spectral radii of specific matrices. In other words, we prove that

$$\lim_{n \to \infty} \|\|g_n\|\|^{\frac{1}{2^n}} = \max\left\{r\left(B^{2t-1} A^{1-2t}\right), r\left(A^{2t-1} B^{1-2t}\right)\right\},$$

where A and B are positive definite matrices and $t \in \left[\frac{1}{2}, 1\right]$.

Hence, we get our main result of this section, which says that

$$\|\|b_t\|\| \leq \|\|A+B\|\| \sqrt{\max\{r\left(B^{2t-1} A^{1-2t}\right), r\left(A^{2t-1} B^{1-2t}\right)\}},$$

where A and B are positive definite matrices and $t \in [0, 1]$ (see [21]).

Consequently, it is shown that if $B \leq A \leq (1+\epsilon)^2 B$, then

$$\|\|b_t\|\| \leq (1+\epsilon) \|\|A+B\|\|$$

for $t \in [0, 1]$, $\epsilon > 0$, and for every unitarily invariant norm. Equivalently, if $\alpha \geq 1$ and if $\mathrm{sp}\big(AB^{-1}\big) \subset [1, \alpha]$, then

$$|||b_t||| \leq \sqrt{\alpha}|||A + B|||.$$

We recall the following well-known facts concerning sequences of nonnegative real numbers:

(1) Let $\{a_n\}_{n\geq 1}$ and $\{b_n\}_{n\geq 1}$ be two real sequences such that $0 \leq a_n \leq b_n$. Then

$$\liminf_{n\to\infty} a_n \leq \liminf_{n\to\infty}\ b_n$$

and

$$\limsup_{n\to\infty} a_n \leq \limsup_{n\to\infty} b_n.$$

(2) Let $\{a_n\}_{n\geq 1}$ and $\{b_n\}_{n\geq 1}$ be two bounded sequences of real numbers such that $\{b_n\}_{n\geq 1}$ converges to $b \geq 0$. Then

$$\limsup_{n\to\infty} a_n b_n = b \limsup_{n\to\infty} a_n$$

and

$$\liminf_{n\to\infty} a_n b_n = b \liminf_{n\to\infty} a_n.$$

Lemma 4.5.1 *Let $\{a_n\}_{n\geq 1}$ and $\{b_n\}_{n\geq 1}$ be two nonnegative sequences of real numbers. If*

$$\lim_{n\to\infty}\ a_n^{\frac{1}{n}} = \lim_{n\to\infty}\ b_n^{\frac{1}{n}} = r \in \mathbb{R},$$

then

$$\lim_{n\to\infty}\ (a_n + b_n)^{\frac{1}{n}} = r.$$

Proof Suppose that

$$\lim_{n\to\infty}\ a_n^{\frac{1}{n}} = \lim_{n\to\infty}\ b_n^{\frac{1}{n}} = r. \tag{4.5.2}$$

Then we have

$$a_n \leq a_n + b_n \leq 2\max\{a_n, b_n\}.$$

Thus,

$$a_n^{\frac{1}{n}} \leq (a_n + b_n)^{\frac{1}{n}} \leq 2^{\frac{1}{n}} \max\left\{a_n^{\frac{1}{n}}, b_n^{\frac{1}{n}}\right\} \text{ for } n = 1, 2, \ldots. \tag{4.5.3}$$

We have

$$\begin{aligned}\lim_{n\to\infty} \max\left\{a_n^{\frac{1}{n}}, b_n^{\frac{1}{n}}\right\} &= \lim_{n\to\infty}\frac{1}{2}\left[a_n^{\frac{1}{n}} + b_n^{\frac{1}{n}} + \left|a_n^{\frac{1}{n}} - b_n^{\frac{1}{n}}\right|\right] \\ &= \frac{1}{2}\left[r + r + |r - r|\right] \text{ (by (4.5.2))} \\ &= r.\end{aligned}$$

Hence,

$$\lim_{n\to\infty} 2^{\frac{1}{n}} \max\left\{a_n^{\frac{1}{n}}, b_n^{\frac{1}{n}}\right\} = r. \tag{4.5.4}$$

Applying the squeeze theorem on (4.5.3), and using (4.5.2) and (4.5.4), we get

$$\lim_{n\to\infty} (a_n + b_n)^{\frac{1}{n}} = r.$$

□

Lemma 4.5.2 *Let T and B be matrices such that B is invertible. Then, for every unitarily invariant norm,*

$$\lim_{m\to\infty} \left|\left|\left|T^m B\right|\right|\right|^{\frac{1}{m}} = r\left(T\right).$$

Proof Since

$$\left|\left|\left|T^m B\right|\right|\right| \leq \left|\left|\left|T^m\right|\right|\right| \; \|B\| \text{ for } m = 1, 2, \ldots,$$

it follows that

$$\left|\left|\left|T^m B\right|\right|\right|^{\frac{1}{m}} \leq \left|\left|\left|T^m\right|\right|\right|^{\frac{1}{m}} \|B\|^{\frac{1}{m}} \text{ for } m = 1, 2, \ldots. \tag{4.5.5}$$

Thus,

$$\begin{aligned}\limsup_{m\to\infty} \left|\left|\left|T^m B\right|\right|\right|^{\frac{1}{m}} &\leq \limsup_{m\to\infty}\left(\left|\left|\left|T^m\right|\right|\right|^{\frac{1}{m}} \|B\|^{\frac{1}{m}}\right) \\ &= \lim_{m\to\infty} \left|\left|\left|T^m\right|\right|\right|^{\frac{1}{m}} \lim_{m\to\infty} \|B\|^{\frac{1}{m}} \\ &= r\left(T\right) \cdot 1 \text{ (by (1.1.6) and since } B \neq 0) \\ &= r\left(T\right).\end{aligned}$$

Therefore,

$$\limsup_{m\to\infty} \left|\left|\left|T^m B\right|\right|\right|^{\frac{1}{m}} \leq r\left(T\right). \tag{4.5.6}$$

We prove that

$$r\left(T\right) \leq \liminf_{m\to\infty} \left|\left|\left|T^m B\right|\right|\right|^{\frac{1}{m}}.$$

We have

$$
\begin{aligned}
r(T) &= r\left(B^{-1}TB\right) \text{ (by (1.1.8))} \\
&= \lim_{m\to\infty} \left|\!\left|\!\left| \left(B^{-1}TB\right)^m \right|\!\right|\!\right|^{\frac{1}{m}} \text{ (by (1.1.6))} \\
&= \lim_{m\to\infty} \left|\!\left|\!\left| B^{-1}T^m B \right|\!\right|\!\right|^{\frac{1}{m}} \\
&\leq \liminf_{m\to\infty} \left\| B^{-1} \right\|^{\frac{1}{m}} \left|\!\left|\!\left| T^m B \right|\!\right|\!\right|^{\frac{1}{m}} \\
&= \liminf_{m\to\infty} \left|\!\left|\!\left| T^m B \right|\!\right|\!\right|^{\frac{1}{m}}.
\end{aligned}
$$

The last equality holds since $\lim_{m\to\infty} \left\| B^{-1} \right\|^{\frac{1}{m}} = 1$ and since the sequence $|||T^m B|||^{\frac{1}{m}}$ is bounded by the inequality (4.5.5). Thus,

$$
r(T) \leq \liminf_{m\to\infty} \left|\!\left|\!\left| T^m B \right|\!\right|\!\right|^{\frac{1}{m}}. \tag{4.5.7}
$$

Using the inequalities (4.5.6) and (4.5.7), we have

$$
\limsup_{m\to\infty} \left|\!\left|\!\left| T^m B \right|\!\right|\!\right|^{\frac{1}{m}} \leq r(T) \leq \liminf_{m\to\infty} \left|\!\left|\!\left| T^m B \right|\!\right|\!\right|^{\frac{1}{m}} \leq \limsup_{m\to\infty} \left|\!\left|\!\left| T^m B \right|\!\right|\!\right|^{\frac{1}{m}}.
$$

Hence,

$$
\lim_{m\to\infty} \left|\!\left|\!\left| T^m B \right|\!\right|\!\right|^{\frac{1}{m}} = r(T).
$$

□

Lemma 4.5.3 *Let S, T be matrices such that $r(T) > r(S) \geq 0$, and let A and B be invertible matrices. Then, for every unitarily invariant norm,*

$$
\lim_{m\to\infty} \frac{|||\, S^m A|||}{|||T^m B|||} = 0.
$$

Proof Let $t_m = \frac{|||S^m A|||}{|||T^m B|||}$. Using Lemma 4.5.2 and the fact that $r(T) > r(S) \geq 0$, we have

$$
\begin{aligned}
\lim_{m\to\infty} t_m^{\frac{1}{m}} &= \lim_{m\to\infty} \frac{|||S^m A|||^{\frac{1}{m}}}{|||T^m B|||^{\frac{1}{m}}} \\
&= \frac{r(S)}{r(T)} < 1.
\end{aligned}
$$

Thus, $\sum_{m=1}^{\infty} t_m$ converges by the root test, and so $\lim_{m\to\infty} t_m = 0$, that is,

$$\lim_{m\to\infty} \frac{|||S^m A|||}{|||T^m B|||} = 0.$$

□

Theorem 4.5.4 *Let S, T be matrices, and let A and B be invertible matrices such that*

$$\max\left\{\left|\left|\left|S^m A\right|\right|\right|, \left|\left|\left|T^m B\right|\right|\right|\right\} \leq \left|\left|\left|S^m A + T^m B\right|\right|\right| \text{ for } m = 1, 2, \ldots.$$

Then

$$\lim_{m\to\infty} \left|\left|\left| S^m A + T^m B\right|\right|\right|^{\frac{1}{m}} = \max\{r(S), r(T)\}.$$

Proof Suppose that $\max\{r(S), r(T)\} = r(T)$. Since $|||T^m B||| \leq |||S^m A + T^m B|||$ for $m = 1, 2, \ldots$, it follows that $\liminf_{m\to\infty} |||T^m B|||^{\frac{1}{m}} \leq \liminf_{m\to\infty} |||S^m A + T^m B|||^{\frac{1}{m}}$. Using Lemma 4.5.2, we have $r(T) \leq \liminf_{m\to\infty} |||S^m A + T^m B|||^{\frac{1}{m}}$. Thus,

$$\max\{r(S), r(T)\} \leq \liminf_{m\to\infty} \left|\left|\left|S^m A + T^m B\right|\right|\right|^{\frac{1}{m}}. \tag{4.5.8}$$

We have the following two cases.

Case 1: $r(T) > r(S) \geq 0$. In this case, since

$$\begin{aligned}\left|\left|\left|S^m A + T^m B\right|\right|\right| &\leq \left|\left|\left|S^m A\right|\right|\right| + \left|\left|\left|T^m B\right|\right|\right| \\ &= \left|\left|\left|T^m B\right|\right|\right| \left(\frac{|||S^m A|||}{|||T^m B|||} + 1\right) \text{ for } m = 1, 2, \ldots,\end{aligned}$$

it follows that

$$\left|\left|\left|S^m A + T^m B\right|\right|\right|^{\frac{1}{m}} \leq \left|\left|\left|T^m B\right|\right|\right|^{\frac{1}{m}} \left(\frac{|||S^m A|||}{|||T^m B|||} + 1\right)^{\frac{1}{m}} \text{ for } m = 1, 2, \ldots.$$

Using Lemma 4.5.3 and the fact that $r(T) > r(S) \geq 0$, we have

$$\lim_{m\to\infty} \left(\frac{|||S^m A|||}{|||T^m B|||} + 1\right)^{\frac{1}{m}} = 1. \tag{4.5.9}$$

Therefore,

$$\begin{aligned}\limsup_{m\to\infty} \left|\left|\left|S^m A + T^m B\right|\right|\right|^{\frac{1}{m}} &\leq \limsup_{m\to\infty} \left|\left|\left|T^m B\right|\right|\right|^{\frac{1}{m}} \text{ (by equality (4.5.9))} \\ &= r(T) \text{ (by Lemma 4.4.7)} \\ &= \max\{r(S), r(T)\}.\end{aligned}$$

Thus,

$$\limsup_{m\to\infty} |||S^m A + T^m B|||^{\frac{1}{m}} \leq \max\{r(S), r(T)\}. \tag{4.5.10}$$

Using inequality (4.5.8) and inequality (4.5.10), we have

$$\lim_{m\to\infty} |||S^m A + T^m B|||^{\frac{1}{m}} = \max\{r(S), r(T)\}.$$

Case 2: $r(S) = r(T) \geq 0$. In this case, since

$$|||S^m A + T^m B||| \leq |||S^m A||| + |||T^m B||| \text{ for } m = 1, 2, \ldots,$$

it follows that

$$|||S^m A + T^m B|||^{\frac{1}{m}} \leq \left(|||S^m A||| + |||T^m B|||\right)^{\frac{1}{m}} \text{ for } m = 1, 2, \ldots. \tag{4.5.11}$$

Using Lemma 4.5.2, we have

$$\lim_{m\to\infty} |||S^m A|||^{\frac{1}{m}} = r(S) = r(T) = \lim_{m\to\infty} |||T^m A|||^{\frac{1}{m}}.$$

Let $a_m = |||S^m A|||$, $b_m = |||T^m B|||$, and $r = r(T)$. Then

$$\lim_{m\to\infty} a_m^{\frac{1}{m}} = r = \lim_{m\to\infty} b_m^{\frac{1}{m}}.$$

Thus, using Lemma 4.5.1, we have

$$\lim_{m\to\infty} (a_m + b_m)^{\frac{1}{m}} = r,$$

that is,

$$\lim_{m\to\infty} \left(|||S^m A||| + |||T^m B|||\right)^{\frac{1}{m}} = r(T).$$

Thus, by inequality (4.5.11), we have

$$\limsup_{m\to\infty} |||S^m A + T^m B|||^{\frac{1}{m}} \leq r(T) = \max\{r(S), r(T)\}. \tag{4.5.12}$$

Using the inequalities (4.5.8) and (4.5.12), we have

$$\lim_{m\to\infty} |||S^m A + T^m B|||^{\frac{1}{m}} = \max\{r(S), r(T)\}.$$

□

We introduce some notations regarding our results in this section. Let A and B be positive definite matrices, and let $t \in \left[\frac{1}{2}, 1\right]$. Let

$$A_0 = A^{1-t} B^{2t-1} A^{1-t} \text{ and } B_0 = B^{1-t} A^{2t-1} B^{1-t}.$$

For $n = 1, 2, \ldots$, we define the following notations:

$$A_n = A_{n-1} A^{-1} A_{n-1} \text{ and } B_n = B_{n-1} B^{-1} B_{n-1}.$$

For $n = 0, 1, 2, \ldots$, we define the following notation:

$$g_n = A_n + B_n.$$

Note that $g_0 = f_t$, as given in (4.5.1).

Theorem 4.5.5 *Let A and B be positive definite matrices. Then*

$$A_n = \left(A_0 A^{-1}\right)^{2^n} A \text{ and } B_n = \left(B_0 B^{-1}\right)^{2^n} B \text{ for } n = 0, 1, 2, \ldots.$$

Proof The statement is true for $n = 1$ since

$$\begin{aligned}\left(A_0 A^{-1}\right)^{2^1} A &= A_0 A^{-1} A_0 A^{-1} A \\ &= A_0 A^{-1} A_0 \\ &= A_1.\end{aligned}$$

Suppose that the statement holds for $n = k - 1$, that is,

$$A_{k-1} = \left(A_0 A^{-1}\right)^{2^{k-1}} A. \tag{4.5.13}$$

Next, we show that the statement holds for $n = k$, that is, $A_k = \left(A_0 A^{-1}\right)^{2^k} A$.

We have

$$\begin{aligned}\left(A_0 A^{-1}\right)^{2^k} A &= \left(\left(A_0 A^{-1}\right)^{2^{k-1}}\right)^2 A \\ &= \left(\left(A_0 A^{-1}\right)^{2^{k-1}} A A^{-1}\right)^2 A \\ &= \left(A_{k-1} A^{-1}\right)^2 A \text{ (by (4.5.13))} \\ &= A_{k-1} A^{-1} A_{k-1} A^{-1} A \\ &= A_k.\end{aligned}$$

Hence, $A_k = \left(A_0 A^{-1}\right)^{2^k} A$.

Thus, by the principle of mathematical induction, we have $A_n = \left(A_0 A^{-1}\right)^{2^n} A$ for $n = 1, 2, \ldots$. Similarly, the statement for B_n can be proved. □

Theorem 4.5.6 *Let A and B be positive definite matrices. Then*

$$\left(A_0A^{-1}\right)^m A > 0 \textit{ and } \left(B_0B^{-1}\right)^m B > 0 \textit{ for } m = 0, 1, \ldots.$$

In particular, $A_n, B_n, g_n > 0$ for $n = 0, 1, \ldots$.

Proof Since $A_0 = A^{1-t}B^{2t-1}A^{1-t} > 0$ and $A^{-1} > 0$, it follows, by Lemma 3.1.13, that $A^{-1}\left(A_0A^{-1}\right)^m > 0$ for $m = 0, 1, \ldots$. Thus, an argument similar to that used in the proof of Lemma 3.1.13 shows that

$$\left(A_0A^{-1}\right)^m A = A\left[A^{-1}\left(A_0A^{-1}\right)^m\right]A > 0. \tag{4.5.14}$$

Similarly, $\left(B_0B^{-1}\right)^m B > 0$.

Now, by Theorem 4.5.5, we have $A_n = \left(A_0A^{-1}\right)^{2^n} A$ and $B_n = \left(B_0B^{-1}\right)^{2^n} B$ for $n = 0, 1, \ldots$. Thus, $A_n, B_n > 0$ by (4.5.14), and so $g_n = A_n + B_n > 0$. □

Using Theorems 4.5.4 and 4.5.5, we have the following theorem.

Theorem 4.5.7 *Let A and B be positive definite matrices, and let $t \in \left[\frac{1}{2}, 1\right]$. Then*

$$\lim_{n\to\infty} |||\, g_n |||^{\frac{1}{2^n}} = \max\left\{r\left(B^{2t-1}A^{1-2t}\right), r\left(A^{2t-1}B^{1-2t}\right)\right\}.$$

Proof Let $S = A_0A^{-1}$, $T = B_0B^{-1}$. Using Theorem 4.5.6, we have

$$S^mA, T^mB > 0 \text{ for } m = 1, 2, \ldots. \tag{4.5.15}$$

We have

$$\begin{aligned} g_n &= A_n + B_n \\ &= \left(A_0A^{-1}\right)^{2^n} A + \left(B_0B^{-1}\right)^{2^n} B \text{ (by Theorem 4.5.5)} \\ &= S^{2^n}A + T^{2^n}B. \end{aligned} \tag{4.5.16}$$

But, by Theorem 4.5.4 and (4.5.15),

$$\lim_{m\to\infty} \left|\left|\left|S^mA + T^mB\right|\right|\right|^{\frac{1}{m}} = \max\left\{r(S), r(T)\right\}.$$

Thus,

$$\lim_{n\to\infty} \left|\left|\left|S^{2^n}A + T^{2^n}B\right|\right|\right|^{\frac{1}{2^n}} = \max\left\{r(S), r(T)\right\}. \tag{4.5.17}$$

We have

$$\begin{aligned}\lim_{n\to\infty} \left\| g_n \right\|^{\frac{1}{2^n}} &= \max\left\{r\left(S\right), r\left(T\right)\right\} \text{ (by (4.5.16) and (4.5.17))}\\ &= \max\left\{r\left(A_0 A^{-1}\right), r\left(B_0 B^{-1}\right)\right\}\\ &= \max\left\{r\left(A^{1-t} B^{2t-1} A^{1-t} A^{-1}\right), r\left(B^{1-t} A^{2t-1} B^{1-t} B^{-1}\right)\right\}\\ &= \max\left\{r\left(B^{2t-1} A^{1-2t}\right), r\left(A^{2t-1} B^{1-2t}\right)\right\}.\end{aligned}$$

□

The following lemma is essential to our main result.

Lemma 4.5.8 *Let A and B be positive definite matrices. Then,*

$$\begin{bmatrix} A+B & g_{n-1} \\ g_{n-1} & g_n \end{bmatrix} \geq 0$$

for $n = 1, 2, \ldots$

Proof We have $A_n, B_n > 0$ for $n = 1, 2, \ldots$ by Theorem 4.5.6. Thus,

$$\begin{aligned}\begin{bmatrix} A+B & g_{n-1} \\ g_{n-1} & g_n \end{bmatrix} &= \begin{bmatrix} A+B & A_{n-1}+B_{n-1} \\ A_{n-1}+B_{n-1} & A_n+B_n \end{bmatrix}\\ &= \begin{bmatrix} A+B & A_{n-1}+B_{n-1} \\ A_{n-1}+B_{n-1} & A_{n-1}A^{-1}A_{n-1}+B_{n-1}B^{-1}B_{n-1} \end{bmatrix}\\ &= \begin{bmatrix} A^{\frac{1}{2}} & B^{\frac{1}{2}} \\ A_{n-1}A^{-\frac{1}{2}} & B_{n-1}B^{-\frac{1}{2}} \end{bmatrix} \begin{bmatrix} A^{\frac{1}{2}} & A^{-\frac{1}{2}}A_{n-1} \\ B^{\frac{1}{2}} & B^{-\frac{1}{2}}B_{n-1} \end{bmatrix}\\ &= \begin{bmatrix} A^{\frac{1}{2}} & B^{\frac{1}{2}} \\ A_{n-1}A^{-\frac{1}{2}} & B_{n-1}B^{-\frac{1}{2}} \end{bmatrix} \begin{bmatrix} A^{\frac{1}{2}} & B^{\frac{1}{2}} \\ A_{n-1}A^{-\frac{1}{2}} & B_{n-1}B^{-\frac{1}{2}} \end{bmatrix}^* \geq 0.\end{aligned}$$

□

Theorem 4.5.9 *Let A and B be positive definite matrices. Then, for every unitarily invariant norm,*

$$\left\| g_{n-1} \right\| \leq \left\| A+B \right\|^{\frac{1}{2}} \left\| g_n \right\|^{\frac{1}{2}} \textit{ for } n = 1, 2, \ldots.$$

Proof By Theorem 4.5.6, Lemma 4.5.8, and Theorem 1.6.1 , for $n = 1, 2, \ldots$, there exists a contraction C_n such that $g_{n-1} = (A+B)^{\frac{1}{2}} C_n g_n^{\frac{1}{2}}$.

Now,

$$\begin{aligned} |||g_{n-1}||| &= \left|\left|\left|(A+B)^{\frac{1}{2}} C_n g_n^{\frac{1}{2}}\right|\right|\right| \\ &\le |||(A+B)\, C_n|||^{\frac{1}{2}} |||C_n g_n|||^{\frac{1}{2}} \text{ (by Lemma 3.3.5)} \\ &\le |||A+B|||^{\frac{1}{2}} \|C_n\|^{\frac{1}{2}} \|C_n\|^{\frac{1}{2}} |||g_n|||^{\frac{1}{2}} \text{ (by inequality (2.3.4))} \\ &\le |||A+B|||^{\frac{1}{2}} |||g_n|||^{\frac{1}{2}}. \end{aligned}$$

□

Theorem 4.5.10 *Let A and B be positive definite matrices, and let* $t \in \left[\frac{1}{2}, 1\right]$. *Then, for every unitarily invariant norm,*

$$|||b_t||| \le |||A+B|||^{\sum_{k=1}^{n+1} \frac{1}{2^k}} |||g_n|||^{\frac{1}{2^{n+1}}} \textit{ for } n = 0, 1, 2, \ldots.$$

Proof The statement is true for $n = 0$ since

$$\begin{aligned} |||b_t||| &\le |||A+B|||^{\frac{1}{2}} \left|\left|\left|A^{1-t} B^{2t-1} A^{1-t} + B^{1-t} A^{2t-1} B^{1-t}\right|\right|\right|^{\frac{1}{2}} \\ &\qquad \text{(by inequality (4.3.1))} \\ &= |||A+B|||^{\frac{1}{2}} |||A_0 + B_0|||^{\frac{1}{2}} \\ &= |||A+B|||^{\frac{1}{2}} |||g_0|||^{\frac{1}{2}}. \end{aligned}$$

Suppose that the statement holds for $n = m - 1$, that is,

$$|||b_t||| \le |||A+B|||^{\sum_{k=1}^{m} \frac{1}{2^k}} |||g_{m-1}|||^{\frac{1}{2^m}}. \tag{4.5.18}$$

We show that the statement holds for $n = m$, that is,

$$|||b_t||| \le |||A+B|||^{\sum_{k=1}^{m+1} \frac{1}{2^k}} |||g_m|||^{\frac{1}{2^{m+1}}}.$$

We have

$$\begin{aligned} |||b_t||| &\le |||A+B|||^{\sum_{k=1}^{m} \frac{1}{2^k}} |||g_{m-1}|||^{\frac{1}{2^m}} \text{ (by inequality (4.5.18))} \\ &\le |||A+B|||^{\sum_{k=1}^{m} \frac{1}{2^k}} \left(|||A+B|||^{\frac{1}{2}} |||g_m|||^{\frac{1}{2}}\right)^{\frac{1}{2^m}} \text{ (by Theorem 4.5.9)} \\ &= |||A+B|||^{\sum_{k=1}^{m+1} \frac{1}{2^k}} |||g_m|||^{\frac{1}{2^{m+1}}}. \end{aligned}$$

Hence, $|||b_t||| \le |||A+B|||^{\sum_{k=1}^{m+1} \frac{1}{2^k}} |||g_m|||^{\frac{1}{2^{m+1}}}$.

Thus, by the principle of mathematical induction, we have

$$|||b_t||| \leq |||A + B|||^{\sum_{k=1}^{m+1} \frac{1}{2^k}} |||g_m|||^{\frac{1}{2^{m+1}}} \text{ for } m = 1, 2, \ldots.$$

□

We are ready to state and prove our main result of this section.

Theorem 4.5.11 *Let A and B be positive definite matrices, and let* $t \in [0, 1]$. *Then, for every unitarily invariant norm,*

$$|||b_t||| \leq |||A + B|||\sqrt{\max\{r\left(B^{2t-1}A^{1-2t}\right), r\left(A^{2t-1}B^{1-2t}\right)\}}.$$

Proof **Case 1:** $t \in \left[\frac{1}{2}, 1\right]$. In this case, since

$$|||b_t||| \leq |||A+|||^{\sum_{k=1}^{n+1} \frac{1}{2^k}} |||g_n|||^{\frac{1}{2^{n+1}}} \text{ for } n = 0, 1, \ldots, \text{ by Theorem 4.5.10,}$$

it follows that

$$|||b_t||| \leq \lim_{n\to\infty} |||A + B|||^{\sum_{k=1}^{n+1} \frac{1}{2^k}} |||g_n|||^{\frac{1}{2^{n+1}}}.$$

Thus, using Theorem 4.5.7, we have

$$\begin{aligned}
|||b_t||| &\leq \lim_{n\to\infty} |||A + B|||^{\sum_{k=1}^{n+1} \frac{1}{2^k}} \lim_{n\to\infty} \sqrt{|||g_n|||^{\frac{1}{2^n}}} \\
&= |||A + B|||\sqrt{\lim_{n\to\infty} |||g_n|||^{\frac{1}{2^n}}} \\
&= |||A + B|||\sqrt{\max\left\{r\left(B^{2t-1}A^{1-2t}\right), r\left(A^{2t-1}B^{1-2t}\right)\right\}}.
\end{aligned}$$

Case 2: $t \in \left[0, \frac{1}{2}\right)$. In this case,

$$\begin{aligned}
|||b_t||| &= \left|\left|\left|b_t^*\right|\right|\right| \\
&= |||b_{1-t}||| \left(\text{because } b_t^* = b_{1-t}\right) \\
&\leq |||A + B|||\sqrt{\max\left\{r\left(B^{2(1-t)-1}A^{1-2(1-t)}\right), r\left(A^{2(1-t)-1}B^{1-2(1-t)}\right)\right\}} \\
&\quad \left(\text{since } 1 - t \in \left[\tfrac{1}{2}, 1\right] \text{ and using Case 1}\right) \\
&= |||A + B|||\sqrt{\max\left\{r\left(B^{1-2t}A^{2t-1}\right), r\left(A^{1-2t}B^{2t-1}\right)\right\}} \\
&= |||A + B|||\sqrt{\max\left\{r\left(A^{2t-1}B^{1-2t}\right), r\left(B^{2t-1}A^{1-2t}\right)\right\}} \text{ (by (1.1.8))}.
\end{aligned}$$

□

Corollary 4.5.12 *Let A and B be positive definite matrices, $t \in [0, 1]$, and let $\epsilon > 0$. If $B \le A \le (1+\epsilon)^2 B$, then, for every unitarily invariant norm,*

$$|||b_t||| \le (1+\epsilon)\, |||A+B|||.$$

Proof **Case 1:** $t \in \left[\frac{1}{2}, 1\right]$. In this case, since

$$B \le A \le (1+\epsilon)^2 B$$

and since $2t-1 \in [0, 1]$, it follows that

$$B^{2t-1} \le A^{2t-1} \le (1+\epsilon)^{2(2t-1)} B^{2t-1}.$$

Let $C = B^{-\frac{2t-1}{2}}$. Then we have

$$C^* B^{2t-1} C \le C^* A^{2t-1} C \le (1+\epsilon)^{2(2t-1)} C^* B^{2t-1} C.$$

Thus,

$$I \le B^{-\frac{2t-1}{2}} A^{2t-1} B^{-\frac{2t-1}{2}} \le (1+\epsilon)^{2(2t-1)} I.$$

Hence,

$$I \le B^{-\frac{2t-1}{2}} A^{2t-1} B^{-\frac{2t-1}{2}} \le (1+\epsilon)^2 I \text{ (since } 2t-1 \in [0, 1]\text{)}. \tag{4.5.19}$$

Let $K = A^{2t-1} B^{1-2t}$. Since

$$I \le B^{-\frac{2t-1}{2}} A^{2t-1} B^{-\frac{2t-1}{2}}, \text{ by the first inequality in (4.5.19)},$$

it follows that

$$r(K) \ge 1 \text{ and } r\left(K^{-1}\right) \le 1.$$

Hence,

$$\max\left\{r(K), r\left(K^{-1}\right)\right\} = r(K). \tag{4.5.20}$$

Since

$$B^{-\frac{2t-1}{2}} A^{2t-1} B^{-\frac{2t-1}{2}} \le (1+\epsilon)^2 I, \text{ by the second inequality in (4.5.19)},$$

it follows that

$$r(K) = r\left(B^{-\frac{2t-1}{2}} A^{2t-1} B^{-\frac{2t-1}{2}}\right) \le (1+\epsilon)^2. \tag{4.5.21}$$

Now, we have

$$\begin{aligned}
\|\!|b_t|\!\| &\le \|\!|A+B|\!\|\sqrt{\max\left\{r\left(K\right), r\left(K^{-1}\right)\right\}} \\
&= \|\!|A+B|\!\|\sqrt{r\left(K\right)} \text{ (by (4.5.20)) }. \\
&\le \left(1+\epsilon\right)\|\!|A+B|\!\| \text{ (by inequality (4.5.21)) }.
\end{aligned}$$

Case 2: $t \in \left[0, \frac{1}{2}\right)$. In this case,

$$\begin{aligned}
\|\!|b_t|\!\| &= \left\|\!\left|b_t^*\right|\!\right\| \\
&= \|\!|b_{1-t}|\!\| \left(\text{ because } b_t^* = b_{1-t}\right) \\
&\le \left(1+\epsilon\right)\|\!|A+B|\!\| \left(\text{ since } 1-t \in \left[\tfrac{1}{2}, 1\right] \text{ and using Case 1}\right).
\end{aligned}$$

□

An equivalent form of Corollary 4.5.12 can be stated as follows.

Corollary 4.5.13 *Let A and B be positive definite matrices, $t \in [0, 1]$, and let $\alpha \ge 1$ such that $sp\left(AB^{-1}\right) \subset [1, \alpha]$. Then, for every unitarily invariant norm,*

$$\|\!|b_t|\!\| \le \sqrt{\alpha}\|\!|A+B|\!\|.$$

4.6 Inequalities Related to Arithmetic–Geometric Mean Inequalities

We begin this section by noting that the double inequality (3.2.1) is equivalent to the inequalities

$$\left(a^{\frac{1}{2}} + b^{\frac{1}{2}}\right)^2 \le \left(a^t + b^t\right)\left(a^{1-t} + b^{1-t}\right) \le 2\left(a+b\right). \tag{4.6.1}$$

In view of (3.2.1) and inequalities (4.6.1), we introduce some notations for their matrix versions: Let A and B be positive semidefinite matrices, $0 \le t \le 1$, and $\|\!|\cdot|\!\|$ represent any unitarily invariant norm. Let

$$h_t = A^t B^{1-t} + A^{1-t} B^t,$$

$$b_t = A^t B^{1-t} + B^t A^{1-t},$$

$$k_t = \left(A^t + B^t\right)\left(A^{1-t} + B^{1-t}\right),$$

and

$$m_t = \left(A^{1-t} + B^{1-t}\right)^{\frac{1}{2}}\left(A^t + B^t\right)\left(A^{1-t} + B^{1-t}\right)^{\frac{1}{2}}.$$

The Heinz inequality (3.2.2) with $X = I$ can be stated as

$$\left|\left|\left| h_{\frac{1}{2}} \right|\right|\right| \leq |||h_t||| \leq |||h_1||| . \tag{4.6.2}$$

In light of the noncommutativity of matrix multiplication, h_t is not the same as b_t. This allows us to perceive the arithmetic–geometric mean inequality in a novel way in terms of b_t, which is equivalent to the following question.

Question 4.6.1 Given $t \in [0, 1]$ and any unitarily invariant norm $|||\cdot|||$, is it true that

$$\left|\left|\left| b_{\frac{1}{2}} \right|\right|\right| \leq |||b_t||| \leq |||b_1||| ? \tag{4.6.3}$$

The second inequality in (4.6.3) is the second question raised in [22], which has been answered positively for the Hilbert–Schmidt norm for $\frac{1}{4} \leq t \leq \frac{3}{4}$.

For the trace norm, this question has a positive answer for $0 \leq t \leq 1$, that is,

$$\|b_t\|_1 \leq \|b_1\|_1 .$$

Moreover, a partial answer to this question in the wider class of unitarily invariant norms is given by proving that

$$|||\text{Re } b_t||| \leq |||b_1|||$$

and

$$|||\text{Im } b_t||| \leq |||b_1||| .$$

Inequalities (4.6.1) have the following matrix versions, which enable us to observe the arithmetic–geometric mean inequality in a different and new way in terms of k_t.

Question 4.6.2 Given $t \in [0, 1]$ and any unitarily invariant norm $|||\cdot|||$, is it true that

$$\left|\left|\left| k_{\frac{1}{2}} \right|\right|\right| \leq |||k_t||| \leq |||k_1||| ? \tag{4.6.4}$$

The first inequality in (4.6.4) can be stated as

$$\left|\left|\left| \left(A^{\frac{1}{2}} + B^{\frac{1}{2}}\right)^2 \right|\right|\right| \leq \left|\left|\left| \left(A^t + B^t\right)\left(A^{1-t} + B^{1-t}\right) \right|\right|\right| . \tag{4.6.5}$$

Inequality (4.6.5) is a special case of the following more general form for commuting positive semidefinite matrices:

$$\left|\left|\left| \left(A_1^{\frac{1}{2}} B_1^{\frac{1}{2}} + A_1^{\frac{1}{2}} B_1^{\frac{1}{2}}\right)^2 \right|\right|\right| \leq |||(A_1 + A_2)(B_1 + B_2)||| , \tag{4.6.6}$$

where A_1, A_2, B_1, B_2 are positive semidefinite matrices such that $A_1B_1 = B_1A_1$ and $A_2B_2 = B_2A_2$. Inequality (4.6.6) is a part of Corollary 4.1.4 for the case $k = 2$. It should be noted that inequality (4.6.5) can be derived from inequality (4.6.6) by substituting A_1, B_1, A_2, and B_2 with A^t, A^{1-t}, B^t, and B^{1-t}, respectively.

Inequalities (4.6.1) have other matrix versions, which enable us to view the arithmetic–geometric mean inequality in a novel manner using m_t.

Question 4.6.3 Let $t \in [0, 1]$, and let $||| \cdot |||$ be any unitarily invariant norm. Is it true that

$$\left|\left|\left| m_{\frac{1}{2}} \right|\right|\right| \leq |||m_t||| \leq |||m_1||| \,? \tag{4.6.7}$$

The second inequality in (4.6.7) was proved in [23]. The first inequality in (4.6.7) can be stated as

$$\left|\left|\left| \left(A^{\frac{1}{2}} + B^{\frac{1}{2}}\right)^2 \right|\right|\right| \leq \left|\left|\left| \left(A^t + B^t\right)^{\frac{1}{2}} \left(A^{1-t} + B^{1-t}\right) \left(A^t + B^t\right)^{\frac{1}{2}} \right|\right|\right|. \tag{4.6.8}$$

Inequality (4.6.8) is a special case of the following more general form for commuting positive semidefinite matrices

$$\left|\left|\left| \left(A_1^{\frac{1}{2}} B_1^{\frac{1}{2}} + A_2^{\frac{1}{2}} B_2^{\frac{1}{2}}\right)^2 \right|\right|\right| \leq \left|\left|\left| (A_1 + A_2)^{\frac{1}{2}} (B_1 + B_2) (A_1 + A_2)^{\frac{1}{2}} \right|\right|\right|, \tag{4.6.9}$$

where A_1, A_2, B_1, B_2 are positive semidefinite matrices such that $A_1B_1 = B_1A_1$ and $A_2B_2 = B_2A_2$. Inequality (4.6.9) is the case $k = 2$ of Theorem 4.1.3. Note that inequality (4.6.8) follows from inequality (4.6.9) by replacing A_1, B_1, A_2, and B_2 with A^t, A^{1-t}, B^t, and B^{1-t}, respectively. This provides an affirmative answer to Question 4.6.3.

Next, we address some of the remaining open questions. The matrices under consideration are in $\mathbb{M}_n$. In this section, we prove the following singular value inequality and majorization relations

$$s_j\left(\left(A^t + B^t\right)\left(A^{1-t} + B^{1-t}\right)\right) \leq 2^{\frac{3}{2}} s_j\left(\left(A^2 + B^2\right)^{\frac{1}{2}}\right) \text{ for } j = 1, 2, \ldots, n,$$

$$s\left(\left(A^t + B^t\right)\left(A^{1-t} + B^{1-t}\right)\right) \prec_w 2\, s\,(A + B),$$

and

$$s\left(A^t B^{1-t} + B^t A^{1-t}\right) \prec_w 2^{\frac{1}{2}} s\left(\left(A^2 + B^2\right)^{\frac{1}{2}}\right),$$

where A and B be positive semidefinite matrices and $0 \leq t \leq 1$.

Notice that these singular value inequalities and majorization relations are sharp, and as a consequence of the second inequality, we prove the second inequality in (4.6.4), that is, the inequality

$$|||k_t||| \leq |||k_1||| \qquad \text{for } t \in [0, 1].$$

This gives an affirmative answer to Question 4.6.2.

Remark 4.6.4 It should be mentioned that Question 4.6.1 remains open.

It is evident that the Heinz inequality and the second main Question 3.2.1 (that is, the second inequality in (4.6.2) and (4.6.3), respectively) are equivalent to the following inequalities:

$$\left|\left|\left| A^p B^q + A^q B^p \right|\right|\right| \leq \left|\left|\left| A^{p+q} + B^{p+q} \right|\right|\right|$$

and

$$\left|\left|\left| A^p B^q + B^p A^q \right|\right|\right| \leq \left|\left|\left| A^{p+q} + B^{p+q} \right|\right|\right|,$$

where A and B are positive semidefinite matrices and p and q are positive real numbers.

Consequently, one can infer from inequality (3.3.14) and the condition (3.3.15) that

$$\|b_t\|_2 \leq \|h_t\|_2 \; for \frac{1}{4} \leq t \leq \frac{3}{4}.$$

The following lemma, together with Lemma 4.4.1, is needed to prove our main results.

Lemma 4.6.5 *Let A and B be Hermitian matrices. Then*

$$(A + B)^2 \leq 2\left(A^2 + B^2\right).$$

Our first main result of this section can be asserted as follows (see [24]).

Theorem 4.6.6 *Let A and B be positive semidefinite matrices, and let* $0 \leq t \leq 1$. *Then*

$$s\left(\left(A^t + B^t\right)\left(A^{1-t} + B^{1-t}\right)\right) \prec_w 2\, s\,(A + B). \tag{4.6.10}$$

Proof Since

$$s\left(\left(A^t + B^t\right)\left(A^{1-t} + B^{1-t}\right)\right) \prec_w s\left(A^t + B^t\right) s\left(A^{1-t} + B^{1-t}\right),$$

it follows that for $k = 1, 2, \ldots, n$, we have

$$\begin{aligned}\sum_{j=1}^{k} s_j\left(\left(A^t + B^t\right)\left(A^{1-t} + B^{1-t}\right)\right) &\leq \sum_{j=1}^{k} \lambda_j\left(A^t + B^t\right)\lambda_j\left(A^{1-t} + B^{1-t}\right)\\ &\leq \sum_{j=1}^{k} 2^{1-t}\lambda_j^t\left(A + B\right)2^t\lambda_j^{1-t}\left(A + B\right)\\ &\qquad \text{(by Lemma 4.4.1)}\\ &= 2\sum_{j=1}^{k} \lambda_j\left(A + B\right).\end{aligned}$$

Thus, we have

$$s\left(\left(A^t + B^t\right)\left(A^{1-t} + B^{1-t}\right)\right) \prec_w 2\, s\left(A + B\right).$$

□

Corollary 4.6.7 *Let A and B be positive semidefinite matrices, and let $0 \leq t \leq 1$. Then, for every unitarily invariant norm,*

$$\left|\!\left|\!\left|\left(A^t + B^t\right)\left(A^{1-t} + B^{1-t}\right)\right|\!\right|\!\right| \leq 2\left|\!\left|\!\left|A + B\right|\!\right|\!\right|. \tag{4.6.11}$$

In other words,

$$|||k_t||| \leq |||k_1|||.$$

Remark 4.6.8 Inequality (4.6.11) is stronger than the second inequality in (4.6.7). To see this, recall that (see inequality (4.1.1)) if X and Y are matrices such that XY is Hermitian, then $|||XY||| \leq |||\mathrm{Re}\,(YX)||| \leq |||YX|||$. Accordingly, $|||m_t||| \leq |||k_t|||$ for $0 \leq t \leq 1$. In the same vein, it is readily seen that inequality (4.6.9) is stronger than the relevant inequality in Corollary 4.1.4.

Our second main result can be stated as follows.

Theorem 4.6.9 *Let A, B be positive semidefinite matrices, and let $0 \leq t \leq 1$. Then*

$$s_j\left(\left(A^t + B^t\right)\left(A^{1-t} + B^{1-t}\right)\right) \leq 2^{\frac{3}{2}} s_j\left(\left(A^2 + B^2\right)^{\frac{1}{2}}\right) \tag{4.6.12}$$

for $j = 1, 2, \ldots, n$.

Proof For $j = 1, 2, \ldots, n$, we have

$$\begin{aligned}
&s_j\left(\left(A^t + B^t\right)\left(A^{1-t} + B^{1-t}\right)\right)\\
&\quad= \lambda_j\left(\left|\left(A^t + B^t\right)\left(A^{1-t} + B^{1-t}\right)\right|\right)\\
&\quad= \lambda_j\left(\left(A^t + B^t\right)\left(A^{1-t} + B^{1-t}\right)^2\left(A^t + B^t\right)\right)^{\frac{1}{2}}\\
&\quad= \lambda_j^{\frac{1}{2}}\left(\left(A^t + B^t\right)\left(A^{1-t} + B^{1-t}\right)^2\left(A^t + B^t\right)\right)\\
&\quad\le 2^{\frac{1}{2}}\lambda_j^{\frac{1}{2}}\left(\left(A^t + B^t\right)\left(A^{2(1-t)} + B^{2(1-t)}\right)\left(A^t + B^t\right)\right) \text{ (by Lemma 4.6.5)}\\
&\quad= 2^{\frac{1}{2}}\lambda_j^{\frac{1}{2}}\left(\left(A^{2(1-t)} + B^{2(1-t)}\right)^{\frac{1}{2}}\left(A^t + B^t\right)^2\left(A^{2(1-t)} + B^{2(1-t)}\right)^{\frac{1}{2}}\right)\\
&\quad\le 2\lambda_j^{\frac{1}{2}}\left(\left(A^{2(1-t)} + B^{2(1-t)}\right)\left(A^{2t} + B^{2t}\right)\right) \text{ (by Lemma 4.6.5)}\\
&\quad\le 2^{\frac{3}{2}}\lambda_j^{\frac{1}{2}}\left(A^2 + B^2\right) \text{ (by Theorem 4.6.6)}\\
&\quad= 2^{\frac{3}{2}}\lambda_j\left(\left(A^2 + B^2\right)^{\frac{1}{2}}\right)\\
&\quad= 2^{\frac{3}{2}}s_j\left(\left(A^2 + B^2\right)^{\frac{1}{2}}\right).
\end{aligned}$$

This proves inequality (4.6.12). □

Using the fact that unitarily invariant norms are increasing functions of singular values, we have the following immediate consequence of Theorem 4.6.9.

Corollary 4.6.10 *Let A and B be positive semidefinite matrices, and let $0 \le t \le 1$. Then for every unitarily invariant norm,*

$$\left|\left|\left|\left(A^t + B^t\right)\left(A^{1-t} + B^{1-t}\right)\right|\right|\right| \le 2^{\frac{3}{2}}\left|\left|\left|\left(A^2 + B^2\right)^{\frac{1}{2}}\right|\right|\right|. \tag{4.6.13}$$

Remark 4.6.11 We remark here that inequality (4.6.13) also follows from Corollary 4.6.7. For Hermitian matrices A and B, we have

$$|||A + B||| \le 2^{\frac{1}{2}}\left|\left|\left|\left(A^2 + B^2\right)^{\frac{1}{2}}\right|\right|\right| \le 2^{\frac{1}{2}}\,|||\,|A| + |B|\,|||. \tag{4.6.14}$$

In fact, considering the Löwner–Heinz theorem 1.4.3, the first inequality in (4.6.14) follows from Lemma 4.6.5, and the second inequality in (4.6.14) can be derived from [25].

Our third main result in this section is as follows.

Theorem 4.6.12 *Let A and B be positive semidefinite matrices, and let $0 \le t \le 1$. Then*

$$s\left(A^tB^{1-t} + B^tA^{1-t}\right) \prec_w 2^{\frac{1}{2}}s\left(\left(A^2 + B^2\right)^{\frac{1}{2}}\right). \tag{4.6.15}$$

Proof Let

$$X = \begin{bmatrix} A^t & B^t \\ 0 & 0 \end{bmatrix}, Y = \begin{bmatrix} B^{1-t} & 0 \\ A^{1-t} & 0 \end{bmatrix}.$$

Then, in view of the fact that $s(T) = \lambda^{\frac{1}{2}}(T^*T) = \lambda^{\frac{1}{2}}(TT^*)$ for every matrix T, we have

$$\begin{aligned} s\left(\left(A^t B^{1-t} + B^t A^{1-t}\right) \oplus 0\right) &= s(XY) \\ &\prec_w s(X)\, s(Y) \\ &= \lambda^{\frac{1}{2}}\left(\left(A^{2t} + B^{2t}\right) \oplus 0\right) \lambda^{\frac{1}{2}}\left(\left(A^{2(1-t)} + B^{2(1-t)}\right) \oplus 0\right). \end{aligned}$$

Thus, for $k = 1, 2, \ldots, n$, we have

$$\begin{aligned} \sum_{j=1}^{k} s_j\left(A^t B^{1-t} + B^t A^{1-t}\right) &\leq \sum_{j=1}^{k} \lambda_j^{\frac{1}{2}}\left(A^{2t} + B^{2t}\right) \lambda_j^{\frac{1}{2}}\left(A^{2(1-t)} + B^{2(1-t)}\right) \\ &\leq \sum_{j=1}^{k} 2^{\frac{1-t}{2}} \lambda_j^{\frac{t}{2}}\left(A^2 + B^2\right) 2^{\frac{1-(1-t)}{2}} \lambda_j^{\frac{1-t}{2}}\left(A^2 + B^2\right) \\ &\qquad \text{(by Lemma 4.4.1)} \\ &= 2^{\frac{1}{2}} \sum_{j=1}^{k} \lambda_j^{\frac{1}{2}}\left(A^2 + B^2\right) \\ &= 2^{\frac{1}{2}} \sum_{j=1}^{k} s_j\left(\left(A^2 + B^2\right)^{\frac{1}{2}}\right). \end{aligned}$$

This proves inequality (4.6.15). □

By the Fan dominance theorem, we have the following corollary.

Corollary 4.6.13 *Let A and B be positive semidefinite matrices, and let $0 \leq t \leq 1$. Then, for every unitarily invariant norm,*

$$\left|\left|\left|A^t B^{1-t} + B^t A^{1-t}\right|\right|\right| \leq 2^{\frac{1}{2}} \left|\left|\left|\left(A^2 + B^2\right)^{\frac{1}{2}}\right|\right|\right|.$$

4.7 Exercises and Problems

Exercise 4.7.1 Prove that if $a, b \geq 0$ and $0 \leq v \leq 1$, then

$$a^v b^{1-v} + r_0\left(\sqrt{a} - \sqrt{b}\right)^2 \leq va + (1 - v)\, b,$$

where $r_0 = \min\{v, 1-v\}$.

Exercise 4.7.2 Let A and B be positive definite matrices, and let $0 \leq v \leq 1$. Prove that

$$\operatorname{tr}\left|A^v B^{1-v}\right| + r_0\left(\sqrt{\operatorname{tr} A} - \sqrt{\operatorname{tr} B}\right) \leq \operatorname{tr}\left(vA + (1-v)B\right)$$

and if A and B are positive definite, then

$$\det\left(A^v B^{1-v}\right) + r_0^n \det\left(A + B - 2A\#B\right) \leq \det\left(vA + (1-v)B\right),$$

where $r_0 = \min\{v, 1-v\}$ and $A\#B = A^{\frac{1}{2}}\left(A^{-\frac{1}{2}}BA^{-\frac{1}{2}}\right)^{\frac{1}{2}}A^{\frac{1}{2}}$ is the geometric mean of A and B.

Exercise 4.7.3 Let A, B, and X be matrices such that A and B are positive semidefinite. Prove that

$$2\left\|A^{\frac{1}{2}}XB^{\frac{1}{2}}\right\|_2 + \left(\sqrt{\|AX\|_2} - \sqrt{\|XB\|_2}\right)^2 \leq \|AX + XB\|_2 .$$

Exercise 4.7.4 Let A, B, and X be matrices such that A and B are positive semidefinite, and let $0 \leq v \leq 1$. Prove that

$$\left\|A^v XB^{1-v} + A^{1-v}XB^v\right\|_2^2 + 2r_0\left\|AX - XB\right\|_2^2 \leq \left\|AX + XB\right\|_2^2,$$

where $r_0 = \min\{v, 1-v\}$.

Exercise 4.7.5 Let A, B, and X be matrices such that A and B are positive semidefinite, and let $0 \leq v \leq 1$. Prove that

$$\left|\left|\left|A^v XB^{1-v}\right|\right|\right| + r_0\left(\sqrt{|||AX|||} - \sqrt{|||XB|||}\right)^2 \leq v\,|||AX||| + (1-v)\,|||XB|||$$

for every unitarily invariant norm, where $r_0 = \min\{v, 1-v\}$.

Exercise 4.7.6 Let A, B, and X be matrices such that A and B are positive semidefinite, and let $0 \leq v \leq 1$. Prove that

$$\left|\left|\left|A^v XB^{1-v} + A^{1-v}XB^v\right|\right|\right| + 2r_0\left(\sqrt{|||AX|||} - \sqrt{|||XB|||}\right)^2 \leq |||AX||| + |||XB|||$$

for every unitarily invariant norm, where $r_0 = \min\{v, 1-v\}$.

Exercise 4.7.7 Let A and B be positive semidefinite matrices, and let $0 \leq v \leq 1$. Prove that

$$\left\|A^v B^{1-v} + A^{1-v}B^v\right\|_1 + 2r_0\left(\sqrt{||A||_1} - \sqrt{||B||_1}\right)^2 \leq \|A + B\|_1 ,$$

where $r_0 = \min\{v, 1-v\}$.

Exercise 4.7.8 Let A and B, and X be matrices such that A and B are positive definite, and let $1 < p < \infty$. Prove that

$$2\left\|A^{\frac{1}{2}}XB^{\frac{1}{2}}\right\|_p = \left\|A^vXB^{1-v} + A^{1-v}XB^v\right\|_p \text{ for some } v, \text{ with } 0 < v < 1 \text{ and } v \neq \frac{1}{2},$$

if and only if $AX = XB$.

Exercise 4.7.9 Let A and B, and X be matrices such that A and B are positive semidefinite, and let $0 \le v \le 1$. Prove that

$$|||A^vXB^v||| \le |||X|||^{1-v}\, |||AXB|||^v$$

for every unitarily invariant norm.

Exercise 4.7.10 Let A and B, and X be matrices such that A and B are positive semidefinite, and let $0 \le v \le 1$. Prove that

$$\left|\left|\left|A^vXB^v + A^{1-v}XB^{1-v}\right|\right|\right| \le |||AXB + X|||$$

and

$$\left|\left|\left|A^vXB^v - A^{1-v}XB^{1-v}\right|\right|\right| \le |2v-1|\, |||AXB - X|||$$

for every unitarily invariant norm.

Problem 4.7.11 Let A, B, and X be matrices such that A and B are positive semidefinite, and let $0 \le v \le 1$. Prove that

$$\left\|A^vXB^{1-v} + A^{1-v}XB^v\right\|_2 + 2r_0\left(\sqrt{\|AX\|_2} - \sqrt{\|XB\|_2}\right)^2 \le \|AX + XB\|_2,$$

where $r_0 = \min\{v, 1-v\}$.

Problem 4.7.12 Let A, B, and X be matrices such that A and B are positive semidefinite, and let $0 \le v \le 1$. Prove that

$$\left|\left|\left|A^vXB^{1-v} + A^{1-v}XB^v\right|\right|\right| \le 4r_0\left|\left|\left|A^{\frac{1}{2}}XB^{\frac{1}{2}}\right|\right|\right| + (1-2r_0)\, |||AX + XB|||$$

for every unitarily invariant norm, where $r_0 = \min\{v, 1-v\}$.

Problem 4.7.13 Let A and B be positive semidefinite matrices, and let $t \in \left[\frac{3}{4}, 1\right]$. Prove that

$$\left|\left|\left|B^{1-t}A^{2t-1}B^{1-t} + A^{1-t}B^{2t-1}A^{1-t}\right|\right|\right| \le 2^{4\left(t-\frac{3}{4}\right)}\, |||A + B|||$$

for every unitarily invariant norm.

Problem 4.7.14 Let A and B be positive semidefinite matrices, and let $t \in \left[\frac{3}{4}, 1\right]$. Prove that

$$\left|\left|\left|A^t B^{1-t} + B^t A^{1-t}\right|\right|\right| \leq 2^{2\left(t-\frac{3}{4}\right)} \left|\left|\left|A + B\right|\right|\right|$$

for every unitarily invariant norm.

Problem 4.7.15 Let A and B be positive semidefinite matrices, and let $t \in \left[0, \frac{1}{4}\right]$. Prove that

$$\left|\left|\left|A^t B^{1-t} + B^t A^{1-t}\right|\right|\right| \leq 2^{2\left(\frac{1}{4}-t\right)} \left|\left|\left|A + B\right|\right|\right|$$

for every unitarily invariant norm.

4.8 Notes, Hints, and References

For a comprehensive account of the first and second main Questions 3.1.1 and 3.2.1 and related trace and norm inequalities, we refer readers to [26–29] and references therein. It is worth noting that Kittaneh and Ricard recently established that $\left|\left|\left|A^t B^{1-t} + B^t A^{1-t}\right|\right|\right| \leq |||A + B|||$ for all $t \in [1/4, 3/4]$ and for all unitarily invariant norms. This considerably expands upon Bhatia's result in [8], where a solution is given for the Hilbert-Schmidt norm. Now, we give some hints and references for the exercises and problems of this section.

Exercise 4.7.1 is a refinement of the classical Young's inequality for nonnegative scalars. This has been proved in [30]. Exercises 4.7.2–4.7.7 and 4.7.11 are noncommutative versions of the refined Young's and Heinz's inequalities. For solutions of these exercises, we refer the reader to [30]. There are also many reverse Young-type inequalities; see [31, Chapter 2].

Exercises 4.7.8 and 4.7.12 are related to the Heinz norm inequalities, and they can be found in [32]. Exercise 4.7.9 is a Heinz–Kato-type inequality, which has been proved in [33].

Exercise 4.7.10 contains two variants of the Heinz norm inequalities, which can be found in [34].

Problems 4.7.13, 4.7.14, and 4.7.15 are related to the second question of Bourin. For solutions of these problems, we refer the reader to [14, 24].

References

1. J. Shao, Y. Han, Some convexity inequalities in noncommutative L_p-spaces. J. Inequal. Appl. **2014**, 385 (2014), 6 pp
2. K.M.R. Audenaert, A norm inequality for pairs of commuting positive semidefinite matrices. Electron. J. Linear Algebra **30**, 80–84 (2015)
3. D.T. Hoa, An inequality for t-geometric means. Math. Inequal. Appl. **19**, 765–768 (2016)
4. M. Lin, Remarks on two recent results of Audenaert. Linear Algebra Appl. **489**, 24–29 (2016)

5. M. Hayajneh, S. Hayajneh, F. Kittaneh, Remarks on some norm inequalities for positive semidefinite matrices and questions of Bourin. Math. Inequal. Appl. **20**, 225–232 (2017)
6. R. Bhatia, F. Kittaneh, Norm inequalities for positive operators. Lett. Math. Phys. **43**, 225–231 (1998)
7. S. Hayajneh, F. Kittaneh, Lieb–Thirring trace inequalities and a question of Bourin. J. Math. Phys. **54**, 033504 (2013), 8 pp
8. R. Bhatia, Trace inequalities for products of positive definite matrices. J. Math. Phys. **55**, 013509 (2014), 3 pp
9. T. Bottazzi, R. Elencwajg, G. Larotonda, A. Varela, Inequalities related to Bourin and Heinz means with complex parameter. J. Math. Anal. Appl. **426**, 765–773 (2015)
10. M. Hayajneh, S. Hayajneh, F. Kittaneh, Norm inequalities for positive semidefinite matrices and a question of Bourin. Int. J. Math. **28**, 1750102 (2017), 7 pp
11. F. Kittaneh, Inequalities for commutators of positive operators. J. Funct. Anal. **250**, 132–143 (2007)
12. R. Bhatia, C. Davis, A Cauchy–Schwarz inequality for operators with applications. Linear Algebra Appl. **223**(224), 119–129 (1995)
13. R. Bhatia, F. Kittaneh, The matrix arithmetic geometric mean inequality revisited. Linear Algebra Appl. **428**, 2177–2191 (2008)
14. M. Hayajneh, S. Hayajneh, F. Kittaneh, *Norm inequalities for positive semidefinite matrices and a question of Bourin II*. Int. J. Math. **32**, 2150043 (2021), 7 pp
15. M. Hayajneh, S. Hayajneh, F. Kittaneh, A Hilbert–Schmidt norm inequality for positive semidefinite matrices related to a question of Bourin. Positivity **26** (2022), Paper No. 60, 9 pp
16. R. Bhatia, *Matrix Analysis* (Springer, New York, 1997)
17. T. Ando, F. Hiai, K. Okubo, Trace inequalities for multiple products of two matrices. Math. Inequal. Appl. **3**, 307–318 (2000)
18. M. Alakhrass, Inequalities related to Heinz mean. Linear Multilinear Algebra **64**, 1562–1569 (2016)
19. B. Simon, *Trace Ideals and Their Applications, Second Edition, Mathematical Surveys and Monographs*, vol. 120 (American Mathematical Society, Providence, RI, 2005)
20. M. Hayajneh, S. Hayajneh, F. Kittaneh, I. Lebaini, Norm inequalities for positive semidefinite matrices and a question of Bourin III. Positivity **26** (2022), Paper No. 23, 13 pp
21. A. Darweesh, M. Hayajneh, S. Hayajneh, F. Kittaneh, Norm Inequalities for positive definite matrices related to a question of Bourin. Linear Multilinear Algebra **71**, 1933–1947 (2023)
22. J.-C. Bourin, A matrix subadditivity inequality for symmetric norms. Proc. Am. Math. Soc. **138**, 495–504 (2010)
23. L. Plevnik, On a matrix trace inequality due to Ando, Hiai and Okubo. Indian J. Pure Appl. Math. **47**, 491–500 (2016)
24. M. Hayajneh, S. Hayajneh, F. Kittaneh, Norm inequalities related to the arithmetic-geometric mean inequalities for positive semidefinite matrices. Positivity **22**, 1311–1324 (2018)
25. T. Ando, X. Zhan, Norm inequalities related to operator monotone functions. Math. Ann. **31**, 771–780 (1999)
26. J.-C. Bourin, Matrix subadditivity inequalities and block-matrices. Int. J. Math. **20**, 979–6910 (2009)
27. M. Hayajneh, S. Hayajneh, F. Kittaneh, Remarks on a question of Bourin for positive semidefinite matrices. Results Math. **78**, 88 (2023), 10 pp
28. M. Hayajneh, S. Hayajneh, F. Kittaneh, A unitarily invariant norm inequality for positive semidefinite matrices and a question of Bourin. Results Math. **78**, 138 (2023), 9 pp
29. J.-T. Liu, Q.-W. Wang, F.-F. Sun, On Hayajneh and Kittaneh's conjecture on unitarily invariant norm. J. Math. Inequal. **11**, 1019–1022 (2017)
30. F. Kittaneh, Y. Manasrah, Improved Young and Heinz inequalities for matrices. J. Math. Anal. Appl. **361**, 262–269 (2010)
31. S. Furuichi, H.R. Moradi, *Advances in Mathematical Inequalities* (De Gruyter, Berlin, 2020)

32. F. Kittaneh, On the convexity of the Heinz means. Integral Equ. Operator Theory **68**, 519–527 (2010)
33. F. Kittaneh, Norm inequalities for fractional powers of positive operators. Lett. Math. Phys. **27**, 279–285 (1993)
34. F. Kittaneh, Some norm inequalities for operators. Can. Math. Bull. **42**, 87–96 (1999)

Chapter 5
Positive Maps and Operator Means

In this chapter, we first study operator convex and operator monotone functions and their fundamental properties. We then explore positive and completely positive maps in a general setting. Finally, we pay attention to the theory of operator means introduced by Kubo and Ando [1] and demonstrate their features, in particular, we seek some important properties of the operator geometric and operator harmonic means. Throughout this chapter, we denote by $\mathbb{B}_{sa}(\mathscr{H})(J)$ the set of all self-adjoint operators in $\mathbb{B}_{sa}(\mathscr{H})$ whose spectra are contained in an interval $J \subseteq \mathbb{R}$.

5.1 Operator Convex and Operator Monotone Functions

We start this section with the following definition.

Definition 5.1.1 A real-valued continuous function f on an interval J is called *operator monotone* if it preserves the usual operator order, that is, $A, B \in \mathbb{B}_{sa}(\mathscr{H})(J)$ and $A \leq B$ entail $f(A) \leq f(B)$. In the framework of matrices, we use the terminology *matrix monotone* for such a function f with Hermitian matrices $A, B \in \mathbb{M}_n$ with eigenvalues in J and for all n.

Clearly a function f is operator monotone on a bounded interval (α, β) if and only if so is $g(t) = f(\frac{1}{2}((\beta - \alpha)t + \beta + \alpha))$ on $(-1, 1)$. So when we deal with functions on bounded intervals, we consider only the case $(-1, 1)$. It is known that a continuous function f on $(-1, 1)$ is operator monotone if and only if there is a finite positive measure m on $[-1, 1]$ such that

$$f(\lambda) = f(0) + \int_{-1}^{1} \frac{\lambda}{1 - \lambda t} dm, \qquad \lambda \in (-1, 1).$$

A. M. Bikchentaev et al., *Trace Inequalities*, Forum for Interdisciplinary Mathematics,
https://doi.org/10.1007/978-981-97-6520-1_5

If f is operator monotone on $(-\infty, \infty)$, then $f(\lambda) = a + \lambda b$ for some real number a and nonnegative real number b. See [2, Chapter 4] for operator inequalities involving operator monotone functions.

Example 5.1.2 (i) It follows from the Löwner–Heinz theorem that $f(t) = t^\alpha$ is operator monotone for any $0 < \alpha < 1$.
(ii) The function $\log t$ is operator monotone on $(0, \infty)$.

Definition 5.1.3 A real-valued continuous function f defined on an interval J is called *operator convex* if

$$f(\lambda A + (1-\lambda)B) \le \lambda f(A) + (1-\lambda) f(B)$$

for all operators $A, B \in \mathbb{B}_{sa}(\mathscr{H})(J)$ and all $\lambda \in [0, 1]$. A function f is said to be *operator concave* if $-f$ is operator convex. In the setting of matrices, we use the terminologies *matrix convex* and *matrix concave* for such a function with Hermitian matrices $A, B \in \mathbb{M}_n$ with eigenvalues in J and for all n.

Example 5.1.4 (i) Let A and B be self-adjoint operators. The inequality $(A - B)^2 \ge 0$ yields that $\left(\frac{A+B}{2}\right)^2 \le \frac{A^2+B^2}{2}$. Therefore, the function $f(t) = t^2$ is operator convex on each interval.
(ii) According to Lemma 4.4.1, the function $f(t) = t^r$, $r \in [0, 1]$ is matrix concave on $[0, \infty)$.
(iii) It follows from Theorem 1.6.1 that $\begin{bmatrix} A_j & X_j \\ X_j^* & X_j^* A_j^{-1} X_j \end{bmatrix} \ge 0$ $(j = 1, 2)$, whence

$$\begin{bmatrix} A_1 + A_2 & X_1 + X_2 \\ X_1^* + X_2^* & X_1^* A_1^{-1} X_1 + X_2^* A_2^{-1} X_2 \end{bmatrix} \ge 0.$$

Theorem 1.6.1 together with the scalar multiplication by $1/2$ gives us the following:

$$\left(\frac{X_1 + X_2}{2}\right)^* \left(\frac{A_1 + A_2}{2}\right)^{-1} \left(\frac{X_1 + X_2}{2}\right) \le \left(\frac{X_1^* A_1^{-1} X_1 + X_2^* A_2^{-1} X_2}{2}\right).$$

Thus, we can say that the map $(A, X) \mapsto X^* A^{-1} X$ from $\mathbb{B}_{++}(\mathscr{H}) \times \mathbb{B}(\mathscr{H})$ to $\mathbb{B}_+(\mathscr{H})$ is *jointly convex*.
(iii) Fixing $X = I$ in (ii) we conclude that the function $f(t) = t^{-1}$ is operator convex on $(0, \infty)$.

A continuous function f on $(-1, 1)$ is operator convex if and only if there exist a finite positive measure m on $[-1, 1]$ and real numbers a and b such that

$$f(\lambda) = a + b\lambda + \int_{-1}^{1} \frac{\lambda^2}{1 - \lambda t} dm, \qquad \lambda \in (-1, 1).$$

There are several characterizations of the operator convexity in the literature. One of them is as follows. A comprehensive proof can be found in [3].

Theorem 5.1.5 *Let f be a real-valued continuous function on an infinite interval J. Then the following assertions are equivalent:*

(i) f is operator convex on J;
(ii) $f(C^*AC) \le C^*f(A)C$ for each Hilbert space $\mathscr{H}$, each $A \in \mathbb{B}_{sa}(\mathscr{H})(J)$ and each isometry $C \in \mathbb{B}(\mathscr{H})$;
(iii) $f(\sum_{j=1}^k C_j^*A_jC_j) \le \sum_{j=1}^k C_j^*f(A_j)C_j$ for each Hilbert space $\mathscr{H}$, each $A_j \in \mathbb{B}_{sa}(\mathscr{H})(J)$ and each $C_j \in \mathbb{B}(\mathscr{H})$ $(j = 1, \dots, k)$ with $\sum_{j=1}^k C_j^*C_j = I$.

The following theorem is known as Hansen's inequality [4].

Theorem 5.1.6 *Let $f : [0, \infty) \to \mathbb{R}$ be an operator monotone function. Then*

$$C^*f(A)C \le f(C^*AC)$$

for each operator $A \in \mathbb{B}_+(\mathscr{H})$ and each contraction $C \in \mathbb{B}(\mathscr{H})$.

Proof It is easy to verify that the operator $U = \begin{bmatrix} C & B \\ D & -C^* \end{bmatrix}$, where $B = (1 - CC^*)^{1/2}$ and $D = (1 - C^*C)^{1/2}$, is unitary. Set $X = \begin{bmatrix} A & 0 \\ 0 & 0 \end{bmatrix}$. Then

$$U^*XU = \begin{bmatrix} C^*AC & C^*AB \\ BAC & -BAB \end{bmatrix}.$$

Let ε and λ be positive real numbers and put

$$Y = \begin{bmatrix} C^*AC + \varepsilon & 0 \\ 0 & 2\lambda \end{bmatrix}.$$

Set $E = -C^*AB$. Then, for $\lambda \ge BAB$ we have

$$Y - U^*XU = \begin{bmatrix} \varepsilon & -C^*AB \\ -BAC & 2\lambda - BAB \end{bmatrix} \ge \begin{bmatrix} \varepsilon & E \\ E^* & \lambda \end{bmatrix}.$$

For any $x, y \in \mathscr{H}$, we have

$$\begin{aligned} \left\langle \begin{bmatrix} \varepsilon & E \\ E^* & \lambda \end{bmatrix} \begin{bmatrix} x \\ y \end{bmatrix}, \begin{bmatrix} x \\ y \end{bmatrix} \right\rangle &= \varepsilon\|x\|^2 + \langle Ey, x\rangle + \langle E^*x, y\rangle + \lambda\|y\|^2 \\ &\ge \varepsilon\|x\|^2 - 2\|E\|\,\|x\|\,\|y\| + \lambda\|y\|^2 \\ &\ge 0 \end{aligned}$$

when $\lambda \ge \|E\|^2/\varepsilon$. Thus, for a sufficiently large λ, we get $U^*XU \le Y$. Using a sequence of polynomials uniformly convergent to A on the spectrum of A (the functional calculus), we conclude that $U^*f(X)U = f(U^*XU)$. Since f is operator monotone, we arrive at $U^*f(X)U = f(U^*XU) \le f(Y)$. Therefore,

$$\begin{bmatrix} C^*f(A)C & C^*f(A)B \\ Bf(A)C & Bf(A)B \end{bmatrix} \le \begin{bmatrix} f(C^*AC+\varepsilon) & 0 \\ 0 & f(2\lambda) \end{bmatrix}.$$

Comparing the (1,1)-entries of the above matrices, we reach $C^*f(A)C \le f(C^*AC+\varepsilon)$. Since f is continuous, letting $\varepsilon \to 0$, we get the desired inequality.

Now we present the *Hansen–Pedersen inequality* that includes the previous theorem in item (ii).

Theorem 5.1.7 (Hansen–Pedersen inequality) ([3]) *Let f be a real-valued continuous function on an interval J containing 0. Then the following assertions are equivalent:*

(i) f is operator convex on J and $f(0) \le 0$;
(ii) $f(C^*AC) \le C^*f(A)C$ for each Hilbert space $\mathscr{H}$, each $A \in \mathbb{B}_{sa}(\mathscr{H})(J)$ and each contraction $C \in \mathbb{B}(\mathscr{H})$;
(iii) $f(\sum_{j=1}^k C_j^*A_jC_j) \le \sum_{j=1}^k C_j^*f(A_j)C_j$ for each Hilbert space $\mathscr{H}$, each $A_j \in \mathbb{B}_{sa}(\mathscr{H})(J)$ and each $C_j \in \mathbb{B}(\mathscr{H})$ $(j=1,\ldots,k)$ with $\sum_{j=1}^k C_j^*C_j \le I$.

Note that $f(t) = \tan t$ is not operator concave on $(-\pi/2, \pi/2)$ while it is operator monotone. However, there is a relationship between the operator concavity and the operator monotonicity when the function is defined on an interval of infinite length; see [5, Theorem 1.15]. The following lemma is due to Uchiyama [6, Theorem 2.4].

Theorem 5.1.8 *Let f be a continuous function on (a,∞), where $a \ge -\infty$, and let $f(\infty) := \lim_{t\to\infty} f(t)$. Then*

(i) f is operator decreasing if and only if f is operator convex and $f(\infty) < \infty$;
(ii) f is operator monotone if and only if f is operator concave and $f(\infty) > -\infty$.

Proof (ii) is a consequence of (i). Hence, it is sufficient to merely prove (i).

Suppose that f is operator convex and $f(\infty) < \infty$. Note that $f(\infty)$ exists since f is convex. We shall prove that $f(B) \le f(A)$ when $a < A \le B$. By considering $B + \delta$ instead of B, we may assume that $B - A \ge \delta > 0$. Then $\frac{\lambda}{1-\lambda}(B-A) > a$ if $0 < \lambda < 1$ and λ is sufficiently close to 1. For such a λ, we have

$$f(\lambda B) = f\left(\lambda A + (1-\lambda)\frac{\lambda}{1-\lambda}(B-A)\right) \le \lambda f(A) + (1-\lambda) f\left(\frac{\lambda}{1-\lambda}(B-A)\right).$$

In the case when $f(\infty) = -\infty$, we have $f(\frac{\lambda}{1-\lambda}(B-A)) \le 0$ for λ sufficiently close to 1. Hence, $f(\lambda B) \le \lambda f(A)$, and hence $f(B) \le f(A)$. In the case when $f(\infty) > -\infty$, we have $(1-\lambda)f(\frac{\lambda}{1-\lambda}(B-A)) \to 0$ as $\lambda \to 1^-$, from which we deduce that $f(B) \le f(A)$.

Next, assume that f is operator decreasing. Then, $f(\infty) < \infty$. For $0 < \lambda < 1$, consider the following unitary operator:

$$U = \begin{bmatrix} \sqrt{\lambda} I & -\sqrt{1-\lambda} I \\ \sqrt{1-\lambda} I & \sqrt{\lambda} I \end{bmatrix}.$$

It follows from

$$U^* \begin{bmatrix} A & 0 \\ 0 & B \end{bmatrix} U = \begin{bmatrix} \lambda A + (1-\lambda)B & \sqrt{\lambda(1-\lambda)}(B-A) \\ \sqrt{\lambda(1-\lambda)}(B-A) & \lambda B + (1-\lambda)A \end{bmatrix},$$

that for an arbitrary $\varepsilon > 0$ there is $\delta > 0$ such that

$$U^* \begin{bmatrix} A & 0 \\ 0 & B \end{bmatrix} U \leq \begin{bmatrix} \lambda A + (1-\lambda)B + \varepsilon I & 0 \\ 0 & \lambda B + (1-\lambda)A + \delta I \end{bmatrix}.$$

In light of that f is operator decreasing, we have

$$\begin{aligned} U^* \begin{bmatrix} f(A) & 0 \\ 0 & f(B) \end{bmatrix} U &= f\left(U^* \begin{bmatrix} A & 0 \\ 0 & B \end{bmatrix} U \right) \\ &\geq f\left(\begin{bmatrix} \lambda A + (1-\lambda)B + \varepsilon I & 0 \\ 0 & \lambda B + (1-\lambda)A + \delta I \end{bmatrix} \right) \\ &= \begin{bmatrix} f(\lambda A + (1-\lambda)B + \varepsilon I) & 0 \\ 0 & f(\lambda B + (1-\lambda)A + \delta I) \end{bmatrix}. \end{aligned}$$

Considering the top-left corner entries, we arrive at

$$\lambda f(A) + (1-\lambda) f(B) \geq f(\lambda A + (1-\lambda)B + \varepsilon I).$$

Letting $\varepsilon \to 0$, we conclude that f is operator convex.

5.2 Positive Linear Maps

Let $\Phi : \mathscr{A} \to \mathscr{B}$ be a not necessarily linear map between C^*-algebras. It is called a *positive map* if $\Phi(A) \geq 0$ whenever $A \geq 0$. We say that Φ is a *unital map* if $\mathscr{A}, \mathscr{B}$ are unital C^*-algebras and Φ preserves the identity. The set of all positive linear maps Φ from $\mathscr{A}$ into $\mathscr{B}$ is denoted by $\mathbb{B}_+(\mathscr{A}, \mathscr{B})$. A positive linear functional is a typical example of a positive map.

The map Φ is called *n-positive* if the map $\Phi_n : \mathbb{M}_n(\mathscr{A}) \to \mathbb{M}_n(\mathscr{B})$ defined by

$$\Phi_n([A_{ij}]_{n\times n}) = [\Phi[A_{ij}]]_{n\times n}$$

is positive. The map Φ is said to be *completely positive* if it is n-positive for every $n \in \mathbb{N}$.

Example 5.2.1 (i) The positive map $\Phi(A) = |A|$ is positive but not linear.
(ii) The map $\Phi(A) = A^*$ is positive and conjugate linear but not linear.
(iii) Any $*$-homomorphism is positive, since $\Phi(A^*A) = \Phi(A)^*\Phi(A) \geq 0$.

Lemma 1 *If $P_1, P_2 \in \mathbb{B}_+(\mathscr{H})$, then $\|P_1 - P_2\| \leq \max\{\|P_1\|, \|P_2\|\}$.*

Proof Since $|t - s| \leq \max\{t, s\}$ for all $t, s \in \mathbb{R}^+$, one has

$$|\langle (P_1 - P_2)x, x\rangle| \leq \max\{\langle P_1 x, x\rangle, \langle P_2 x, x\rangle\} \leq \max\{\|P_1\|, \|P_2\|\}.$$

It follows that

$$\|P_1 - P_2\| \leq \max\{\|P_1\|, \|P_2\|\}.$$

Lemma 2 *Every positive linear map $\Phi : \mathscr{A} \to \mathscr{B}$ preserves the adjoint as well as the usual operator order. In addition, if $\mathscr{A}$ is unital, then Φ is bounded.*

Proof First note that if $A \in \mathscr{A}$ is self-adjoint and $A = A_+ - A_-$ is the Jordan decomposition of A, where A_+, A_- are positive, then $\Phi(A) = \Phi(A_+) - \Phi(A_-)$ is self-adjoint.
Let $A \in \mathscr{A}$. If $A = B + \mathrm{i}C$ is the Cartesian decomposition of A into its self-adjoint parts, then

$$\Phi(A^*) = \Phi(B - \mathrm{i}C) = \Phi(B) - \mathrm{i}\Phi(C) = (\Phi(B) + \mathrm{i}\Phi(C))^* = \Phi(A)^*.$$

Let $A, B \in \mathscr{A}$ and $A \leq B$. Then $B - A \geq 0$, so $\Phi(B) - \Phi(A) = \Phi(B - A) \geq 0$, which gives $\Phi(A) \leq \Phi(B)$.

Finally, let $A \in \mathscr{A}$ be positive. Functional calculus gives $0 \leq A \leq \|A\|I$. Hence, $\Phi(A) \leq \|A\|\Phi(I)$ whence $\|\Phi(A)\| \leq \|A\|\,\|\Phi(I)\|$. Next, using the Jordan decomposition, we get

$$\|\Phi(A)\| = \|\Phi(A_+) - \Phi(A_-)\| \leq \max\{\|\Phi(A_+)\|, \|\Phi(A_-)\|\} \leq \|A\|\,\|\Phi(I)\|$$

for all self-adjoint elements $A \in \mathscr{A}$. The first inequality is true by Lemma 1, and the second inequality is due to $\|A_+\|, \|A_-\| \leq \|A\|$. If $A \in \mathscr{A}$ is arbitrary, using the Cartesian decomposition $A = B + iC$, we get

$$\|\Phi(A)\| \leq \|\Phi(B)\| + \|\Phi(C)\| \leq 2\|A\|\,\|\Phi(I)\|$$

since $\|B\|, \|C\| \leq \|A\|$.

Theorem 5.2.2 *Every positive linear map $\Phi : \mathscr{A} \to \mathscr{B}$ between C^*-algebras is bounded.*

Proof Recall that the vector space direct sum $\widetilde{\mathscr{D}} = \mathscr{D} \oplus \mathbb{C}$ of a C^*-algebra $\mathscr{D}$ and $\mathbb{C}$ equipped with the multiplication $(D, \lambda)(D', \lambda') = (DD' + \lambda D' + \lambda' D, \lambda\lambda')$ and the norm $\|(D, \lambda)\| = \sup\{\|DC + \lambda C\| : \|C\| \leq 1\}$ is a unitization of $\mathscr{D}$, in the sense that it is a unital C^*-algebra containing $\mathscr{D}$ as an essential ideal. Note that $(0, 1)$ is the identity of $\widetilde{\mathscr{D}}$. If $\Phi : \mathscr{A} \to \mathscr{B}$ is a nonunital positive linear map, then $\widetilde{\Phi} : \widetilde{\mathscr{A}} \to \widetilde{\mathscr{B}}$ defined by $\widetilde{\Phi}(A, \lambda) = \Phi(A) + \lambda$ is a unital positive linear map extending Φ. By Lemma 2, $\widetilde{\Phi}$ is bounded and so is its restriction Φ.

The following theorem plays a key role in establishing several significant results.

Theorem 5.2.3 (Stinespring theorem [7]) *Suppose that Φ is a completely positive unital linear map from a C^*-algebra $\mathscr{A}$ into $\mathbb{B}(\mathscr{K})$. Then there exist a representation π of $\mathscr{A}$ on a Hilbert space $\mathscr{L}$ and an isometry V from $\mathscr{K}$ into $\mathscr{L}$ such that $\Phi(X) = V^*\pi(X)V$ for all $X \in \mathscr{A}$.*

The following two theorems are known.

Theorem 5.2.4 ([7–9]) *If $\Phi \in \mathbb{B}_+(\mathscr{A}, \mathscr{B})$ is a unital map and $\mathscr{A}$ or $\mathscr{B}$ is commutative, then Φ is completely positive.*

Theorem 5.2.5 (Choi Inequality) *Suppose that $\Phi \in \mathbb{B}_+(\mathbb{B}(\mathscr{H}), \mathbb{B}(\mathscr{K}))$ is unital. Then*

$$\Phi(A^*A) \geq \Phi(A)^*\Phi(A)$$

for all normal operators $A \in \mathbb{B}(\mathscr{H})$.

Proof The restriction of Φ to the commutative C^*-algebra $\mathscr{B}$ generated by A and I is completely positive. It follows from Theorem 5.2.3 that $\Phi(X) = W^*\pi(X)W$ $(X \in \mathscr{B})$, where π is a representation of $\mathscr{B}$ on a Hilbert space $\mathscr{L}$ and W is an isometry from $\mathscr{K}$ into $\mathscr{L}$. From $W^*W = I$, we conclude that $\|WW^*\| = \|W^*W\| = 1$, whence $WW^* \leq I$. Therefore,

$$\Phi(A)^*\Phi(A) = W^*\pi(A^*)W\,W^*\pi(A)W \leq W^*\pi(A^*)\pi(A)W = W^*\pi(A^*A)W = \Phi(A^*A).$$

As a consequence of Choi's inequality we immediately get Kadison's Schwarz inequality.

Theorem 5.2.6 (Kadison inequality') *Suppose that $\Phi \in \mathbb{B}_+(\mathbb{B}(\mathscr{H}), \mathbb{B}(\mathscr{K}))$ is unital. Then*

$$\Phi(A^2) \geq \Phi(A)^2 \tag{5.2.1}$$

for all self-adjoint operators $A \in \mathbb{B}(\mathscr{H})$.

Theorem 5.2.7 *Let $\Phi \in \mathbb{B}_+(\mathbb{B}(\mathscr{H}), \mathbb{B}(\mathscr{K}))$ with $\Phi(I) > 0$. Then*

$$\Phi(BA^{-1}B) \geq \Phi(B)\Phi(A)^{-1}\Phi(B)$$

for all $B \in \mathbb{B}_{sa}(\mathscr{H})$ and all $A \in \mathbb{B}_{++}(\mathscr{H})$.

Proof It follows from $\Phi(I) > 0$ that $\Phi(A)$ is invertible. Set

$$\Psi(X) := \Phi(A)^{-1/2}\Phi(A^{1/2}XA^{1/2})\Phi(A)^{-1/2}.$$

It is easy to verify that $\Psi \in \mathbb{B}_+(\mathbb{B}(\mathscr{H}), \mathbb{B}(\mathscr{K}))$ and

$$\Psi(I) = \Phi(A)^{-1/2}\Phi(A)\Phi(A)^{-1/2} = I.$$

It follows from the Kadison inequality (5.2.1) used for Ψ and $X = A^{-1/2}BA^{-1/2}$ that

$$\Phi(A)^{-1/2}\Phi(BA^{-1}B)\Phi(A)^{-1/2} = \Psi(X^2) \geq \Psi(X)^2 = \Phi(A)^{-1/2}\Phi(B)\Phi(A)^{-1}\Phi(B)\Phi(A)^{-1/2},$$

from which we get the desired inequality.

Putting $B = I$ in Theorem 5.2.7, we get the next result.

Corollary 5.2.8 *Let $\Phi \in \mathbb{B}_+(\mathbb{B}(\mathscr{H}), \mathbb{B}(\mathscr{K}))$ be unital. Then*

$$\Phi(A^{-1}) \geq \Phi(A)^{-1}$$

for all $A \in \mathbb{B}_{++}(\mathscr{H})$.

The next theorem reads as follows; see [3, 10, 11].

Theorem 5.2.9 (Choi–Davis–Jensen theorem) *Suppose that f is a real-valued continuous function defined on an interval J. Then the following assertions are equivalent:*

(i) f is operator convex on J;
(ii) $f(\Phi(A)) \leq \Phi(f(A))$ for every unital $\Phi \in \mathbb{B}_+(\mathbb{B}(\mathscr{H}), \mathbb{B}(\mathscr{K}))$ and every $A \in \mathbb{B}_{sa}(\mathscr{H})(J)$.

Proof (i) $\Rightarrow$ (ii). It is a consequence of the spectral representation for self-adjoint operators that A can be approximated uniformly by operators of the form $A' = \sum_j \lambda_j E_j$ where $\sum_j E_j = I_{\mathscr{H}}$ and E_j 's are orthogonal. Since Φ is unital, $\sum_j \Phi(E_j) = I_{\mathscr{K}}$. It follows from Theorem 5.1.5 (iii) with $C_j = \Phi(E_j)^{1/2}$ that

$$\begin{aligned} f(\Phi(A')) &= f\left(\sum_j \lambda_j \Phi(E_j)\right) = f\left(\sum_j C_j^* \lambda_j C_j\right) \\ &\leq \sum_j C_j f(\lambda_j) C_j = \sum_j f(\lambda_j)\Phi(E_j) \\ &= \Phi\left(\sum_j f(\lambda_j)E_j\right) = \Phi(f(A')). \end{aligned}$$

From the continuity of Φ we infer (ii).
(ii) $\Rightarrow$ (i). Let C be an isometry and put $\Phi(X) = C^*XC$. Evidently, $\Phi \in \mathbb{B}_+(\mathbb{B}(\mathscr{H}), \mathbb{B}(\mathscr{K}))$. Employing (ii) $\Rightarrow$ (i) of Theorem 5.1.5, we deduce (i) from (ii).

Let us state some historical note: Brown and Kosaki [12] proved that if f is a continuous convex function on the interval $[0, \infty)$ with $f(0) = 0$, then

$$f(\langle C^*ACx, x\rangle) \leq \langle C^*f(A)Cx, x\rangle$$

for all $x \in \mathscr{H}$ with $\|x\| \leq 1$, all positive operators A, and all contractions C. It is known that if $\mathscr{A}$ is a C^*-algebra, φ is a state on $\mathscr{A}$, and f is a convex function on an interval J, then

$$f(\varphi(A)) \leq \varphi(f(A)) \tag{5.2.2}$$

is valid for each self-adjoint element $A \in \mathscr{A}$ with spectrum contained in J; see [13, Theorem 2.10]. In the special case, it holds that

$$f\left(\frac{1}{n}\operatorname{tr}(A)\right) \leq \frac{1}{n}\operatorname{tr}(f(A))$$

for any convex function f on an interval J and any Hermitian matrix $A \in \mathbb{M}_n$ with eigenvalues in J. But it is not generally true if the state φ is replaced with an arbitrary positive linear map between C^*-algebras. However, Davis [14] showed that if Φ is a unital completely positive linear map from a C^*-algebra into $\mathbb{B}(\mathscr{H})$ for some Hilbert space $\mathscr{H}$ and if f is an operator convex function on an interval J, then

$$f(\Phi(A)) \leq \Phi(f(A)) \tag{5.2.3}$$

holds for every self-adjoint operator A, whose spectrum is contained in J. Subsequently, Choi [15] proved that it is sufficient to assume that Φ is a unital positive map. Ando [16] gave an alternative proof for this inequality by using the integral representation of operator convex functions. Various types of inequality (5.2.3), when f is not necessarily operator convex, have been explored by some researchers. For instance, Antezana, Massey, and Stojanoff [13] showed that if $\Phi : \mathscr{A} \to \mathscr{B}$ is a positive unital map between unital C^*-algebras $\mathscr{A}$ and $\mathscr{B}$, f is a convex function, $A \in \mathscr{A}$ is such that $\Phi(f(A))$ and $\Phi(A)$ commute, then $f(\Phi(A)) \leq \Phi(f(A))$.

As a consequence of (5.2.3), we get an extension of the Hölder–McCarthy inequality (see Lemma 1.4.7).

Corollary 5.2.10 *Let $\Phi \in \mathbb{B}_+(\mathbb{B}(\mathscr{H}), \mathbb{B}(\mathscr{K}))$ and $A \in \mathbb{B}_+(\mathscr{H})$. Then*

(i) $\Phi(A^r) \leq \Phi(A)^r$ *for* $0 \leq r \leq 1$;
(ii) $\Phi(A) \leq \Phi(A^r)^{\frac{1}{r}}$ *for* $1 \leq r < \infty$.

Proof The operator convexity of $f(t) = -t^r$, the linearity of Φ, and Theorem 5.2.9 give us (i). Applying (i) to $0 < 1/r \le 1$ we get $\Phi(A^{1/r}) \le \Phi(A)^{1/r}$. Replacing A with A^r we reach (ii).

5.3 Operator Means

The operator version of the arithmetic mean of two positive numbers is $(A + B)/2$, where $A, B \in \mathbb{B}_+(\mathscr{H})$. Now recall that the geometric mean of two positive numbers a and b is the unique positive solution to the equation $x^2 = ab$. Since operators do not commute in general, an operator version of this equation can be $XA^{-1}X = B$, where $A, B \in \mathbb{B}_{++}(\mathscr{H})$. Assume that we intend to solve this operator equation. We can write it in the form $Y(B^{1/2}A^{-1}B^{1/2})Y = I$, where $Y = B^{-1/2}XB^{-1/2}$. We can see that the solution to $Y(B^{1/2}A^{-1}B^{1/2})Y = I$ is the solution to $B^{1/2}A^{-1}B^{1/2} = Y^{-2}$, which is $Y = (B^{-1/2}AB^{-1/2})^{1/2}$. Thus, $X = B^{1/2}YB^{1/2} = B^{1/2}(B^{-1/2}AB^{-1/2})^{1/2}B^{1/2}$. This is a motivation for defining the operator geometric mean $B\sharp A$ as $B^{1/2}(B^{-1/2}AB^{-1/2})^{1/2}B^{1/2}$.

Taking inverses of both sides of $XA^{-1}X = B$ we get the equivalent equation $XB^{-1}X = A$, whose solution is $B\sharp A$. The uniqueness ensures that $A\sharp B = B\sharp A$. Also, if $AB = BA$, then every polynomial in A commutes with B and so, by using the functional calculus, $A^{1/2}$ commutes with B. A similar statement holds by changing the roles of A and B. Hence, $A\sharp B = (AB)^{1/2}$.

If $A, B \in \mathbb{B}_{++}(\mathscr{H})$ with $A \le B$, $p, q \in [0, 1]$, $A^p \le B^p$, and $A^q \le B^q$, then Theorem 1.4.2 (iv) and (x) ensures that

$$\begin{aligned}
A^{(p+q)/2} &= A^p \sharp A^q \\
&= A^{q/2}(A^{-q/2}A^pA^{-q/2})^{1/2}A^{q/2} \\
&\le A^{q/2}(A^{-q/2}B^pA^{-q/2})^{1/2}A^{q/2} \\
&\qquad\qquad \text{(by Theorem 1.42 (iv) and noting } A^p \le B^p\text{)} \\
&= B^p \sharp A^q \\
&= A^q \sharp B^p \\
&\le B^q \sharp B^p \qquad \text{(since } A^q \le B^q\text{)} \\
&= B^{(p+q)/2}\,.
\end{aligned}$$

This and $0, 1 \in \Delta = \{r \in [0, 1] : A^r \le B^r\}$ show that Δ is convex, and so $\Delta = [0, 1]$. Hence, we reach the Löwner–Heinz theorem.

The axiomatic theory for the operator mean of two positive operators has been developed by Kubo and Ando [1]. An operator mean is a binary operation σ defined on the set of positive operators, if the following conditions are satisfied:

(i) $A \le C, B \le D$ imply $A\sigma B \le C\sigma D$;
(ii) $A_n \downarrow A$, $B_n \downarrow B$ imply $A_n\sigma B_n \downarrow A\sigma B$, where $A_n \downarrow A$ means that $A_1 \ge A_2 \ge \cdots$ and $A_n \to A$ as $n \to \infty$ in the strong operator topology;
(iii) $T^*(A\sigma B)T \le (T^*AT)\sigma(T^*BT) \quad (T \in \mathbb{B}(\mathscr{H}))$;
(iv) $I\sigma I = I$.

There exists an affine order isomorphism between the class of operator means and the class of positive operator monotone functions f defined on $(0, \infty)$ with $f(1) = 1$ via $f(t)I = I\sigma(tI)$ $(t > 0)$. In addition, $A\sigma B = A^{1/2} f(A^{-1/2}BA^{-1/2})A^{1/2}$ for all positive invertible operators A and B. The operator monotone function f is called the *representing function* of σ.

Writing $A_\varepsilon = A + \varepsilon I$ with $\varepsilon > 0$, one can have the limit expression for general positive operators as

$$A\sigma B = \lim_{\varepsilon \to 0} A_\varepsilon \sigma B_\varepsilon = \lim_{\varepsilon \to 0} A_\varepsilon^{1/2} f\left(A_\varepsilon^{-1/2} B_\varepsilon A_\varepsilon^{-1/2}\right) A_\varepsilon^{1/2}.$$

Let σ be an operator mean with the representing function f. The dual $\sigma^\perp$, the transpose σ', and the adjoint σ^* of the mean σ are the operator means corresponding to the representing function $tf(t)^{-1}$, $tf(t^{-1})$, and $f(t^{-1})^{-1}$, respectively. Thus,

$$A\sigma^\perp B = A^{1/2}(A^{-1/2}BA^{-1/2}) f(A^{-1/2}BA^{-1/2})^{-1} A^{1/2},$$

$$A\sigma' B = A^{1/2}(A^{-1/2}BA^{-1/2}) f(A^{1/2}B^{-1}A^{1/2}) A^{1/2},$$

and

$$A\sigma^* B = A^{1/2} f(A^{1/2}B^{-1}A^{1/2})^{-1} A^{1/2}.$$

For $\alpha \in [0, 1]$, the operator means corresponding to the operator monotone functions $f(t) = t^\alpha$, $f(t) = \alpha + (1 - \alpha)t$, and $f(t) = \frac{t}{(1-\alpha)+\alpha t}$ are called the *weighted geometric operator mean*, the *weighted arithmetic operator mean*, and the *weighted harmonic operator mean*, respectively, and are denoted by

$$A\sharp_\alpha B = A^{1/2}(A^{-1/2}BA^{-1/2})^\alpha A^{1/2}.$$

$$A\nabla_\alpha B = (1 - \alpha)A + \alpha B.$$

$$A!_\alpha B = ((1 - \alpha)A^{-1} + \alpha B^{-1})^{-1}.$$

If $\alpha = 1/2$, then we get the usual *geometric mean* $A\sharp B = A^{1/2}(A^{-1/2}BA^{-1/2})^{1/2}A^{1/2}$, *arithmetic mean* $A\nabla B = \frac{A+B}{2}$, and *harmonic mean* $A!B = 2(A^{-1} + B^{-1})^{-1}$.

For convenience, we put $A : B = (A^{-1} + B^{-1})^{-1}$. We have

$$\begin{aligned} A : B &= (A^{-1} + B^{-1})^{-1} = (A^{-1}(A + B)B^{-1})^{-1} = B(A + B)^{-1}A \\ &= B(A + B)^{-1}A + B(A + B)^{-1}B - B(A + B)^{-1}B \\ &= B - B(A + B)^{-1}B\,. \end{aligned}$$

By symmetry, we also have $A : B = A - A(A + B)^{-1}A$.
Example 5.1.4(ii), together with the implication $0 < C \le D \Rightarrow D^{-1} \le C^{-1}$, shows that the map $(A, B) \mapsto A - A(A + B)^{-1}$ is jointly concave and jointly increasing. This means that $A_1 \le A_2$ and $B_1 \le B_2$ imply that $A_1 - A_1(A_1 + B_1)^{-1} \le A_2 - A_2(A_2 + B_2)^{-1}$. Thus, we get the following.

Proposition 5.3.1 *The operator harmonic mean ! is jointly increasing and jointly concave on* $\mathbb{B}_{++}(\mathscr{H}) \times \mathbb{B}_{++}(\mathscr{H})$.

Theorem 5.3.2 *If* $A, B \in \mathbb{B}_{++}(\mathscr{H})$, *then*

$$A : B = \max\left\{X : X \ge 0 \textit{ and } \begin{bmatrix} A + B & A \\ A & A \end{bmatrix} \ge \begin{bmatrix} 0 & 0 \\ 0 & X \end{bmatrix}\right\}.$$

Proof It follows from Theorem 1.6.1 that

$$A(A + B)^{-1}A = \min\left\{Y \in \mathbb{B}_{+}(\mathscr{H}) : \begin{bmatrix} A + B & A \\ A & Y \end{bmatrix} \ge 0\right\}.$$

Hence,

$$\begin{aligned} A : B &= A - A(A + B)^{-1}A \\ &= \max\left\{A - Y : Y \in \mathbb{B}_{+}(\mathscr{H}), Y \le A, \begin{bmatrix} A + B & A \\ A & Y \end{bmatrix} \ge 0\right\} \\ &= \max\left\{X : X \ge 0 \text{ and } \begin{bmatrix} A + B & A \\ A & A \end{bmatrix} \ge \begin{bmatrix} 0 & 0 \\ 0 & X \end{bmatrix}\right\}. \end{aligned}$$

The following characterization of the operator geometric mean is interesting.

Theorem 5.3.3 *If* $A, B \in \mathbb{B}_{+}(\mathscr{H})$, *then*

$$A \sharp B = \max\left\{X : X = X^* \textit{ and } \begin{bmatrix} A & X \\ X & B \end{bmatrix} \ge 0\right\}.$$

Proof By continuity, we may assume that $A, B > 0$. First,

$$\begin{aligned} &(A\sharp B)A^{-1}(A\sharp B) \\ &\quad = (A^{1/2}(A^{-1/2}BA^{-1/2})^{1/2}A^{1/2})\,A^{-1}\,(A^{1/2}(A^{-1/2}BA^{-1/2})^{1/2}A^{1/2}) \\ &\quad = B\,. \end{aligned}$$

Second, if $X = X^*$ and $\begin{bmatrix} A & X \\ X & B \end{bmatrix} \geq 0$, then by Theorem 1.6.1 it follows that $B \geq XA^{-1}X$. Hence,

$$A^{-1/2}BA^{-1/2} \geq (A^{-1/2}XA^{-1/2})\,(A^{-1/2}XA^{-1/2}) = (A^{-1/2}XA^{-1/2})^2\,.$$

Taking the square roots of both sides of the above inequality, we reach $(A^{-1/2}BA^{-1/2})^{1/2} \geq A^{-1/2}XA^{-1/2}$, whence

$$A\sharp B = B^{1/2}(B^{-1/2}AB^{-1/2})^{1/2}B^{1/2} \geq X\,.$$

It follows from $\begin{bmatrix} A & A\sharp B \\ A\sharp B & B \end{bmatrix} \geq 0$ and Exercise 5.4.11 that if $\Phi \in \mathbb{B}_+(\mathbb{B}(\mathscr{H}), \mathbb{B}(\mathscr{K}))$, then

$$\begin{bmatrix} \Phi(A) & \Phi(A\sharp B) \\ \Phi(A\sharp B) & \Phi(B) \end{bmatrix} \geq 0\,.$$

Now, Theorem 5.3.3 yields that $\Phi(A\sharp B) \leq \Phi(A)\sharp\Phi(B)$. We can extend this property to any operator mean.

Theorem 5.3.4 (Ando theorem) *Let* $\Phi \in \mathbb{B}_+(\mathbb{B}(\mathscr{H}), \mathbb{B}(\mathscr{K}))$. *If* σ *is an operator mean, then*

$$\Phi(A\sigma B) \leq \Phi(A)\sigma\Phi(B) \tag{5.3.1}$$

for all $A, B \in \mathbb{B}_+(\mathscr{H})$.

Proof By a continuity argument, we may assume that A is invertible, and thus there is a number m such that $0 < mI \leq A$. By a limit argument, we can assume that $\Phi(I) > 0$. Hence, $0 < m\Phi(I) \leq \Phi(A)$ showing that $\Phi(A)$ is invertible. Set

$$\Psi(X) = \Phi(A)^{-1/2}\Phi(A^{1/2}XA^{1/2})\Phi(A)^{-1/2}\,.$$

Then Ψ is a unital positive linear map. If f is the representing function of σ, then f is operator monotone and so operator concave, by Theorem 5.1.8. Putting $X = A^{-1/2}BA^{-1/2}$ and utilizing the Choi–Davis–Jensen Theorem 5.2.9 to the operator convex function $-f$ and Ψ we have

$$\begin{aligned}
\Phi(A\sigma B) &= \Phi(A^{1/2}f(X)A^{1/2}) \\
&= \Phi(A)^{1/2}\Psi(f(X))\Phi(A)^{1/2} \\
&\leq \Phi(A)^{1/2}f(\Psi(X))\Phi(A)^{1/2} \\
&= \Phi(A)^{1/2}f(\Phi(A)^{-1/2}\Phi(B)\Phi(A)^{-1/2})\Phi(A)^{1/2} \\
&= \Phi(A)\sigma\Phi(B).
\end{aligned}$$

Remark 5.3.5 Let f be a continuous real function on $(0, \infty)$, and let $A\sigma B$ be defined as $A^{1/2}f(A^{-1/2}BA^{-1/2})A^{1/2}$. If inequality (5.3.1) holds for all $A, B \in \mathbb{B}_{++}(\mathscr{H})$ and all $\Phi \in \mathbb{B}_{+}(\mathbb{B}(\mathscr{H}), \mathbb{B}(\mathscr{K}))$, then f is operator monotone. Indeed, the hypothesis implies that $\Phi(I\sigma A) \leq \Phi(I)\sigma\Phi(A)$, which yields $\Phi(f(A)) \leq f(\Phi(A))$. This, by the Choi–Davis–Jensen Theorem 5.2.9, yields that $-f$ is operator convex, and so f is operator concave, or, equivalently, operator monotone.

If $C \geq 0$ and $\Phi(X) = \operatorname{tr}(XC)$, then we derive the following assertion.

Corollary 5.3.6 *Let Φ be a positive linear map on $\mathbb{M}_n$, and let σ be an operator mean. Then*

$$\operatorname{tr}(C(A\sigma B)) \leq \operatorname{tr}(CA)\sigma \operatorname{tr}(CB)$$

for all positive semidefinite matrices $A, B, C \in \mathbb{M}_n$.

The next result provides an operator arithmetic–geometric–harmonic mean inequality.

Theorem 5.3.7 *If $A, B \in \mathbb{B}_{++}(\mathscr{H})$, then*

$$A!B \leq A\sharp B \leq A\nabla B\,.$$

Proof Applying functional calculus to the positive operator $B^{-1/2}AB$ and the numerical arithmetic–geometric–harmonic inequality $\frac{2t}{1+t} \leq t^{1/2} \leq \frac{1+t}{2}$ for $t > 0$, we get

$$2A^{-1/2}BA^{-1/2}(1 + A^{-1/2}BA^{-1/2})^{-1} \leq (A^{-1/2}BA^{-1/2})^{1/2} \leq \frac{1 + A^{-1/2}BA^{-1/2}}{2}\,.$$

Multiplying every term by $A^{1/2}$ from both sides, we get

$$2(B + BA^{-1}B) \leq A^{1/2}(A^{-1/2}BA^{-1/2})^{1/2}A^{1/2} \leq \frac{A+B}{2}\,.$$

The left-hand side is $2(A^{-1} + B^{-1})^{-1}$, whence we obtain $A!B \leq A\sharp B \leq A\nabla B$.

The first inequality can be deduced from the second. In fact, by the arithmetic–geometric mean inequality, it holds that $A^{-1}\sharp B^{-1} \leq A^{-1}\nabla B^{-1}$. Taking the inverse, we get

$$\begin{aligned} A!B = (A^{-1}\nabla B^{-1})^{-1} &\leq (A^{-1}\sharp B^{-1})^{-1} \\ &= (A^{-1/2}(A^{1/2}B^{-1}A^{1/2})^{1/2}A^{-1/2})^{-1} = A\sharp B\,. \end{aligned}$$

5.4 Exercises and Problems

Exercise 5.4.1 For $A, B \in \mathbb{B}_{++}(\mathscr{H})$, prove that $XA^{-1}X = B$ has a unique solution $X > 0$.

Exercise 5.4.2 Let $A, B \in \mathbb{B}_{++}(\mathscr{H})$. Show that there is a unitary operator $U \in \mathbb{B}(\mathscr{H})$ such that $A \sharp B = A^{1/2} U B^{1/2}$. Conversely, if $A^{1/2} U B^{1/2}$ is positive for some unitary operator U, then $A \sharp B = A^{1/2} U B^{1/2}$.

Exercise 5.4.3 If $A, B \in \mathbb{B}_{++}(\mathscr{H})$, show that

$$A!B = \max\left\{X;\, X \geq 0 \text{ and } \begin{bmatrix} 2A & 0 \\ 0 & 2B \end{bmatrix} \geq \begin{bmatrix} X & X \\ X & X \end{bmatrix}\right\}.$$

Exercise 5.4.4 If $A, B \in \mathbb{B}_{++}(\mathscr{H})$, prove that $(A \nabla B) \sharp (A!B) = A \sharp B$.

Exercise 5.4.5 Prove that the function $f(t) = t^3$ on $[0, \infty)$ is convex but not operator convex.

Exercise 5.4.6 [Russo–Dye theorem] Suppose that $\Phi \in \mathbb{B}_+(\mathbb{B}(\mathscr{H}), \mathbb{B}(\mathscr{K}))$ is unital. Then $\|\Phi\| = 1$.

Exercise 5.4.7 Give an example of a linear map Φ on $\mathbb{M}_n$ for some $n \geq 2$ such that $\|\Phi\| = \|\Phi(I)\| = 1$ but Φ is not positive.

Exercise 5.4.8 Suppose that Φ is a positive unital linear map on $\mathbb{B}(\mathscr{H})$ and $A \in \mathbb{B}(\mathscr{H})$ is a self-adjoint operator such that $m \leq A \leq M$ for some positive numbers M and m. Show that

$$\Phi(A^2) - \Phi(A)^2 \leq \frac{1}{4}(M - m)^2.$$

Exercise 5.4.9 (*Kantorovich inequality*) Suppose that $A \in \mathbb{B}(\mathscr{H})$ is a positive operator such that $0 < m \leq A \leq M$ for some positive numbers M and m. Show that

$$\langle Ax, x\rangle\langle A^{-1}x, x\rangle \leq \frac{(M + m)^2}{4mM}$$

for all unit vectors $x \in \mathscr{H}$.

Exercise 5.4.10 Suppose that $\Phi : \mathbb{B}(\mathscr{H}) \to \mathbb{B}(\mathscr{K})$ is a unital positive linear map and $A \in \mathbb{B}(\mathscr{H})$ is a positive operator satisfying $0 < m \leq A \leq M$ for some scalars m and M. Prove that

$$\Phi(A) \sharp \Phi(A^{-1}) \leq \frac{M + m}{2\sqrt{Mm}}.$$

Exercise 5.4.11 Let $\Phi \in \mathbb{B}_+(\mathbb{B}(\mathscr{H}), \mathbb{B}(\mathscr{K}))$.

(i) If $\begin{bmatrix} A & B \\ B & C \end{bmatrix} \geq 0$, then $\begin{bmatrix} \Phi(A) & \Phi(B) \\ \Phi(B) & \Phi(C) \end{bmatrix} \geq 0$;

(ii) If $\begin{bmatrix} A & B \\ B^* & A \end{bmatrix} \geq 0$, then $\begin{bmatrix} \Phi(A) & \Phi(B) \\ \Phi(B^*) & \Phi(A) \end{bmatrix} \geq 0$.

Exercise 5.4.12 Prove that

(i) the map $\Phi : \mathbb{M}_3 \to \mathbb{M}_3$ defined by $\Phi(A) = 2\operatorname{tr}(A)I - A$ is 2-positive but not 3-positive,
(ii) the map $\Psi : \mathbb{M}_2 \to \mathbb{M}_2$ defined by $\Phi(A) = A^*A$ is positive but not 2-positive.

Exercise 5.4.13 Suppose that σ is an operator mean on $\mathbb{M}_n$ and $A, B, C \in \mathbb{M}_n$ are positive semidefinite. Show that

$$|||A\sigma B||| \leq |||A|||\,\sigma\,|||B|||$$

for all unitarily invariant norms $|||\cdot|||$ on $\mathbb{M}_n$.

Problem 5.4.14 A map $\Phi : \mathbb{M}_n \to \mathbb{M}_k$ is called *trace preserving* if $\operatorname{tr}(\Phi(A)) = \operatorname{tr} A$ for all $A \in \mathbb{M}_n$.

(i) Show that a linear map $\Phi : \mathbb{M}_n \to \mathbb{M}_k$ is trace preserving if and only if its adjoint $\Phi^* : \mathbb{M}_k \to \mathbb{M}_n$ with respect to the Hilbert–Schmidt inner product $\langle A, B\rangle = \operatorname{tr}(B^*A)$ is unital.
(ii) Prove that a positive trace preserving linear map $\Phi : \mathbb{M}_n \to \mathbb{M}_n$ satisfies $\|\Phi(A)\|_p \leq \|A\|_p$ for all $1 \leq p \leq \infty$ if and only if Φ is unital; see [17].
(iii) Characterize all completely positive linear maps $\Phi : \mathbb{M}_n \to \mathbb{M}_n$ that preserve the density matrices. In matrix theory, a *density matrix* is a positive semidefinite matrix A with $\operatorname{tr} A = 1$; see [18].
(iv) Give a nontrivial example of a nonlinear completely positive trace preserving map $\Phi : \mathbb{M}_n \to \mathbb{M}_k$.

Problem 5.4.15 Let

$$\Phi(A_1, \ldots, A_n) = \operatorname{tr}\left(\left(\sum_{j=1}^{n} A_j^p\right)^{\frac{1}{p}}\right),$$

where A_j $(1 \leq j \leq n)$ is positive semidefinite and $p > 0$. Discuss the (joint) concavity and the (joint) convexity of Φ for various values of p; see [19].

Problem 5.4.16 Let A and B be positive semidefinite matrices, and let f be a nonnegative operator monotone function on $[0, \infty)$. Prove that

$$|||f(A) - f(B)||| \leq |||f(|A - B|)|||$$

for every unitarily invariant norm.

5.5 Notes, Hints, and References

This chapter includes some results on operator convex functions, operator concave functions, positive linear maps, and operator means. We refer readers who are looking for more detailed explanations on these topics to the books [2, 5, 20]. Let us provide some hints and references for the exercises and problems.

For Exercise 5.4.1, one can use the fact that for any positive invertible operator A, there exists a unique positive invertible operator X such that $X^2 = A$ (Theorem 1.4.1). For Exercise 5.4.2, verify that $U = (A^{-1/2}BA^{-1/2})^{1/2}A^{1/2}B^{-1/2}$ is unitary. For the reverse statement; see [21, Proposition 4.1.8]. To solve Exercise 5.4.3, note that for each $x \in \mathscr{H}$, $\langle (A : B)x, x\rangle = \inf\{\langle Ay, y\rangle + \langle Bz, z\rangle : y, z \in \mathscr{H}, y + z = x\}$, where $A : B = B - B(A + B)^{-1}B$.

To solve Exercise 5.4.4, use the representing functions of !, $\sharp$, and ∇ which are $\frac{2t}{1+t}$, $t^{1/2}$, and $\frac{1+t}{2}$, respectively. For Exercise 5.4.5, one can find a counterexample of 2×2 matrices; see [22, Example 1.5]. Exercise 5.4.6 is the same as [21, Theorem 2.3.7].

A hint for Exercise 5.4.7 is that $\|\Phi\| = \|\Phi(I)\| = 1$ and $\Phi(I) > 0$ for the linear map on $\mathbb{M}_2$ defined by $\Phi\left(\begin{pmatrix} a_{11} & a_{12} \\ a_{21} & a_{22} \end{pmatrix}\right) = \begin{pmatrix} a_{11} & a_{12} \\ 0 & 0 \end{pmatrix}$. For Exercise 5.4.8, first show that $\Phi(A^2) - \Phi(A)^2 \le (M - \Phi(A))(\Phi(A) - m)$.

To solve Exercise 5.4.9, first show that $M + m \ge MmA^{-1} + A$ and then employ the arithmetic–geometric mean inequality; see [22, Theorem 1.23]. For Exercise 5.4.10, one can use Exercise 5.4.9 and Theorem 5.3.7. See [22, Theorem 1.26]. For Exercise 5.4.11, one can use Theorem 1.6.1 and Theorem 5.2.7. Similar results to Exercise 5.4.12 (i) can be found in [23]. For (ii) see [21, Exercise 1.6.6].

Exercise 5.4.13 appears in [24]. For (i) in Problem 5.4.14, notice that if Φ is a linear map on $\mathbb{M}_n$, then we define its adjoint Φ^* as the linear map that satisfies the relation $\langle \Phi(A), B\rangle = \langle A, \Phi^*(B)\rangle$ for all A and B. For (ii), we refer readers to [17]. For (iii); see [18]. For Problem 5.4.15, consult [19]. Finally, Problem 5.4.16 is a perturbation norm inequality, which has been established in [25].

References

1. F. Kubo, T. Ando, Means of positive linear operators. Math. Ann. **246**, 205–224 (1980)
2. M.B. Ghaemi, N. Gharakhanlu, T.M. Rassias, R. Saadati, *Advances in Matrix Inequalities, Springer Optimization and Its Applications*, vol. 176 (Springer, Cham, 2021)
3. F. Hansen, K.G. Pedersen, Jensen's operator inequality. Bull. Lond. Math. Soc. **35**, 553–564 (2003)
4. F. Hansen, An operator inequality. Math. Ann. **246**, 249–250 (1979/80)
5. M. Fujii, J. Mićić Hot, J. Pečarić and Y. Seo, *Recent Developments of Mond–Pečarić method in operator inequalities*, Monographs in Inequalities 4, Element, Zagreb (2012)
6. M. Uchiyama, Operator monotone functions, positive definite kernel and majorization. Proc. Am. Math. Soc. **138**, 3985–3996 (2010)
7. W.F. Stinespring, Positive functions on C^*-algebras. Proc. Am. Math. Soc. **6**, 211–216 (1955)

8. E. Størmer, Positive linear maps of operator algebras. Acta Math. **110**, 233–278 (1963)
9. W.B. Arveson, Subalgebras of C^*-algebras. Acta Math. **123**, 141–224 (1969)
10. F. Hansen, K.G. Pedersen, Jensen's inequality for operators and Löwner's theorem. Math. Ann. **258**, 229–241 (1982)
11. J.I. Fujii, M. Fujii, *Jensen's Inequalities on Any Interval for Operators, Nonlinear Analysis and Convex Analysis* (Yokohama Publ, Yokohama, 2004), pp.29–30
12. L.G. Brown, H. Kosaki, Jensen's inequality in semi-finite von Neuman algebras. J. Operator Theory **23**, 3–19 (1990)
13. J. Antezana, P. Massey, D. Stojanoff, Jensen's inequality for spectral order and submajorization. J. Math. Anal. Appl. **331**, 297–307 (2007)
14. C. Davis, A Schwarz inequality for convex operator functions. Proc. Am. Math. Soc. **8**, 42–44 (1957)
15. M.D. Choi, A Schwarz inequality for positive linear maps on C^*-algebras. Illinois J. Math. **18**, 565–574 (1974)
16. T. Ando, *Topics on Operator Inequalities* (Hokkaido University, Sapporo, 1978)
17. D. Pérez-García, M.M. Wolf, D. Petz, M.B. Ruskai, Contractivity of positive and trace-preserving maps under L_p norms. J. Math. Phys. **47**, 083506 (2006), 5 pp
18. M.B. Ruskai, S. Szarek, E. Werner, An analysis of completely positive trace-preserving maps on $\mathscr{M}_2$. Linear Algebra Appl. **347**, 159–187 (2002)
19. E.A. Carlen, E.H. Lieb, A Minkowski type trace inequality and strong subadditivity of quantum entropy. II. Convexity and concavity. Lett. Math. Phys. **83**, 107–126 (2008)
20. R. Bhatia, *Matrix Analysis* (Springer, New York, 1997)
21. R. Bhatia, *Positive Definite Matrices* Princeton Ser. Appl. Math. (Princeton University Press, Princeton, NJ, 2007)
22. T. Furuta, J. Mićić Hot, J. Pečarić, Y. Seo, *Mond–Pečarić Method in Operator Inequalities*, Monographs in Inequalities 1, Element, Zagreb (2005)
23. M.D. Choi, Positive linear maps on C^*-algebras. Canad. J. Math. **24**, 520–529 (1972)
24. T. Ando, Majorizations and inequalities in matrix theory. Linear Algebra Appl. **199**, 17–67 (1994)
25. T. Ando, Comparison norms $|||f(A) - f(B)|||$ and $|||f(|A - B|)|||$. Math. Z. **197**, 403–409 (1988)

Chapter 6
Golden–Thompson Trace Inequalities

In this chapter, we first study the Golden–Thompson trace inequality and its generalizations. We then explore log-majorization and unitarily invariant norm inequalities, which have evolved from considering the Golden–Thompson trace inequality. Next, we study the reverse inequalities of the Golden–Thompson trace inequality in terms of the generalized Kantorovich constant and the Specht ratio, determined by the maximum and minimum eigenvalues of given matrices. Finally, we study the Hadamard product versions and the multivariable versions as further generalizations of the Goldem–Thompson trace inequality.

6.1 Golden–Thompson Trace Inequalities

In the commutative case, if A and B are commuting Hermitian matrices, then $e^{A+B} = e^A e^B$. However, in the noncommutative case, there is no relation between e^{A+B} and $e^A e^B$ under the usual order. It is known that the equality $\mathrm{tr}(e^{A+B}) = \mathrm{tr}(e^A e^B)$ holds if and only if $AB = BA$ [1, 2]. The celebrated Golden–Thompson trace inequality, independently proved by Golden [3], Symanzik [4], and Thompson [5], is given by the following theorem.

Theorem 6.1.1 *If A and B are Hermitian matrices in $\mathbb{M}_n$, then*

$$\mathrm{tr}(e^{A+B}) \le \mathrm{tr}(e^A e^B). \tag{6.1.1}$$

To prove this theorem, we need certain preliminaries. A useful tool in proving inequalities involving the matrix exponential is the following Lie–Trotter formula that can be found in [6, Theorem IX.1.3] and [7, Theorem 3.8, Corollary 3.9].

A. M. Bikchentaev et al., *Trace Inequalities*, Forum for Interdisciplinary Mathematics,
https://doi.org/10.1007/978-981-97-6520-1_6

Theorem 6.1.2 (Lie–Trotter formula) *For any matrices A and B in $\mathbb{M}_n$*

$$e^{A+B} = \lim_{m\to\infty} \left(e^{A/m}e^{B/m}\right)^m . \tag{6.1.2}$$

If A, $B \in \mathbb{M}_n$ are Hermitian, then

$$e^{A+B} = \lim_{m\to\infty} \left(e^{B/2m}e^{A/m}e^{B/2m}\right)^m . \tag{6.1.3}$$

Proof Put $X_m = e^{(A+B)/m}$ and $Y_m = e^{A/m}e^{B/m}$. Then, with the convention $O\left(\frac{1}{m^2}\right)$ as $m \to \infty$, we have

$$\begin{aligned} X_m - Y_m &= \left[I + \frac{A+B}{m} + \frac{1}{2}\left(\frac{A+B}{m}\right)^2 + \cdots\right] \\ &\quad - \left[I + \frac{A}{m} + \frac{1}{2}\left(\frac{A}{m}\right)^2 + \cdots\right]\left[I + \frac{B}{m} + \frac{1}{2}\left(\frac{B}{m}\right)^2 + \cdots\right] \\ &= O\left(\frac{1}{m^2}\right) \qquad \text{for large } m. \end{aligned}$$

Since

$$X_m^m - Y_m^m = \sum_{j=0}^{m-1} X_m^{m-1-j}(X_m - Y_m)Y_m^j,$$

it follows that

$$\begin{aligned} \left\|X_m^m - Y_m^m\right\| &\le m\,\|X_m - Y_m\| \max\{\|X_m\|, \|Y_m\|\}^{m-1} \\ &\le m e^{\|A\|+\|B\|} O\left(\frac{1}{m^2}\right) \to 0 \end{aligned}$$

as $m \to \infty$. Since $X_m^m = e^{A+B}$ for all m, $\lim_{m\to\infty}(e^{A/m}e^{B/m})^m = \lim_{m\to\infty} Y_m^m = e^{A+B}$, and we have (6.1.2).

For the second part, we have

$$\left(e^{A/2m}e^{B/m}e^{A/2m}\right)^m = e^{-A/2m}\left(e^{A/m}e^{B/m}\right)^m e^{A/2m}$$

and we deduce (6.1.3). □

Remark 6.1.3 The Lie–Trotter formula $e^{A+B} = \lim_{m\to\infty}\left(e^{B/2m}e^{A/m}e^{B/2m}\right)^m$ holds for bounded linear operators A and B on a Hilbert space.

From Theorem 2.1.12, we obtain $\mathrm{tr}(AB)^2) \le \mathrm{tr}(A^2B^2)$ for all Hermitian matrices A and B. The following lemma is a generalization of it. This lemma follows from Lemma 3.2.4 and we provide a proof for the case when $m = 2^k$.

Lemma 6.1.4 *For any two Hermitian matrices A and B in $\mathbb{M}_n$,*

$$|\operatorname{tr}((AB)^{2^k})| \le \operatorname{tr}(A^{2^k} B^{2^k})$$

for any positive integer k.

Proof If we put $X = AB$ and $|X| = (BA^2B)^{1/2}$ in Lemma 3.1.4, then we have

$$\begin{aligned}|\operatorname{tr}((AB)^{2m})| &\le \operatorname{tr}((BA^2B)^m) = \operatorname{tr}(BA^2B^2\cdots B^2A^2B)\\ &= \operatorname{tr}((A^2B^2)^m)\end{aligned} \tag{6.1.4}$$

for any positive integers m.

We utilize induction. If $k = 1$, then it is just the case of $m = 1$ in (6.1.4). Assume that the statement is true for $k = \ell$. We show that it is true for $k = \ell + 1$.

$$\begin{aligned}|\operatorname{tr}((AB)^{2^{\ell+1}})| &= |\operatorname{tr}((AB)^{2\cdot 2^{\ell}})|\\ &\le \operatorname{tr}((A^2B^2)^{2^{\ell}}) \qquad \text{(by (6.1.4))}\\ &\le \operatorname{tr}(A^{2\cdot 2^{\ell}} B^{2\cdot 2^{\ell}}) \qquad \text{(by assumption for } k = \ell)\\ &= \operatorname{tr}(A^{2^{\ell+1}} B^{2^{\ell+1}}).\end{aligned}$$

□

Proof of Theorem 6.1.1 By Lemma 6.1.4, we have

$$\operatorname{tr}((e^{\frac{1}{2^k}A}e^{\frac{1}{2^k}B})^{2^k}) \le \operatorname{tr}(e^Ae^B) \qquad \text{for any positive integer } k.$$

As $k \to \infty$, the left-hand side of this inequality converges to $\operatorname{tr}(e^{A+B})$ according to (6.1.2) of Theorem 6.1.2 (Lie–Trotter formula). Therefore, we have the Golden–Thompson trace inequality $\operatorname{tr}(e^{A+B}) \le \operatorname{tr}(e^Ae^B)$. □

Remark 6.1.5 The obvious generalization of Lemma 6.1.4, namely,

$$|\operatorname{tr}((ABC)^{2^k})| \le \operatorname{tr}(A^{2^k} B^{2^k} C^{2^k}) \tag{6.1.5}$$

is not true, so that Theorem 6.1.1 has no obvious generalization. A counterexample to (6.1.5) (for $k = 1$) is

$$A = \begin{bmatrix} 2&0&0\\0&1&0\\0&0&1\end{bmatrix}, \quad B = \begin{bmatrix} 2&0&-2\\0&1&0\\-2&0&1\end{bmatrix}, \text{ and } \quad C = \begin{bmatrix} 1&0&1\\0&1&0\\1&0&2\end{bmatrix},$$

for which $\operatorname{tr}(A^2B^2C^2) = \operatorname{tr}(B^2A^2C^2) = 0$ and $\operatorname{tr}((ABC)^2) = 9$.

If A and B and C are Hermitian, and B and C commute, then we have

$$\operatorname{tr}(e^{A+B+C}) \le \operatorname{tr}(e^{A}e^{B+C}) = \operatorname{tr}(e^{A}e^{B}e^{C}) \qquad \text{by (6.1.1)}.$$

As mentioned above, the 3-variable version of the Golden–Thompson trace inequality

$$\operatorname{tr}(e^{A+B+C}) \le |\operatorname{tr}(e^{A}e^{B}e^{C})|$$

does not hold in general. To see it, put

$$S_1 = \begin{bmatrix} 0 & 1 \\ 1 & 0 \end{bmatrix}, \; S_2 = \begin{bmatrix} 0 & -\mathrm{i} \\ \mathrm{i} & 0 \end{bmatrix}, \; \text{and} \quad S_3 = \begin{bmatrix} 1 & 0 \\ 0 & -1 \end{bmatrix}$$

and

$$A = tS_1,\; B = tS_2 \;\; \text{and} \;\; C = t(S_3 - S_2 - S_1) \qquad \text{for } t \in \mathbb{R}.$$

Then for small t, we have $\operatorname{tr}(e^{A+B+C}) > |\operatorname{tr}(e^{A}e^{B}e^{C})|$; see [6, p. 280] for details. In the last section, we consider a k-variable version of the Golden–Thompson trace inequality (6.1.1).

6.2 Strengthened Variant of the Golden–Thompson Trace Inequality

In this section, we show a strengthened variant of the Golden–Thompson trace inequality. For any matrix X and $k = 1, 2, \ldots, n$, we denote by $C_k(X)$ the k-fold antisymmetric tensor power (or the k-th compound) of X; refer to [8] for details. Then (i)–(iii) below are basic facts, and (iv) is easily seen from (ii) and (iii).

Lemma 6.2.1 *(i)* $C_k(X^*) = C_k(X)^*$;
(ii) $C_k(XY) = C_k(X)C_k(Y)$ *for every pair of matrices* X, Y;
(iii) $C_k(X^{-1}) = C_k(X)^{-1}$ *for invertible* X;
(iv) $C_k(A^p) = C_k(A)^p$ *for every positive definite* $A > 0$ *and all* $p \in \mathbb{R} \setminus \{0\}$.

Moreover, by the Binet–Cauchy theorem [9, p. 123],

$$\prod_{i=1}^{k} \lambda_i(A) = \lambda_1(C_k(A)) = \|C_k(A)\| \qquad \text{for } k = 1, 2, \ldots, n$$

for every $A \ge 0$. Hence, we have the following equivalent condition for the log-majorization.

Lemma 6.2.2 *Let A and B be two positive definite matrices in* $\mathbb{M}_n$. *Then* $A \prec_{\log} B$ *if and only if* $\lambda_1(C_k(A)) \le \lambda_1(C_k(B))$ *for all* $k = 1, \ldots, n-1$ *and* $\det(A) = \det(B)$.

The following lemma, clearly true by the definition of the log-majorization, is a basic result that we frequently apply in this chapter.

Lemma 6.2.3 *For positive definite matrices A and B,*

$$A \prec_{\log} B \quad \textit{implies} \quad A^p \prec_{\log} B^p \quad \textit{for all } p > 0.$$

Proof If $A \prec_{\log} B$, then it follows from Lemma 6.2.1 that

$$\lambda_1(C_k(A^p)) = \lambda_1(C_k(A)^p) = \lambda_1(C_k(A))^p \leq \lambda_1(C_k(B))^p = \lambda_1(C_k(B^p))$$

for $k = 1, \ldots, n-1$ and

$$\det(A^p) = \det(A)^p = \det(B)^p = \det(B^p) \qquad \text{for all } p > 0.$$

Hence, we have $A^p \prec_{\log} B^p$ for all $p > 0$ by Lemma 6.2.2. □

Now, we show a strengthened variant of the Golden–Thompson trace inequality. To show it, we need the following Araki's log-majorization [10].

Theorem 6.2.4 *Let A and B be two positive semidefinite matrices in $\mathbb{M}_n$. Then*

$$(A^{1/2}BA^{1/2})^r \prec_{\log} A^{r/2}B^rA^{r/2} \qquad \textit{for } r \geq 1 \tag{6.2.1}$$

or equivalently

$$A^{r/2}B^rA^{r/2} \prec_{\log} (A^{1/2}BA^{1/2})^r \qquad \textit{for } 0 < r \leq 1 \tag{6.2.2}$$

or equivalently

$$(A^{q/2}B^qA^{q/2})^{1/q} \prec_{\log} (A^{p/2}B^pA^{p/2})^{1/p} \qquad \textit{for } 0 < q \leq p. \tag{6.2.3}$$

Proof Suppose that A and B are positive definite. It follows from Lemma 6.2.1 that

$$C_k((A^{1/2}BA^{1/2})^r) = (C_k(A)^{1/2}C_k(B)C_k(A)^{1/2})^r$$

and

$$C_k(A^{r/2}B^rA^{r/2}) = C_k(A)^{r/2}C_k(B)^rC_k(A)^{r/2}$$

for $k = 1, 2, \ldots, n-1$. Also, we have

$$\begin{aligned}\det((A^{1/2}BA^{1/2})^r) &= \det(A^{1/2}BA^{1/2})^r = \det(A)^{r/2}\det(B)^r\det(A)^{r/2}\\ &= \det(A^{r/2}B^rA^{r/2}).\end{aligned}$$

By Lemma 6.2.2 it suffices to show that

$$\lambda_1((A^{1/2}BA^{1/2})^r) \le \lambda_1(A^{r/2}B^rA^{r/2}) \qquad \text{for } r \ge 1. \tag{6.2.4}$$

To do so, we prove that

$$A^{r/2}B^rA^{r/2} \le I \quad \text{implies that} \quad (A^{1/2}BA^{1/2})^r \le I \quad \text{for all } r \ge 1. \tag{6.2.5}$$

In fact, since $A^{r/2}B^rA^{r/2} \le \|A^{r/2}B^rA^{r/2}\|$ holds in general, we have

$$A^{r/2}(B/\|A^{r/2}B^rA^{r/2}\|^{1/r})^rA^{r/2} \le I.$$

Hence, it follows from (6.2.5) that

$$\left(A^{1/2}\frac{B}{\|A^{r/2}B^rA^{r/2}\|^{1/r}}A^{1/2}\right)^r \le I,$$

and so $(A^{1/2}BA^{1/2})^r \le \|A^{r/2}B^rA^{r/2}\|$. Hence, we have

$$\|(A^{1/2}BA^{1/2})^r\| \le \|A^{r/2}B^rA^{r/2}\|,$$

that is, it follows that $\lambda_1(A^{1/2}BA^{1/2})^r) \le \lambda_1(A^{r/2}B^rA^{r/2})$ for $r \ge 1$.

Let us prove (6.2.5). If $A^{r/2}B^rA^{r/2} \le I$, then we have $B^r \le A^{-r}$ for $r \ge 1$. Since $0 < 1/r \le 1$, it follows from the Löwner–Heinz theorem (Theorem 1.4.3) that $B \le A^{-1}$, and hence $A^{1/2}BA^{1/2} \le I$. Hence, we have $(A^{1/2}BA^{1/2})^r \le I$ for $r \ge 1$.

Assume that A and B are positive semidefinite. Since $A + \varepsilon I$, $B + \varepsilon I$ are positive definite for all $\varepsilon > 0$, we can use (6.2.4) to deduce that

$$\lambda_1((A+\varepsilon I)^{1/2}(B+\varepsilon I)(A+\varepsilon I)^{1/2})^r) \le \lambda_1((A+\varepsilon I)^{r/2}(B+\varepsilon I)^r(A+\varepsilon I)^{r/2})$$

and as $\varepsilon \to 0$ we have $\lambda_1((A^{1/2}BA^{1/2})^r) \le \lambda_1(A^{r/2}B^rA^{r/2})$. Hence, we have (6.2.1).

By Lemma 6.2.3, (6.2.1) implies (6.2.2) and (6.2.3). □

Remark 6.2.5 As in the proof of Theorem 6.2.4, we have the following operator version corresponding to Theorem 6.2.4: If A and B are positive operators on a Hilbert space, then it follows that

$$\|(A^{1/2}BA^{1/2})^r\| \le \|A^{r/2}B^rA^{r/2}\| \qquad \text{for all } r \ge 1$$

and

$$\|(A^{1/2}BA^{1/2})^r\| \ge \|A^{r/2}B^rA^{r/2}\| \qquad \text{for all } 0 < r \le 1.$$

By Theorem 6.2.4, we have the following inequality of Araki–Lieb–Thirring [10], also compare it with Lemma 3.2.4.

Corollary 6.2.6 *Let $A, B \in \mathbb{M}_n$ be positive semidefinite, and let s and t be positive real numbers with $t \geq 1$. Then*

$$\operatorname{tr}((A^{1/2}BA^{1/2})^{st}) \leq \operatorname{tr}((A^{t/2}B^tA^{t/2})^s).$$

Proof By (2.2.5), it suffices to prove that

$$(A^{t/2}B^tA^{t/2})^s \leq I \quad \text{implies that} \quad (A^{1/2}BA^{1/2})^{st} \leq I.$$

Since $s > 0$ and $t \geq 1$, this corollary follows from the proof of Theorem 6.2.4. □

The next result reads as follows; see Lemma 3.1.3.

Lemma 6.2.7 *Let $A, B \in \mathbb{M}_n$ be Hermitian matrices. Then*

$$|||e^{A/2}e^Be^{A/2}||| \leq |||e^Ae^B|||$$

for any unitarily invariant norm $|||\cdot|||$.

Proof 1. Since $e^{A/2}e^Be^{A/2}$ is positive definite, it follows that

$$\begin{aligned}\|e^{A/2}e^Be^{A/2}\|_{(k)} &= \sum_{i=1}^k s_i(e^{A/2}e^Be^{A/2}) = \sum_{i=1}^k \lambda_i(e^{A/2}e^Be^{A/2}) \\ &= \sum_{i=1}^k \lambda_i(e^Ae^B) \leq \sum_{i=1}^k s_i(e^Ae^B) = \|e^Ae^B\|_{(k)}\end{aligned}$$

for $k = 1, \ldots, n$, where the last inequality follows from the Weyl majorant theorem [6, Theorem II.3.6]. Hence, by Theorem 2.3.1 (Fan dominance theorem), we have the desired inequality $|||e^{A/2}e^Be^{A/2}||| \leq |||e^Ae^B|||$.

Proof 2. Alternatively, by Theorem 6.2.4, we have

$$\left(e^{A/2}e^Be^{A/2}\right)^2 \prec_{\log} e^Ae^{2B}e^A = |e^Be^A|^2$$

and it follows from Lemma 6.2.3 that $e^{A/2}e^Be^{A/2} \prec_{\log} |e^Be^A|$. Hence, we have

$$|||e^{A/2}e^Be^{A/2}||| \leq |||\,|e^Be^A|\,||| = |||U|e^Be^A|\,||| = |||e^Be^A||| = |||(e^Be^A)^*||| = |||e^Ae^B|||,$$

where $e^Be^A = U|e^Be^A|$ is the polar decomposition of e^Be^A. □

The following theorem is due to Ando and Hiai [11], which is a strengthened variant of the Golden–Thompson trace inequality.

Theorem 6.2.8 *Let A and B be Hermitian matrices in $\mathbb{M}_n$. Then,*

$$|||e^{A+B}||| \leq |||(e^{pA/2}e^{pB}e^{pA/2})^{1/p}||| \quad \text{for all } p > 0 \tag{6.2.6}$$

for any unitarily invariant norm $|||\cdot|||$. *The right-hand side decreases to* $|||e^{A+B}|||$ *as* $p \downarrow 0$.

In particular,

$$|||e^{A+B}||| \le |||(e^{pA/2}e^{pB}e^{pA/2})^{1/p}||| \le |||e^{A/2}e^{B}e^{A/2}||| \le |||e^{A}e^{B}|||$$

for all $0 < p \le 1$.

Proof Replace A and B with e^A and e^B, respectively, in (6.2.3) of Theorem 6.2.4, and conclude that

$$\left(e^{qA/2}e^{qB}e^{qA/2}\right)^{1/q} \prec_{\log} \left(e^{pA/2}e^{pB}e^{pA/2}\right)^{1/p} \qquad \text{for } 0 < q \le p.$$

Hence, it follows from (2.2.5) that

$$\left|\left|\left|\left(e^{qA/2}e^{qB}e^{qA/2}\right)^{1/q}\right|\right|\right| \le \left|\left|\left|\left(e^{pA/2}e^{pB}e^{pA/2}\right)^{1/p}\right|\right|\right| \qquad \text{for } 0 < q \le p.$$

By Theorem 6.1.2 (Lie–Trotter formula), the left-hand side of this relation converges to $|||e^{A+B}|||$ as $q \downarrow 0$, and so we have the desired inequality (6.2.6).

It follows from (6.2.1) in Theorem 6.2.4 and Lemma 6.2.7 that

$$|||(e^{pA/2}e^{pB}e^{pA/2})^{1/p}||| \le |||e^{A/2}e^{B}e^{A/2}||| \le |||e^{A}e^{B}|||$$

for $0 < p \le 1$. □

Remark 6.2.9 By Remarks 6.1.3 and 6.2.5, we have the operator version of the inequality corresponding to Theorem 6.2.8. If A and B are self-adjoint operators on a Hilbert space, then

$$\|e^{A+B}\| \le \|(e^{pA/2}e^{pB}e^{pA/2})^{1/p}\| \qquad \text{for all } p > 0.$$

The following theorem represents a 1-variable extension of the Golden–Thompson trace inequality.

Theorem 6.2.10 *Let* A *and* B *be Hermitian matrices in* $\mathbb{M}_n$. *Then*

$$\operatorname{tr} e^{A+B} \le \operatorname{tr}((e^{pA/2}e^{pB}e^{pA/2})^{1/p}) \le \operatorname{tr}(e^{A}e^{B}) \le \operatorname{tr}((e^{qA/2}e^{qB}e^{qA/2})^{1/q}) \qquad (6.2.7)$$

for $0 < p \le 1 \le q$.

Proof Since $|||A||| = \operatorname{tr}(|A|)$ for $A \in \mathbb{M}_n$ is a unitarily invariant norm, it follows from Theorem 6.2.8 and Corollary 6.2.6 that inequality (6.2.7) holds for all $0 < p \le 1 \le q$. □

6.3 Complementary Golden–Thompson Trace Inequality

The Golden–Thompson trace inequality says that

$$\mathrm{tr}(e^{A+B}) \leq \mathrm{tr}(e^A e^B)$$

for all Hermitian matrices A and B.

In this section, we estimate lower bounds of the Golden–Thompson trace inequality. To do this, we first recall the definition of the weighted geometric matrix mean: Let A and B be positive definite matrices and $\alpha \in [0, 1]$. Then the weighted geometric matrix mean is defined by

$$A \sharp_\alpha B = A^{1/2}(A^{-1/2}BA^{-1/2})^\alpha A^{1/2}.$$

In particular, when $\alpha = \frac{1}{2}$, we obtain the usual geometric mean $A \sharp B = A \sharp_{1/2} B$, whicis also discussed in Sect. 5.3.

To prove this, we need to use the Ando–Hiai inequality [11].

Theorem 6.3.1 (Ando–Hiai inequality) *Let A and B be two positive semidefinite matrices in $\mathbb{M}_n$ and $\alpha \in [0, 1]$. Then*

$$A^r \sharp_\alpha B^r \prec_{\log} (A \sharp_\alpha B)^r \quad \textit{for } r \geq 1 \tag{6.3.1}$$

or equivalently

$$(A \sharp_\alpha B)^r \prec_{\log} A^r \sharp_\alpha B^r \quad \textit{for } 0 < r \leq 1, \tag{6.3.2}$$

$$(A^p \sharp_\alpha B^p)^{1/p} \prec_{\log} (A^q \sharp_\alpha B^q)^{1/q} \quad \textit{for } 0 < q \leq p. \tag{6.3.3}$$

Proof The equivalence of (6.3.1)–(6.3.3) is immediate. For (6.3.1), as in the proof of Theorem 6.2.4, it suffices to show that

$$A \sharp_\alpha B \leq I \quad \text{implies that} \quad A^r \sharp_\alpha B^r \leq I \quad \text{for } r \geq 1$$

for positive definite matrices A and B and $\alpha \in [0, 1]$.

Assume that $1 \leq r \leq 2$ and write $r = 2 - \varepsilon$ with $0 \leq \varepsilon \leq 1$. Let $C = A^{-1/2}BA^{-1/2}$. Then we have $A \sharp_\alpha B = A^{1/2}C^\alpha A^{1/2}$. If $A \sharp_\alpha B \leq I$, then $C^\alpha \leq A^{-1}$, so that $A \leq C^{-\alpha}$. Since $0 \leq 1 - \varepsilon \leq 1$, it follows from the Löwner–Heinz theorem (Theorem 1.4.3) that

$$A^{1-\varepsilon} \leq C^{-\alpha(1-\varepsilon)},$$

and so

$$\begin{aligned} A^r \natural_\alpha B^r &= A^{1-\varepsilon/2}(A^{-1+\varepsilon/2}B^{2-\varepsilon}A^{-1+\varepsilon/2})^\alpha A^{1-\varepsilon/2} \\ &= A^{1/2}\{A^{1-\varepsilon} \natural_\alpha [C(A \natural_\varepsilon C^{-1})C]\}A^{1/2} \\ &\le A^{1/2}\{C^{-\alpha(1-\varepsilon)} \natural_\alpha [C(C^{-\alpha} \natural_\varepsilon C^{-1})C]\}A^{1/2} \\ &= A^{1/2}C^\alpha A^{1/2} = A \natural_\alpha B \le I. \end{aligned}$$

For $r > 2$, by writing $r = 2^k(2-\varepsilon)$ with $k \in \mathbb{N}$ and $0 \le \varepsilon \le 1$, we can proceed by induction. □

In particular, we have the following unitarily invariant norm versions.

Corollary 6.3.2 *Let A and B be two positive semidefinite matrices in $\mathbb{M}_n$ and $\alpha \in [0, 1]$. Then for each unitarily invariant norm $|||\cdot|||$,*

$$|||A^r \natural_\alpha B^r||| \le |||(A \natural_\alpha B)^r||| \quad \text{for } r \ge 1$$

or equivalently

$$|||(A \natural_\alpha B)^r||| \le |||A^r \natural_\alpha B^r||| \quad \text{for } 0 < r \le 1,$$

and

$$|||(A^p \natural_\alpha B^p)^{1/p}||| \le |||(A^q \natural_\alpha B^q)^{1/q}||| \quad \text{for } 0 < q \le p.$$

Remark 6.3.3 As in the proof of Theorem 6.3.1, the operator version of the Ando–Hiai inequality holds. If A and B are positive operators on a Hilbert space and $\alpha \in [0, 1]$, then

$$A \natural_\alpha B \le I \quad \text{implies that} \quad A^r \natural_\alpha B^r \le I \quad \text{for } r \ge 1.$$

Hence, we have

$$\|A^r \natural_\alpha B^r\| \le \|(A \natural_\alpha B)^r\| \quad \text{for } r \ge 1.$$

We present another proof of the Ando–Hiai inequality (Theorem 6.3.1): To show it, we need the following three lemmas.

Lemma 6.3.4 *Let A and B be positive definite matrices. If $X = A \natural_\alpha B$ for some $\alpha \in [0, 1]$, then X is the unique solution to the equation*

$$(1-\alpha)\log X^{-1/2}AX^{-1/2} + \alpha \log X^{-1/2}BX^{-1/2} = 0. \tag{6.3.4}$$

Proof In the case of $\alpha = 0, 1$, $X = A, B$ are the solution to the equation (6.3.4). In the case of $0 < \alpha < 1$, $X = A \natural_\alpha B$ implies that $(A^{-1/2}BA^{-1/2})^\alpha = A^{-1/2}XA^{-1/2}$ and so

$$\begin{aligned}A^{-1/2}BA^{-1/2} &= (A^{-1/2}XA^{-1/2})^{-\frac{1}{\alpha}} \\ &= A^{-1/2}X^{1/2}(X^{-1/2}AX^{-1/2})^{1-\frac{1}{\alpha}}X^{1/2}A^{-1/2}.\end{aligned}$$

Hence, we have $B = X^{1/2}(X^{-1/2}AX^{-1/2})^{1-\frac{1}{\alpha}}X^{1/2}$, and so

$$\log(X^{-1/2}AX^{-1/2})^{\frac{1-\alpha}{-\alpha}} = \log X^{-1/2}BX^{-1/2}.$$

Therefore, X is a solution to the equation (6.3.4).

Next, let Y be a solution to equation (6.3.4) for some $\alpha \in [0, 1]$. Then by the agument above, we have $Y = A \sharp_\alpha B$. □

Definition 6.3.5 For positive definite $A, B \in \mathbb{M}_n$, the chaotic order $A \gg B$ holds if $\log A \geq \log B$. Clearly, if $A \geq B$, then $A \gg B$.

The next lemma is due to Fujii, Furuta, and Kamei [12], see also [13, Theorem 3,15]. The following proof is due to Uchiyama [14].

Lemma 6.3.6 (Chaotic Furuta inequality) *Let A and B be positive definite matrices. If $A \gg B$, then*

$$A^r \geq (A^{r/2}B^pA^{r/2})^{\frac{r}{p+r}} \quad \textit{for } p, r \geq 0.$$

Proof Since

$$A_n = I + \log A/n \geq B_n = I + \log B/n > 0$$

for sufficiently large n, the Furuta inequality (Theorem 1.4.4) ensures that for given $p, r > 0$

$$A_n^{1+nr} \geq (A_n^{nr/2}B_n^{np}A_n^{nr/2})^{\frac{1+nr}{n(p+r)}},$$

or, equivalently,

$$A_n^{n(\frac{1}{n}+r)} \geq (A_n^{nr/2}B_n^{np}A_n^{nr/2})^{\frac{1}{n(p+r)}+\frac{r}{p+r}}.$$

Since $(I + \log X/n)^n \to X$ for $X > 0$, we make $n \to \infty$ in both sides, and achieve the desired inequality. □

Lemma 6.3.7 *Let A and B be positive definite matrices and $\alpha \in [0, 1]$. If $(1 - \alpha)\log A + \alpha \log B \leq 0$, then $A \sharp_\alpha B \leq I$.*

Proof We can assume that $\alpha \in (0, 1)$. Since $\log B^\alpha \leq \log A^{\alpha-1}$, then it follows from Lemma 6.3.6 that for $r = \frac{1}{1-\alpha}$ and $p = \frac{1}{\alpha}$

$$(A^{\alpha-1})^{\frac{1}{1-\alpha}} \geq \left((A^{\alpha-1})^{\frac{1}{2(1-\alpha)}}(B^\alpha)^{\frac{1}{\alpha}}(A^{\alpha-1})^{\frac{1}{2(1-\alpha)}}\right)^\alpha,$$

and so

$$A^{-1} \geq (A^{-1/2}BA^{-1/2})^\alpha.$$

Hence, we have $A \sharp_\alpha B \leq I$. □

The second proof of Theorem 6.3.1: We show that the relation

$$A \,\sharp_\alpha\, B \le I \qquad \text{yields the inequality} \quad A^r \,\sharp_\alpha\, B^r \le I \qquad \text{for } r \ge 1.$$

Let $X = A \,\sharp_\alpha\, B$ for $\alpha \in [0, 1]$. Then it follows from Lemma 6.3.4 that

$$(1-\alpha)\log X^{-1/2}AX^{-1/2} + \alpha \log X^{-1/2}BX^{-1/2} = 0.$$

For $r \in [1, 2]$, by the Hansen–Pedersen inequality (Theorem 5.1.7) for $X \le I$ and the matrix monotonicity of $\log t$, we have

$$\begin{aligned} 0 &= r((1-\alpha)\log X^{1/2}A^{-1}X^{1/2} + \alpha \log X^{1/2}B^{-1}X^{1/2}) \\ &= (1-\alpha)\log(X^{1/2}A^{-1}X^{1/2})^r + \alpha \log(X^{1/2}B^{-1}X^{1/2})^r \\ &\le (1-\alpha)\log(X^{1/2}A^{-r}X^{1/2}) + \alpha \log(X^{1/2}B^{-r}X^{1/2}). \end{aligned}$$

Hence,

$$(1-\alpha)\log(X^{-1/2}A^rX^{-1/2}) + \alpha \log(X^{-1/2}B^rX^{-1/2}) \le 0$$

and it follows form Lemma 6.3.7 that

$$(X^{-1/2}A^rX^{-1/2}) \,\sharp_\alpha\, (X^{-1/2}B^rX^{-1/2}) \le I.$$

By the property (iii) of operator means in Sect. 5.3, we have

$$A^r \,\sharp_\alpha\, B^r \le X \le I.$$

To establish this, put $Z = A^{r/2}X^{-1/2}$. Then,

$$\begin{aligned} (X^{-1/2}A^rX^{-1/2}) \,\sharp_\alpha\, (X^{-1/2}B^rX^{-1/2}) &= Z^*Z \,\sharp_\alpha\, (X^{-1/2}B^rX^{-1/2}) \\ &= |Z|(|Z|^{-1}X^{-1/2}B^rX^{1/2}|Z|^{-1})^\alpha|Z| \\ &= |Z|(|Z|^{-1}Z^*A^{-r/2}B^rA^{-r/2}Z|Z|^{-1})^\alpha|Z| \\ &= Z^*(A^{-r/2}B^rA^{-r/2})^\alpha Z \\ &= X^{-/2}(A^r \,\sharp_\alpha\, B^r)X^{-1/2}, \end{aligned}$$

where the fourth equality follows from the fact that $|Z|^{-1}Z^*$ is a unitary matrix. Therefore, we have

$$X^{-/2}(A^r \,\sharp_\alpha\, B^r)X^{-1/2} = (X^{-1/2}A^rX^{-1/2}) \,\sharp_\alpha\, (X^{-1/2}B^rX^{-1/2}) \le I.$$

By repeating this procedure for $A^r \,\sharp_\alpha\, B^r \le I$, we complete the second proof of the Ando–Hiai inequality (Theorem 6.3.1). □

Here, we show a slight extension of the Ando–Hiai inequality (Theorem 6.3.1).

Theorem 6.3.8 *Let A and B be positive definite matrices and $\alpha \in [0, 1]$. Then*

$$\lambda_{\min}(A \natural_\alpha B)^{r-1} A \natural_\alpha B \le A^r \natural_\alpha B^r \le \|A \natural_\alpha B\|^{r-1} A \natural_\alpha B \tag{6.3.5}$$

for all $r \ge 1$, where $\lambda_{\min}(X)$ is the minimum of the spectrum of X, that is, $\lambda_{\min}(X) = \left\|X^{-1}\right\|^{-1}$.

Proof Since $A \natural_\alpha B \le \|A \natural_\alpha B\|$, by putting $A_1 = \frac{A}{\|A \natural_\alpha B\|}$ and $B_1 = \frac{B}{\|A \natural_\alpha B\|}$, we obtain $A_1 \natural_\alpha B_1 \le I$. By the proof of the Ando–Hiai inequality (Theorem 6.3.1), we have $A_1^r \natural_\alpha B_1^r \le A_1 \natural_\alpha B_1$, and so

$$A^r \natural_\alpha B^r \le \|A \natural_\alpha B\|^{r-1} A \natural_\alpha B.$$

Next, we show the first inequality of (6.3.5). By applying the second inequality of (6.3.5) to A^{-1}, B^{-1}, we achieve

$$A^{-r} \natural_\alpha B^{-r} \le \left\|A^{-1} \natural_\alpha B^{-1}\right\|^{r-1} A^{-1} \natural_\alpha B^{-1} \qquad \text{for all } r \ge 1,$$

and so

$$\left(A^r \natural_\alpha B^r\right)^{-1} \le \left\|(A \natural_\alpha B)^{-1}\right\|^{r-1} (A \natural_\alpha B)^{-1}.$$

By taking the inverse of both sides, we have

$$\begin{aligned} A^r \natural_\alpha B^r &\ge \left\|(A \natural_\alpha B)^{-1}\right\|^{1-r} A \natural_\alpha B \\ &= \lambda_{\min}(A \natural_\alpha B)^{r-1} A \natural_\alpha B \end{aligned}$$

as desired. □

Remark 6.3.9 As in the proof of Theorem 6.3.8, we have the operator version of Theorem 6.3.8.

The following lemma is due to Hiai and Petz [15, Lemma 3.3]. By virtue of the reverse inequality in Sect. 6.5, we give another proof.

Lemma 6.3.10 *Let A and B be positive definite matrices in $\mathbb{M}_n$, and $p > 0$ and $\alpha \in [0, 1]$. Then*

$$\lim_{p \to 0} (A^p \natural_\alpha B^p)^{1/p} = e^{(1-\alpha)\log A + \alpha \log B}.$$

Proof We can assume that there exist scalars $0 < m \le M$ such that $m \le A, B \le M$. Put $h = \frac{M}{m}$. It follows from Theorem 6.5.8 that

$$0 \le \log((1-\alpha)A + \alpha B) - ((1-\alpha)\log A + \alpha \log B) \le \log S(h)$$

where $S(h)$ is the Specht ratio defined by (6.5.9). By replacing A and B with A^p and B^p for $p > 0$, respectively, since $m^p \leq A^p, B^p \leq M^p$, we conclude that

$$0 \leq \log((1-\alpha)A^p + \alpha B^p) - ((1-\alpha)\log A^p + \alpha \log B^p) \leq \log S(h^p),$$

and hence

$$0 \leq \log((1-\alpha)A^p + \alpha B^p)^{1/p} - ((1-\alpha)\log A + \alpha \log B) \leq \log S(h^p)^{1/p}.$$

Since $S(h^p)^{1/p} \to 1$ as $p \to 0$, by (iv) of Lemma 6.5.5, it follows that

$$\lim_{p \downarrow 0}((1-\alpha)A^p + \alpha B^p)^{1/p} = e^{(1-\alpha)\log A + \alpha \log B}.$$

On the other hand, since $M^{-1} \leq A^{-1}, B^{-1} \leq m^{-1}$, the generalized condition number of A^{-1} and B^{-1} is $\frac{m^{-1}}{M^{-1}} = \frac{M}{m} = h$ and

$$0 \leq \log((1-\alpha)A^{-1} + \alpha B^{-1}) - ((1-\alpha)\log A^{-1} + \alpha \log B^{-1}) \leq \log S(h).$$

Therefore,

$$0 \geq \log((1-\alpha)A^{-1} + \alpha B^{-1})^{-1} - ((1-\alpha)\log A + \alpha \log B) \geq -\log S(h),$$

and this implies that

$$0 \geq \log((1-\alpha)A^{-p} + \alpha B^{-p})^{-1/p} - ((1-\alpha)\log A + \alpha \log B) \geq -\log S(h^p)^{1/p}$$

for $p > 0$. Hence, we have

$$\lim_{p \downarrow 0}((1-\alpha)A^{-p} + \alpha B^{-p})^{-1/p} = e^{(1-\alpha)\log A + \alpha \log B}.$$

Since

$$\log((1-\alpha)A^{-p} + \alpha B^{-p})^{-1/p} \leq \log(A^p \sharp_\alpha B^p)^{1/p} \leq \log((1-\alpha)A^p + \alpha B^p)^{1/p}$$

for $p > 0$, we achieve the result. □

By Lemma 6.3.10, we have the following matrix geometric mean version of the Lie–Trotter formula.

Theorem 6.3.11 *If A and B are Hermitian matrices in $\mathbb{M}_n$ and $\alpha \in [0, 1]$, then*

$$\lim_{p \to 0}(e^{pA} \sharp_\alpha e^{pB})^{1/p} = e^{(1-\alpha)A + \alpha B}.$$

The following log-majorization is due to Ando and Hiai [11, Corollary 2.3].

Theorem 6.3.12 *If A and B are Hermitian matrices in $\mathbb{M}_n$ and $\alpha \in [0, 1]$, then*

$$(e^{pA} \natural_\alpha e^{pB})^{1/p} \prec_{\log} e^{(1-\alpha)A+\alpha B} \prec_{\log} (e^{(1-\alpha)qA/2} e^{\alpha qB} e^{(1-\alpha)qA/2})^{1/q}$$

for all $p, q > 0$.

Proof By (6.3.3) of Theorem 6.3.1, we have

$$(e^{pA} \natural_\alpha e^{pB})^{1/p} \prec_{\log} (e^{qA} \natural_\alpha e^{qB})^{1/q}$$

for $0 < q < p$. By Theorem 6.3.11, it follows that

$$\prod_{i=1}^{k} \lambda_i((e^{pA} \natural_\alpha e^{pB})^{1/p}) \le \prod_{i=1}^{k} \lambda_i((e^{qA} \natural_\alpha e^{qB})^{1/q}) \nearrow \prod_{i=1}^{k} \lambda_i(e^{(1-\alpha)A+\alpha B}) \qquad \text{as } q \downarrow 0$$

for $k = 1, \ldots, n$, and so

$$(e^{pA} \natural_\alpha e^{pB})^{1/p} \prec_{\log} e^{(1-\alpha)A+\alpha B}.$$

For the second inequality, it follows from (6.2.3) of Theorem 6.2.4 that

$$(e^{(1-\alpha)rA/2} e^{\alpha rB} e^{(1-\alpha)rA/2})^{1/r} \prec_{\log} (e^{(1-\alpha)qA/2} e^{\alpha qB} e^{(1-\alpha)qA/2})^{1/q}$$

for $0 < r < q$, and as $r \to 0$ by (6.1.3) of Theorem 6.1.2, we have

$$e^{(1-\alpha)A+\alpha B} \prec_{\log} (e^{(1-\alpha)qA/2} e^{\alpha qB} e^{(1-\alpha)qA/2})^{1/q}.$$

□

By Theorem 6.3.12, we have the following complemented Golden–Thompson inequalities in [11, Corollary 2.4] and [15, Theorem 3.4].

Theorem 6.3.13 *If A and B are Hermitian matrices in $\mathbb{M}_n$ and $\alpha \in [0, 1]$, then*

$$\left|\!\left|\!\left| (e^{pA} \natural_\alpha e^{pB})^{1/p} \right|\!\right|\!\right| \le \left|\!\left|\!\left| e^{(1-\alpha)A+\alpha B} \right|\!\right|\!\right| \tag{6.3.6}$$

for all $p > 0$ and the left-hand side increases to $|||e^{(1-\alpha)A+\alpha B}|||$ as $p \downarrow 0$ for any unitarily invariant norm $|||\cdot|||$. In particular,

$$\operatorname{tr}((e^{pA} \natural_\alpha e^{pB})^{1/p}) \le \operatorname{tr}(e^{(1-\alpha)A+\alpha B}) \qquad \textit{for all } p > 0.$$

By Theorem 6.3.12, we have

$$e^{2A} \sharp e^{2B} \prec_{\log} e^{A+B},$$

and hence

$$\left|\!\left|\!\left| e^{2A} \sharp e^{2B} \right|\!\right|\!\right| \leq \left|\!\left|\!\left| e^{A+B} \right|\!\right|\!\right|$$

for any unitarily invariant norm $|||\cdot|||$.

In particular,

$$\operatorname{tr}(e^{2A} \sharp e^{2B}) \leq \operatorname{tr}(e^{A+B}) \leq \operatorname{tr}(e^A e^B).$$

This estimates the lower bound of the Golden–Thompson trace inequality.

Bhatia and Grover [16] showed the following series of norm inequalities related to the complemented Golden–Thompson inequalities.

Theorem 6.3.14 *If A and B are Hermitian matrices in $\mathbb{M}_n$ and $\alpha \in [0, 1]$, then*

$$\begin{aligned}
\left|\!\left|\!\left| e^A \sharp_\alpha e^B \right|\!\right|\!\right| &\leq \left|\!\left|\!\left| e^{(1-\alpha)A+\alpha B} \right|\!\right|\!\right| \leq \left|\!\left|\!\left| e^{(1-\alpha)A/2} e^{\alpha B} e^{(1-\alpha)A/2} \right|\!\right|\!\right| \\
&\leq \left|\!\left|\!\left| \frac{1}{2}(e^{(1-\alpha)A} e^{\alpha B} + e^{\alpha B} e^{(1-\alpha)A}) \right|\!\right|\!\right| \leq \left|\!\left|\!\left| e^{(1-\alpha)A} e^{\alpha B} \right|\!\right|\!\right| \\
&\leq \left|\!\left|\!\left| (1-\alpha)e^A + \alpha e^B \right|\!\right|\!\right|
\end{aligned}$$

for any unitarily invariant norm $|||\cdot|||$.

We want to estimate the upper bound of (6.3.6):

$$\left|\!\left|\!\left| (e^{pA} \sharp_\alpha e^{pB})^{1/p} \right|\!\right|\!\right| \leq \left|\!\left|\!\left| e^{(1-\alpha)A+\alpha B} \right|\!\right|\!\right| \leq \boxed{\;?\;}.$$

For this, we extend the notion of matrix geometric mean for $\alpha \in [0, 1]$.

We apply the notation $\natural_\alpha$ to represent the binary operation defined as follows:

$$A \natural_\alpha B = A^{1/2}(A^{-1/2} B A^{-1/2})^\alpha A^{1/2} \qquad \text{for } \alpha \notin [0, 1]. \tag{6.3.7}$$

This formula is the same as $\sharp_\alpha$. Although matrices $A \natural_\alpha B$ for $\alpha \notin [0, 1]$ are not considered matrix means in the sense of the Kubo–Ando theory [17], they possess properties similar to those of operator means when A and B are positive definite matrices. Thus, we call (6.3.7) the *quasi α-geometric mean* for $\alpha \notin [0, 1]$, as mentioned in [18].

Lemma 6.3.15 *Let A, B, C and D be positive definite matrices in $\mathbb{M}_n$, and $\alpha \in [-1, 0)$. Then the following properties similar to those of the operator means hold:*

(i) **consistency with scalars:** *If A and B commute, then $A \natural_\alpha B = A^{1-\alpha} B^\alpha$;*
(ii) **right reverse monotonicity:** *$B \leq C$ implies that $A \natural_\alpha B \geq A \natural_\alpha C$;*
(iii) **super-additivity:** *$A \natural_\alpha B + C \natural_\alpha D \geq (A + C) \natural_\alpha (B + D)$;*
(iv) **homogeneity:** *$(\alpha A) \natural_\alpha (\alpha B) = \alpha(A \natural_t B)$ for all $\alpha > 0$;*

(v) **joint convexity:** *For* $t \in [0,1]$

$$((1-t)A+tC) \natural_\alpha ((1-t)B+tD) \le (1-t)A \natural_\alpha B + t\, C \natural_\alpha D;$$

(vi) **information monotonicity:** *For a positive linear map* Φ

$$\Phi(A \natural_\alpha B) \ge \Phi(A) \natural_\alpha \Phi(B);$$

(vii) **AG mean inequality:** $(1-\alpha)A + \alpha B \le A \natural_\alpha B.$

Proof (i) and (iv): They follow from the definition of the quasi α-geometric means for $\alpha \in [-1, 0)$.

(ii): Since t^α is matrix monotone decreasing for $\alpha \in [-1, 0)$ and $B \le C$, we have

$$A \natural_\alpha B = A^{1/2}(A^{-1/2}BA^{-1/2})^\alpha A^{1/2} \ge A^{1/2}(A^{-1/2}CA^{-1/2})^\alpha A^{1/2} = A \natural_\alpha C.$$

(iii): Put $X = A^{1/2}(A+C)^{-1/2}$ and $Y = C^{1/2}(A+C)^{-1/2}$. Since t^α is matrix convex for $\alpha \in [-1, 0)$ and $X^*X + Y^*Y = I$, we have

$$\begin{aligned}&X^*(A^{-1/2}BA^{-1/2})^\alpha X + Y^*(C^{-1/2}DC^{-1/2})^\alpha Y\\ &\ge (X^*A^{-1/2}BA^{-1/2}X + Y(C^{-1/2}DC^{-1/2}Y)^\alpha = ((A+C)^{-1/2}(B+D)(A+C)^{-1/2})^\alpha,\end{aligned}$$

and so

$$A \natural_\alpha B + C \natural_\alpha D \ge (A+C) \natural_\alpha (B+D).$$

(v): It follows from (iii) and (iv).

(vi): For $n \in \mathbb{N}$, put $\Phi_n(X) = \Phi(X) + \frac{1}{n}\phi(X)I$, where ϕ is a faithful state. Then the linear map Φ_n is positive and invertible for all $n \in \mathbb{N}$, that is, $X > 0$ implies that $\Phi_n(X) > 0$ for all $n \in \mathbb{N}$. Put

$$\Psi_n(X) = \Phi_n(B)^{-1/2}\Phi_n(B^{1/2}XB^{1/2})\Phi_n(B)^{-1/2} \quad \text{for } n \in \mathbb{N}.$$

Then Ψ_n is a unital positive linear map for all $n \in \mathbb{N}$ and the Jensen operator inequality yields the relation

$$\Psi_n(X^{1-\alpha}) \ge \Psi(X)^{1-\alpha} \qquad \text{for } X > 0,$$

since $t^{1-\alpha}$ is matrix convex for $1 < 1-\alpha < 2$. Hence, we have

$$\begin{aligned}\Phi_n(A \natural_\alpha B) &= \Phi_n(B \natural_{1-\alpha} A) = \Phi_n(B)^{1/2}\Psi_n((B^{-1/2}AB^{-1/2})^{1-\alpha})\Phi_n(B)^{1/2}\\ &\ge \Phi_n(B)^{1/2}\Psi_n(B^{-1/2}AB^{-1/2})^{1-\alpha}\Phi_n(B)^{1/2}\\ &= \Phi_n(B) \natural_{1-\alpha} \Phi_n(A)\\ &= \Phi_n(A) \natural_\alpha \Phi_n(B).\end{aligned}$$

As $n \to \infty$, we have the desired inequality.

(vii): For $\alpha \in [-1, 0)$, since $t^\alpha \geq \alpha(t-1)+1$ for $t > 0$. we have

$$\begin{aligned} A \natural_\alpha B &= A^{1/2}(A^{-1/2}BA^{-1/2})^\alpha A^{1/2} \\ &\geq A^{1/2}(\alpha A^{-1/2}BA^{-1/2} + (1-\alpha)I)A^{1/2} \\ &= (1-\alpha)A + \alpha B. \end{aligned}$$

□

To demonstrate the norm inequality associated with the quasi α-geometric mean for $\alpha \in [-1, 0)$, we require the following Ando–Hiai log-majorization of negative power $\alpha \in [-1, 0)$ in [19].

Theorem 6.3.16 *For every positive definite matrices $A, B > 0$ and $\alpha \in [-1, 0)$,*

$$A^r \natural_\alpha B^r \prec_{\log} (A \natural_\alpha B)^r \quad \textit{for all } 0 < r \leq 1 \tag{6.3.8}$$

or equivalently

$$(A \natural_\alpha B)^r \prec_{\log} (A^r \natural_\alpha B^r) \quad \textit{for all } r \geq 1, \tag{6.3.9}$$

$$(A^q \natural_\alpha B^q)^{\frac{1}{q}} \prec_{\log} (A^p \natural_\alpha B^p)^{\frac{1}{p}} \quad \textit{for all } 0 < q \leq p. \tag{6.3.10}$$

Proof The equivalence of (6.3.8)–(6.3.10) is immediate. It follows from Lemma 6.2.1 that for $k = 1, \ldots, n$

$$C_k(A^r \natural_\alpha B^r) = C_k(A)^r \natural_\alpha C_k(B)^r$$

and

$$C_k((A \natural_\beta B)^r) = (C_k(A) \natural_\beta C_k(B))^r.$$

Also,

$$\det(A^r \natural_\alpha B^r) = (\det A)^{r(1-\alpha)}(\det B)^{r\alpha} = \det(A \natural_\alpha B)^r.$$

Hence, in order to prove (6.3.8), it suffices to show that

$$\lambda_1(A^r \natural_\alpha B^r) \leq \lambda_1(A \natural_\alpha B)^r \quad \text{for all } 0 < r \leq 1. \tag{6.3.11}$$

For this purpose, we prove that the relation $A \natural_\alpha B \leq I$ yields the inequality $A^r \natural_\alpha B^r \leq I$. This is because both sides of (6.3.11) have the same order of homogeneity for A and B. Thus, we can multiply A and B by a positive constant.

First assume that $\frac{1}{2} \leq r \leq 1$ and write $r = 1 - \varepsilon$ with $0 \leq \varepsilon \leq \frac{1}{2}$. Let $C = A^{\frac{1}{2}}B^{-1}A^{\frac{1}{2}}$. Then $B^{-1} = A^{-\frac{1}{2}}CA^{-\frac{1}{2}}$ and $A \natural_\alpha B = A^{\frac{1}{2}}C^{-\alpha}A^{\frac{1}{2}}$. If $A \natural_\alpha B \leq I$, then $C^{-\alpha} \leq A^{-1}$ so that $A \leq C^\alpha$ and $A^\varepsilon \leq C^{\alpha\varepsilon}$ for $0 \leq \varepsilon \leq \frac{1}{2}$ by Löwner–Heinz inequality. Since $-\alpha \in (0, 1]$ and $1 - \varepsilon \in [\frac{1}{2}, 1]$, we obtain

$$\begin{aligned}
A^r \natural_\alpha B^r &= A^{\frac{1-\varepsilon}{2}}(A^{\frac{\varepsilon-1}{2}} B^{1-\varepsilon} A^{\frac{\varepsilon-1}{2}})^\alpha A^{\frac{1-\varepsilon}{2}} \\
&= A^{\frac{1-\varepsilon}{2}}(A^{\frac{1-\varepsilon}{2}} (B^{-1})^{1-\varepsilon} A^{\frac{1-\varepsilon}{2}})^{-\alpha} A^{\frac{1-\varepsilon}{2}} \\
&= A^{\frac{1-\varepsilon}{2}}(A^{\frac{1-\varepsilon}{2}} (A^{-\frac{1}{2}} C A^{-\frac{1}{2}})^{1-\varepsilon} A^{\frac{1-\varepsilon}{2}})^{-\alpha} A^{\frac{1-\varepsilon}{2}} \\
&= A^{\frac{1-\varepsilon}{2}}(A^{-\frac{\varepsilon}{2}} [A \sharp_{1-\varepsilon} C] A^{-\frac{\varepsilon}{2}})^{-\alpha} A^{\frac{1-\varepsilon}{2}} \\
&= A^{\frac{1}{2}-\varepsilon}[A^\varepsilon \sharp_{-\alpha} (A \sharp_{1-\varepsilon} C)] A^{\frac{1}{2}-\varepsilon} \\
&\le A^{\frac{1}{2}-\varepsilon}[C^{\alpha\varepsilon} \sharp_{-\alpha} (C^\alpha \sharp_{1-\varepsilon} C)] A^{\frac{1}{2}-\varepsilon},
\end{aligned}$$

by applying the joint monotonicity of matrix geometric means. A direct computation yields

$$C^{\alpha\varepsilon} \sharp_{-\alpha} (C^\alpha \sharp_{1-\varepsilon} C) = C^{\alpha(2\varepsilon-1)}.$$

It follows from the Löwner–Heinz inequality and $0 \le 1 - 2\varepsilon \le 1$ that by $C^{-\alpha} \le A^{-1}$ we have $C^{-\alpha(1-2\varepsilon)} \le A^{-(1-2\varepsilon)}$. Therefore,

$$A^r \natural_\alpha B^r \le A^{\frac{1}{2}-\varepsilon} C^{\alpha(2\varepsilon-1)} A^{\frac{1}{2}-\varepsilon} \le A^{\frac{1}{2}-\varepsilon} A^{-1+2\varepsilon} A^{\frac{1}{2}-\varepsilon} = I.$$

Thus, (6.3.11) is proved in the case of $\frac{1}{2} \le r \le 1$.

When $0 < r < \frac{1}{2}$, by writing $r = 2^{-k}(1-\varepsilon)$ with $k \in \mathbb{N}$ and $0 \le \varepsilon \le \frac{1}{2}$, and repeating the argument above, we have

$$\begin{aligned}
\lambda_1(A^r \natural_\alpha B^r) &\le \lambda_1(A^{2^{-(k-1)}(1-\varepsilon)} \natural_\alpha B^{2^{-(k-1)}(1-\varepsilon)})^{\frac{1}{2}} \\
&\;\;\vdots \\
&\le \lambda_1(A^{1-\varepsilon} \natural_\alpha B^{1-\varepsilon})^{2^{-k}} \\
&\le \lambda_1(A \natural_\alpha B)^r.
\end{aligned}$$

□

By Theorem 6.3.16, we have the following norm inequality for quasi α-geometric means in [19, Theorem 3.2].

Theorem 6.3.17 *Let A and B be positive definite matrices and $|||\cdot|||$ any unitarily invariant norm, and $\alpha \in [-1, 0)$. If f is a continuous nondecreasing function on $[0, \infty)$ such that $f(0) \ge 0$ and $f(e^t)$ is convex, then*

$$|||f(A^r \natural_\alpha B^r)||| \le |||f((A \natural_\alpha B)^r)||| \quad \text{for all } 0 < r \le 1.$$

In particular,

$$|||A^r \natural_\alpha B^r||| \le |||(A \natural_\alpha B)^r||| \quad \text{for all } 0 < r \le 1$$

or equivalently

$$|||(A \natural_\alpha B)^r||| \le |||A^r \natural_\alpha B^r||| \qquad \text{for all } r \ge 1,$$

$$\left|\left|\left|(A^q \natural_\alpha B^q)^{\frac{1}{q}}\right|\right|\right| \le \left|\left|\left|(A^p \natural_\alpha B^p)^{\frac{1}{p}}\right|\right|\right| \qquad \text{for all } 0 < q \le p.$$

Proof By [20, Proposition 4.4.13], if $A \prec_{w\log} B$ for positive definite matrices A and B and f is a continuous nondecreasing function on $[0, \infty)$ such that $f(0) \ge 0$ and $f(e^t)$ is convex, then $f(A) \prec_w f(B)$. Therefore, $|||f(A)||| \le |||f(B)|||$. Hence, Theorem 6.3.17 follows from Theorem 6.3.16. □

Corollary 6.3.18 *For every positive definite matrices $A, B > 0$ and $\alpha \in [-1, 0)$,*

$$A \natural_\alpha B \le I \quad \text{implies that} \quad A^r \natural_\alpha B^r \le I \quad \text{for all } 0 < r \le 1.$$

Remark 6.3.19 Lemma 6.3.15 and Corollary 6.3.18 hold for positive operators on a Hilbert space.

6.4 Norm Inequalities for Quasi α-Geometric Means

In this section, we establish some norm inequalities related to the quasi α-geometric mean for positive definite matrices. In light of [13, Lemma 5.5], we have the following quasi α-geometric mean version of the Lie–Trotter formula (Theorem 6.1.2).

Lemma 6.4.1 *If A and B are Hermitian matrices in $\mathbb{M}_n$, then*

$$e^{(1-\alpha)A+\alpha B} = \lim_{r\to+0} \left(e^{rA} \natural_\alpha e^{rB}\right)^{1/r} \tag{6.4.1}$$

for all $\alpha \in \mathbb{R}$.

Proof For $0 < r < 1$ and $\alpha \in \mathbb{R}$, let $X(r) = e^{rA} \natural_\alpha e^{rB}$, $Y(r) = e^{r((1-\alpha)A+\alpha B)}$ and $r^{-1} = m + s$, where $m \in \mathbb{N}$ and $s \in [0, 1)$. It suffices to prove that $\|X(r)^m - Y(r)^m\| \to 0$. Since, with the convention $o(r)/r \to 0$ as $r \to 0$,

$$\begin{aligned}
X(r) &= e^{rA/2}\left(\sum_{k=0}^{\infty}\frac{1}{k!}\left(-\frac{rA}{2}\right)^k \sum_{k=0}^{\infty}\frac{(rB)^k}{k!} \sum_{k=0}^{\infty}\frac{1}{k!}\left(-\frac{rA}{2}\right)^k\right)^\alpha e^{rA/2} \\
&= e^{rA/2}\left(I + r(B-A) + o(r)\right)^\alpha e^{rA/2} \\
&= \left(I + \frac{rA}{2} + o(r)\right)\left(I + r(B-A) + o(r)\right)\left(I + \frac{rA}{2} + o(r)\right) \\
&= I + r((1-\alpha)A + \alpha B) + o(r),
\end{aligned}$$

we have $X(r) - Y(r) = o(r)$. Since

$$X(r)^m - Y(r)^m = \sum_{j=0}^{m-1} X(r)^{m-1-j}(X(r) - Y(r))Y(r)^j,$$

it follows that

$$\begin{aligned} \left\|X(r)^m - Y(r)^m\right\| &\le m \,\|X(r) - Y(r)\| \max\{\|X(r)\|, \|Y(r)\|\}^{m-1} \\ &\le \frac{1}{r}\,\|X(r) - Y(r)\|\, e^{(1+|\alpha|)\|A\|+|\alpha|\|B\|} \to 0 \quad \text{as } r \to +0, \end{aligned}$$

which entails (6.4.1). □

By Lemma 6.4.1, we have the following quasi α-geometric mean version of Theorem 6.3.13, see also [19, Theorem 4.1].

Theorem 6.4.2 *If A and B are Hermitian matrices in $\mathbb{M}_n$, then for each $\alpha \in [-1, 0)$ and any unitarily invariant norm $|||\cdot|||$,*

$$\left|\left|\left|e^{(1-\alpha)A+\alpha B}\right|\right|\right| \le \left|\left|\left|(e^{pA} \natural_\alpha e^{pB})^{1/p}\right|\right|\right| \quad \textit{for all } p > 0 \tag{6.4.2}$$

and the left-hand side decreases to $\left|\left|\left|e^{(1-\alpha)A+\alpha B}\right|\right|\right|$ as $p \downarrow 0$. In particular,

$$\left|\left|\left|e^{(1-\alpha)A+\alpha B}\right|\right|\right| \le \left|\left|\left|e^{A} \natural_\alpha e^{B}\right|\right|\right| \quad \textit{for all } \alpha \in [-1, 0).$$

Proof It follows from Theorem 6.3.17 that

$$\left|\left|\left|(e^{qA} \natural_\alpha e^{qB})^{\frac{1}{q}}\right|\right|\right| \le \left|\left|\left|(e^{pA} \natural_\alpha e^{pB})^{\frac{1}{p}}\right|\right|\right| \qquad \text{for all } 0 < q < p$$

and as $q \to 0$ we have (6.4.2) by Lemma 6.4.1. □

Theorem 6.4.3 *Let A and B be positive definite matrices in $\mathbb{M}_n$. Then for every unitarily invariant norm $|||\cdot|||$,*

$$|||A \natural_\alpha B||| \le \left|\left|\left|A^{1-\alpha}B^{\alpha}\right|\right|\right| \qquad \textit{for all } \alpha \in [-1, -\tfrac{1}{2}]. \tag{6.4.3}$$

Proof For the operator norm, by Araki's log-majorization (Theorem 6.2.4) and Remark 6.2.5, we have $\|A^pB^pA^p\| \le \|(ABA)^p\|$ for all $p \in [0, 1]$. Hence, for $-1 \le \alpha \le -\frac{1}{2}$, we have

$$\begin{aligned}\|A \natural_\alpha B\| &= \left\|A^{1/2}(A^{-1/2}BA^{-1/2})^\alpha A^{1/2}\right\| \\ &= \left\|A^{1/2}(A^{1/2}B^{-1}A^{1/2})^{-\alpha}A^{1/2}\right\| \\ &\le \left\|A^{-1/2\alpha}A^{1/2}B^{-1}A^{1/2}A^{-1/2\alpha}\right\|^{-\alpha} \qquad (\text{by } \tfrac{1}{2} \le -\alpha \le 1) \\ &= \left\|A^{(\alpha-1)/2\alpha}B^{-1}A^{(\alpha-1)/2\alpha}\right\|^{-\alpha} \\ &\le \left\|(A^{1-\alpha}B^{2\alpha}A^{1-\alpha})^{1/2}\right\| \qquad (\text{for } \tfrac{1}{2} \le -\tfrac{1}{2\alpha} \le 1) \\ &= \left\||B^\alpha A^{1-\alpha}|\right\|,\end{aligned}$$

and this ensures

$$\lambda_1(A \natural_\alpha B) \le \lambda_1(|B^\alpha A^{1-\alpha}|)$$

and

$$\det(A \natural_\alpha B) = \det(|B^\alpha A^{1-\alpha}|).$$

By Lemma 6.2.2, we have the log-majorization $A \natural_\alpha B \prec_{\log} \left|B^\alpha A^{1-\alpha}\right|$, and this yields

$$|||A \natural_\alpha B||| \le \left|\left|\left|\,|B^\alpha A^{1-\alpha}|\,\right|\right|\right| = \left|\left|\left|B^\alpha A^{1-\alpha}\right|\right|\right| = \left|\left|\left|A^{1-\alpha}B^\alpha\right|\right|\right|$$

for every unitarily invariant norm, and so we have the desired inequality (6.4.3). □

Remark 6.4.4 In Theorem 6.4.3, the inequality $|||A \natural_\alpha B||| \le \left|\left|\left|A^{1-\alpha}B^\alpha\right|\right|\right|$ does not always hold for $-1/2 < \alpha < 0$. In fact, if we put $\alpha = -\frac{1}{3}$, $A = \begin{bmatrix} 2 & 1 \\ 1 & 2 \end{bmatrix}$ and $B = \begin{bmatrix} 2 & 1 \\ 1 & 1 \end{bmatrix}$, we have $\left\|A \natural_{-\frac{1}{3}} B\right\| = 3.385$ and $\left\|A^{4/3}B^{-1/3}\right\| = 3.375$, and so $\|A \natural_\alpha B\| > \left\|A^{1-\alpha}B^\alpha\right\|$.

Since the trace norm $\|\cdot\|_1$ is a unitarily invariant norm, we have the following trace inequality by Theorem 6.4.3.

Corollary 6.4.5 *Let A and B be positive definite matrices in $\mathbb{M}_n$. Then*

$$\operatorname{tr}(A \natural_\alpha B) \le \operatorname{tr}((A^{1-\alpha}B^{2\alpha}A^{1-\alpha})^{1/2})$$

for all $\alpha \in [-1, -1/2]$.

The following norm inequalities are due to [19, Theorem 4.4].

Theorem 6.4.6 *Let A and B be positive definite matrices in $\mathbb{M}_n$. Then for every unitarily invariant norm $|||\cdot|||$,*

$$\left|\left|\left|e^{(1-\alpha)\log A + \alpha \log B}\right|\right|\right| \le \left|\left|\left|\left(A^{(1-\alpha)p/2}B^{\alpha p}A^{(1-\alpha)p/2}\right)^{\frac{1}{p}}\right|\right|\right| \le |||A \natural_\alpha B||| \qquad (6.4.4)$$

for $0 < p \le \frac{1}{2}$ *and* $\alpha \in [-1, 0)$, *and*

$$|||A \natural_\alpha B||| \le \left|\left|\left|A^{1-\alpha} B^\alpha\right|\right|\right| \le \left|\left|\left|\left(B^{\alpha p/2} A^{(1-\alpha)p} B^{\alpha p/2}\right)^{\frac{1}{p}}\right|\right|\right| \tag{6.4.5}$$

for $p \ge 2$ *and* $\alpha \in [-1, -\frac{1}{2}]$.

Proof Since the first inequality in (6.4.4) follows from Theorem 6.2.8, we show the second inequality in (6.4.4). By Theorem 6.3.17, we have

$$\left|\left|\left|\left(A^r \natural_\alpha B^r\right)^{\frac{1}{r}}\right|\right|\right| \le |||A \natural_\alpha B||| \qquad \text{for all } 0 < r \le 1.$$

For the operator norm, by the Araki log-majorization (Theorem 6.2.4) and Remark 6.2.5, $\|(BAB)^p\| \le \|B^p A^p B^p\|$ for all $p \ge 1$, and so we have for $0 < r \le 1$,

$$\begin{aligned}
\|A \natural_\alpha B\| &\ge \left\|A^r \natural_\alpha B^r\right\|^{\frac{1}{r}} \\
&= \left\|B^r \natural_{1-\alpha} A^r\right\|^{\frac{1}{r}} \\
&= \left\|B^{r/2}(B^{-r/2} A^r B^{-r/2})^{1-\alpha} B^{r/2}\right\|^{\frac{1}{r}} \\
&\ge \left\|B^{\alpha r/2(1-\alpha)} A^r B^{\alpha r/2(1-\alpha)}\right\|^{\frac{1-\alpha}{r}} \quad (\text{by } 1 < 1-\alpha \le 2) \\
&\ge \left\|B^{\alpha r/4} A^{(1-\alpha)r/2} B^{\alpha r/4}\right\|^{\frac{2}{r}} \quad (\text{by } \tfrac{1}{2} < \tfrac{1-\alpha}{2} \le 1).
\end{aligned}$$

If we put $r = 2p$, we have $\left\|\left(B^{\alpha p/2} A^{(1-\alpha)p} B^{\alpha p/2}\right)^{\frac{1}{p}}\right\| \le \|A \natural_\alpha B\|$ for $0 < p \le \frac{1}{2}$, and this ensures that

$$\lambda_1(\left(B^{\alpha p/2} A^{(1-\alpha)p} B^{\alpha p/2}\right)^{\frac{1}{p}}) \le \lambda_1(A \natural_\alpha B)$$

and

$$\det(\left(B^{\alpha p/2} A^{(1-\alpha)p} B^{\alpha p/2}\right)^{\frac{1}{p}}) = \det(A \natural_\alpha B).$$

By Lemma 6.2.2, we have the second inequality of (6.4.4).

Next, for $s \ge 1$, it follows from Theorem 6.3.17 that $|||A \natural_\alpha B||| \le \left|\left|\left|(A^s \natural_\alpha B^s)^{\frac{1}{s}}\right|\right|\right|$. For the operator norm, we have

$$\begin{aligned}
\|A \natural_\alpha B\| \le \left\|A^s \natural_\alpha B^s\right\|^{\frac{1}{s}} &= \left\|A^{\frac{s}{2}}(A^{\frac{s}{2}} B^{-s} A^{\frac{s}{2}})^{-\alpha} A^{\frac{s}{2}}\right\|^{\frac{1}{s}} \\
&\le \left\|A^{\frac{-(1-\alpha)s}{2\alpha}} B^{-s} A^{\frac{-(1-\alpha)s}{2\alpha}}\right\|^{-\frac{\alpha}{s}} \quad (\text{by } \tfrac{1}{2} \le -\alpha \le 1 \text{ and Remark 6.2.5}) \\
&\le \left\|A^{(1-\alpha)s} B^{2\alpha s} A^{(1-\alpha)s}\right\|^{\frac{1}{2s}} \quad (\text{by } \tfrac{1}{2} \le -\tfrac{1}{2\alpha} \le 1 \text{ and Remark 6.2.5}).
\end{aligned}$$

If we put $s = \frac{p}{2}$, we have

$$\begin{aligned}\|A \natural_\alpha B\| &\le \left\| A^{\frac{(1-\alpha)p}{2}} B^{\alpha p} A^{\frac{(1-\alpha)p}{2}} \right\|^{\frac{1}{p}} \\ &= \left\| B^{\frac{\alpha p}{2}} A^{(1-\alpha)p} B^{\frac{\alpha p}{2}} \right\|^{\frac{1}{p}} \\ &= \left\| \left(B^{\frac{\alpha p}{2}} A^{(1-\alpha)p} B^{\frac{\alpha p}{2}} \right)^{\frac{1}{p}} \right\|\end{aligned}$$

for $p \ge 2$, where the first equality follows from $\|X^*X\| = \|XX^*\|$ for $X \in \mathbb{M}_n$. By an argument similar to the above, we achieve inequality (6.4.5). □

Let A and B be positive definite matrices in $\mathbb{M}_n$ and $\alpha \in [-1, 0)$. Since it may happen that $(1-\alpha)A + \alpha B$ is not positive semidefinite, we have no relation between $|||(1-\alpha)A + \alpha B|||$ and $|||A \natural_\alpha B|||$ although $(1-\alpha)A + \alpha B \le A \natural_\alpha B$. Assume that $A \ge B$. Then $(1-\alpha)A^p + \alpha B^p$ is positive definite for all $p \in (0, 1]$. In particular, $0 < (1-\alpha)A + \alpha B \le A \natural_\alpha B$ and so $|||(1-\alpha)A + \alpha B||| \le |||A \natural_\alpha B|||$ for every unitarily invariant norm. Thus, under the assumption of $A \ge B$, we consider the refinement of this norm inequality. For this, we need the following result due to J. I. Fujii [21]: A real-valued continuous function f on an interval J is matrix concave if and only if

$$f((1-\alpha)H + \alpha K) \le (1-\alpha)f(H) + \alpha f(K) \tag{6.4.6}$$

for all Hermitian matrices H and K with $\mathrm{sp}(H)$, $\mathrm{sp}(K)$ and $\mathrm{sp}((1-\alpha)H + \alpha K) \subset J$ and $\alpha \in [-1, 0)$. Compare with Definition 5.1.3.

Let $0 < q < p \le 1$. Then the function $f(t) = t^{\frac{q}{p}}$ on $[0, \infty)$ is matrix monotone and by Theorem 5.1.8 is also matrix concave. Therefore, it follows from (6.4.6) that

$$\left((1-\alpha)A^p + \alpha B^p\right)^{\frac{q}{p}} \le (1-\alpha)A^q + \alpha B^q. \tag{6.4.7}$$

Note that $(1-\alpha)A^p + \alpha B^p = A^p - \alpha(A^p - B^p) > 0$ for all $p \in (0, 1]$ since $A \ge B$. This implies that

$$\lambda_i \left((1-\alpha)A^p + \alpha B^p\right)^{\frac{q}{p}} \le \lambda_i \left((1-\alpha)A^q + \alpha B^q\right) \quad \text{for all } i = 1, \ldots, n.$$

Taking q-th roots of both sides, we obtain

$$\lambda_i \left((1-\alpha)A^p + \alpha B^p\right)^{\frac{1}{p}} \le \lambda_i \left((1-\alpha)A^q + \alpha B^q\right)^{\frac{1}{q}} \quad \text{for all } i = 1, \ldots, n,$$

and so $\left|\left|\left| ((1-\alpha)A^p + \alpha B^p)^{\frac{1}{p}} \right|\right|\right|$ is a decreasing function of p.

On the other hand, by taking the logarithm of both sides in (6.4.7) and applying (6.4.6), we obtain

$$\log\left((1-\alpha)A^p+\alpha B^p\right)^{\frac{1}{p}} \le \frac{1}{q}\log\left((1-\alpha)A^q+\alpha B^q\right)$$
$$\le (1-\alpha)\log A+\alpha\log B,$$

and this entails

$$\lambda_i\left(\log((1-\alpha)A^p+\alpha B^p)^{\frac{1}{p}}\right) \le \lambda_i\left((1-\alpha)\log A+\alpha\log B\right) \quad \text{for all } i=1,\ldots,n.$$

By taking the exponent of both sides, we obtain

$$\lambda_i\left((1-\alpha)A^p+\alpha B^p\right)^{\frac{1}{p}} \le \lambda_i\left(e^{(1-\alpha)\log A+\alpha\log B}\right) \quad \text{for all } i=1,\ldots,n,$$

and so

$$\left|\!\left|\!\left|\left((1-\alpha)A^p+\alpha B^p\right)^{\frac{1}{p}}\right|\!\right|\!\right| \le \left|\!\left|\!\left|e^{(1-\alpha)\log A+\alpha\log B}\right|\!\right|\!\right|$$

for all $p\in(0,1]$. Summing up, we achieve the following norm inequalities.

Theorem 6.4.7 *Let A and B be positive definite matrices in $\mathbb{M}_n$ such that $A\ge B$ and $\alpha\in[-1,0)$. Then for every unitarily invariant norm $|\!|\!|\cdot|\!|\!|$,*

$$|\!|\!|(1-\alpha)A+\alpha B|\!|\!| \le \left|\!\left|\!\left|\left((1-\alpha)A^p+\alpha B^p\right)^{\frac{1}{p}}\right|\!\right|\!\right| \le \left|\!\left|\!\left|e^{(1-\alpha)\log A+\alpha\log B}\right|\!\right|\!\right| \le |\!|\!|A\natural_\alpha B|\!|\!|$$

for all $p\in(0,1]$.

Finally, as an application, we refine the generalized Golden–Thompson inequality in terms of the quasi α-geometric means. Let A and B be Hermitian matrices. The Golden–Thompson trace inequality is

$$\mathrm{tr}[e^{A+B}] \le \mathrm{tr}[e^Ae^B].$$

Hiai–Petz [15] proved the complemented Golden–Thompson inequality:

$$\left|\!\left|\!\left|e^A\sharp_\alpha e^B\right|\!\right|\!\right| \le \left|\!\left|\!\left|e^{(1-\alpha)A+\alpha B}\right|\!\right|\!\right| \quad \text{for all } \alpha\in[0,1]$$

for every unitarily invariant norm. By Theorems 6.4.6 and 6.4.3, we have a refinement of the Golden–Thompson inequality in terms of the quasi α-geometric means:

$$\left|\!\left|\!\left|e^{(1-\alpha)A+\alpha B}\right|\!\right|\!\right| \le \left|\!\left|\!\left|e^A\natural_\alpha e^B\right|\!\right|\!\right| \le \left|\!\left|\!\left|e^{\alpha B}e^{(1-\alpha)A}\right|\!\right|\!\right|$$

for all $\alpha\in[-1,-\frac{1}{2}]$, and so

$$\mathrm{tr}[e^{A+B}] \le \mathrm{tr}[e^{\frac{1}{1-\alpha}A}\natural_\alpha e^{\frac{1}{\alpha}B}] \le \mathrm{tr}[(e^Ae^{2B}e^A)^{1/2}].$$

In particular, we have

$$\operatorname{tr}[e^{2A} \sharp e^{2B}] \le \operatorname{tr}[e^{A+B}] \le \operatorname{tr}[e^{\frac{2}{3}A} \natural_{-\frac{1}{2}} e^{-2B}] \le \operatorname{tr}[(e^{A}e^{2B}e^{A})^{1/2}].$$

6.5 Reverse Inequality to Golden–Thompson Inequality

In this section, we present reverse inequalities to the Golden–Thompson trace inequality, by using the Specht ratio and the generalized Kantorovich constant. For a positive definite matrix $A \in \mathbb{M}_n$, we consider the following condition:

$$0 < m \le A \le M$$

for some scalars $m < M$. Recall that the constant $h = h(A) = \frac{M}{m}$ is the *generalized condition number* of A in the sense of Turing [22].

First, let us recall the Kantorovich inequality. Suppose $a_i > 0$ for $i = 1, \dots, k$ are positive real numbers and $\omega = (\omega_1, \dots, \omega_k)$ is a weight vector. The arithmetic–harmonic mean inequality states that

$$\left(\sum_{i=1}^{k} \omega_i a_i^{-1}\right)^{-1} \le \sum_{i=1}^{k} \omega_i a_i.$$

Then, Kantorovich [23] showed that if $m \le a_i \le M$ for $i = 1, \dots, k$ and some scalars $0 < m \le M$, then

$$\sum_{i=1}^{k} \omega_i a_i \le \frac{(M+m)^2}{4Mm} \left(\sum_{i=1}^{k} \omega_i a_i^{-1}\right)^{-1}. \tag{6.5.1}$$

The constant $\frac{(M+m)^2}{4Mm}$ is called the *Kantorovich constant*. In particular, we have the matrix version of (6.5.1): If $A_1, \dots, A_k$ are positive definite and $m \le A_i \le M$ for $i = 1, \dots, k$ and some scalars $0 < m \le M$, then

$$\sum_{i=1}^{k} \omega_i A_i \le \frac{(M+m)^2}{4Mm} \left(\sum_{i=1}^{k} \omega_i A_i^{-1}\right)^{-1}. \tag{6.5.2}$$

In fact, since $MmA_i^{-1} + A_i \le M + m$ for $i = 1, \dots, k$, we have

$$Mm \sum_{i=1}^{k} \omega_i A_i^{-1} + \sum_{i=1}^{k} \omega_i A_i \le M + m.$$

Since

$$\frac{(M+m)^2}{4Mm}X^{-1} - (M+m) + MmX = \left(\frac{M+m}{2\sqrt{Mm}}X^{-1/2} - \sqrt{Mm}X^{1/2}\right)^2 \geq 0$$

for any positive definite X, we obtain the desired inequality:

$$\sum_{i=1}^{k} \omega_i A_i \leq M + m - Mm \sum_{i=1}^{k} \omega_i A_i^{-1} \leq \frac{(M+m)^2}{4Mm}\left(\sum_{i=1}^{k} \omega_i A_i^{-1}\right)^{-1}.$$

Moreover, Greub and Rheinboldt [24] showed the following inequality as a generalization of the Kantorovich inequality: If A is a positive definite matrix such that $m \leq A \leq M$ for some scalars $0 < m \leq M$, then

$$\langle Ax, x\rangle \leq \frac{(M+m)^2}{4Mm}\langle A^{-1}x, x\rangle^{-1} \tag{6.5.3}$$

for every unit vector $x \in \mathbb{C}^n$. The following theorem is a generalization of the Kantorovich inequality (6.5.3), see also [13].

Theorem 6.5.1 *Let $A \in \mathbb{M}_n$ be a positive definite matrix such that $m \leq A \leq M$ for some scalars $0 < m \leq M$. If f is a convex function on $[m, M]$ such that $f > 0$ on $[m, M]$, then*

$$f(\langle Ax, x\rangle) \leq \langle f(A)x, x\rangle \leq K(m, M, f)f(\langle Ax, x\rangle) \tag{6.5.4}$$

for every unit vector $x \in \mathbb{C}^n$, where

$$K(m, M, f) = \max\left\{\frac{1}{f(t)}\left(\frac{f(M)-f(m)}{M-m}(t-m) + f(m)\right) : m \leq t \leq M\right\}.$$

If f is a concave function on $[m, M]$ such that $f > 0$ on $[m, M]$, then

$$\tilde{K}(m, M, f)f(\langle Ax, x\rangle) \leq \langle f(A)x, x\rangle \leq f(\langle Ax, x\rangle)$$

for every unit vector $x \in \mathbb{C}^n$, where

$$\tilde{K}(m, M, f) = \min\left\{\frac{1}{f(t)}\left(\frac{f(M)-f(m)}{M-m}(t-m) + f(m)\right) : m \leq t \leq M\right\}.$$

Proof We only prove the case where f is convex on $[m, M]$. By the spectral decomposition theorem, there exists a unitary matrix U such that $A = UDU^*$ where $D = \text{diag}(\lambda_1, \ldots, \lambda_n)$ and $\lambda_1, \ldots, \lambda_n$ are eigenvalues of A. For a unit vector x, let $Ux = (x_1, \ldots, x_n)^t$ with $\sum_{i=1}^n |x_i|^2 = 1$. Then it follows from the convexity of f that

$$f(\langle Ax, x\rangle) = f\left(\sum_{i=1}^{n} \lambda_i |x_i|^2\right) \leq \sum_{i=1}^{n} f(\lambda_i)|x_i|^2 = \langle f(A)x, x\rangle.$$

Next, since $f(t)$ is convex on $[m, M]$, we have

$$f(t) \leq \frac{f(M) - f(M)}{M - m}(t - m) + f(m) \qquad \text{for all } t \in [m, M].$$

By the functional calculus, it follows that

$$f(A) \leq \frac{f(M) - f(M)}{M - m}(A - mI) + f(m)I,$$

and hence

$$\langle f(A)x, x\rangle \leq \frac{f(M) - f(M)}{M - m}(\langle Ax, x\rangle - m) + f(m)$$

for every unit vector x. So, we have

$$\langle f(A)x, x\rangle \leq \frac{f(M) - f(M)}{M - m}(\langle Ax, x\rangle - m) + f(m) \leq K(m, M, f) f\left(\langle Ax, x\rangle\right),$$

since $m \leq \langle Ax, x\rangle \leq M$. □

In particular, in the case of the power function $f(t) = t^\alpha$ for $\alpha \in \mathbb{R}$, we have the following precise estimate:

$$K(m, M, \alpha) := K(m, M, t^\alpha) = \frac{mM^p - Mm^p}{(\alpha - 1)(M - m)}\left(\frac{\alpha - 1}{\alpha}\frac{M^\alpha - m^\alpha}{mM^\alpha - Mm^\alpha}\right)^\alpha$$

for any real number $\alpha \in \mathbb{R}$. Putting $h = \frac{M}{m} \geq 1$, it follows that $K(m, M, \alpha)$ coincides with $K(h, \alpha)$:

$$K(h, \alpha) := \frac{h^\alpha - h}{(\alpha - 1)(h - 1)}\left(\frac{\alpha - 1}{\alpha}\frac{h^\alpha - 1}{h^\alpha - h}\right)^\alpha \qquad \text{for all } \alpha \in \mathbb{R} \tag{6.5.5}$$

and we call $K(h, \alpha)$ the *generalized Kantorovich constant*.

If we put $f(t) = t^\alpha$ for $\alpha \in \mathbb{R}$ in Theorem 6.5.1, we have the Hölder–McCarthy inequality and its reverses due to Furuta [25], see also [13, Theorem 2.11].

Theorem 6.5.2 *Let A be a positive definite matrix such that $m \leq A \leq M$ for some scalars $0 < m \leq M$, and $h = \frac{M}{m}$. Then*

$$\langle Ax, x\rangle^\alpha \leq \langle A^\alpha x, x\rangle \leq K(h, \alpha)\langle Ax, x\rangle^\alpha \qquad \textit{for } \alpha \notin [0, 1] \tag{6.5.6}$$

and

$$K(h,\alpha)\langle Ax,x\rangle^{\alpha} \le \langle A^{\alpha}x,x\rangle \le \langle Ax,x\rangle^{\alpha} \quad \text{for } \alpha \in [0,1] \tag{6.5.7}$$

for every unit vector $x \in \mathbb{C}^n$, *where* $K(h,\alpha)$ *is defined as in* (6.5.5). *The constant* $K(h,\alpha)$ *is sharp in the sense that for any* $\alpha \in \mathbb{R}$ *there exists a unit vector* $z \in \mathbb{C}^n$ *such that*

$$\langle A^{\alpha}z,z\rangle = K(h,\alpha)\langle Az,z\rangle^{\alpha}.$$

Proof We only show the sharpness of $K(h,\alpha)$ for $\alpha > 1$. Let $Ax = mx$, $Ay = My$, and $z = sx + ty$, where $\|x\| = \|y\| = 1$, $|s|^2 + |t|^2 = 1$, and $h = \frac{M}{m}$. Then we have

$$\langle A^{\alpha}z,z\rangle = \langle sm^{\alpha}x + tM^{\alpha}y, sx+ty\rangle = |s|^2 m^{\alpha} + |t|^2 M^{\alpha}$$

and

$$\langle Az,z\rangle^{\alpha} = (|s|^2 m + |t|^2 M)^{\alpha}.$$

Therefore, we want to obtain the unit vector z satisfying the following equality:

$$|s|^2 m^{\alpha} + |t|^2 M^{\alpha} = K(h,\alpha)(|s|^2 m + |t|^2 M)^{\alpha},$$

that is,

$$1 + |t|^2(h^{\alpha} - 1) = K(h,\alpha)(1 + |t|^2(h-1))^{\alpha}.$$

We construct a solution β of the above equation as

$$\beta = \left(\frac{h^{\alpha} - 1 - \alpha(h-1)}{(\alpha-1)(h-1)((h^{\alpha}-1)}\right)^{1/2} < 1.$$

For example, we have $z = \frac{1}{\sqrt{M}+\sqrt{m}}(\sqrt{M}x + \sqrt{m}y)$ for $\alpha = 2$. □

We mention basic properties of $K(h,\alpha)$ in [13, Theorem 2.12, Theorem 2.13] and [26, Lemma 2.4].

Lemma 6.5.3 *Let* $h > 0$ *be given. Then the generalized Kantorovich constant* $K(h,\alpha)$ *has the following properties:*

(i) $K(h,\alpha) = K(h^{-1},\alpha)$ *for all* $\alpha \in \mathbb{R}$;
(ii) $K(h,\alpha)$ *is symmetric with respect to* $\alpha = 1/2$;
(iii) $K(h,0) = K(h,1) = 1$ *and* $K(1,\alpha) = 1$ *for all* $\alpha \in \mathbb{R}$;
(iv) $K(h,\alpha)$ *is an increasing function of* α *for* $\alpha > 1/2$ *and a decreasing function of* α *for* $\alpha < 1/2$;
(v) $K(h,\alpha) \ge 1$ *for* $\alpha \notin (0,1)$ *and* $K(h,\alpha) \le 1$ *for* $\alpha \in [0,1]$;
(vi) $K(h^r, \frac{p}{r})^{1/p} = K(h^p, \frac{r}{p})^{-1/r}$ *for* $pr \neq 0$;
(vii) Let $h > 1$. *If* $\alpha > 1$ *(respectively,* $0 < \alpha < 1$*), then* $K(h^x,\alpha)^{1/x}$ *is increasing for* $x > 0$ *(respectively, decreasing for* $x > 0$*).*

Moreover, we present an important constant due to Specht. In [27] he estimated the upper bound of the arithmetic mean by the geometric one for positive numbers: For $x_1, \ldots, x_n \in [m, M]$

$$\sqrt[n]{x_1 x_2 \cdots x_n} \le \frac{x_1 + x_2 + \cdots + x_n}{n} \le S(h)\sqrt[n]{x_1 x_2 \cdots x_n}, \tag{6.5.8}$$

where $h = \frac{M}{m}$ and the Specht ratio is defined by

$$S(h) = \frac{(h-1)h^{\frac{1}{h-1}}}{e \log h} \quad (h \neq 1) \quad \text{and} \quad S(1) = 1. \tag{6.5.9}$$

We note that Specht inequality (6.5.8) means a ratio-type reverse inequality of the arithmetic–geometric mean inequality.

Now, let us show a noncommutative version of the Specht inequality (6.5.8) due to [28].

Theorem 6.5.4 *Let A be a positive definite matrix in $\mathbb{M}_n$ such that $0 < m \le A \le M$ for some scalars $0 < m < M$ and put $h = \frac{M}{m}$. Then*

$$e^{\langle \log A\, x, x\rangle} \le \langle Ax, x\rangle \le S(h) e^{\langle \log A\, x, x\rangle} \tag{6.5.10}$$

holds for every unit vector $x \in \mathbb{C}^n$.

Proof For the first inequality in (6.5.10), since $y = e^t$ is convex, it follows from Lemma 1.4.1.

For the second inequality in (6.5.10), put $a = \frac{M-m}{\log M - \log m}$ and $b = \frac{m \log M - M \log m}{\log M - \log m}$. Since $y = e^t$ is a convex function, then, for the line $at + b$ crossing e^t at $t = \log m$ and $t = \log M$, we have

$$e^t \le at + b \le a e^{\frac{b-a}{a}} e^t$$

on $[\log m, \log M]$. In fact, $F(t) = a e^{\frac{b-a}{a}} e^t - at - b$ is a convex function with the minimum zero at $t = \frac{a-b}{a}$ by $F'(\frac{a-b}{a}) = 0$ and $\frac{a-b}{a} \in [\log m, \log M]$ is guaranteed by the inequalities $m \le a = \frac{M-m}{\log M - \log m} \le M$. Thus, we have the latter inequality. Put $S = \log A$, then

$$\langle e^S x, x\rangle \le \langle (aS + b)x, x\rangle = a\langle Sx, x\rangle + b \le a e^{\frac{b-a}{a}} e^{\langle Sx, x\rangle}.$$

The number $a e^{\frac{b-a}{a}}$ is exactly the Specht ratio. In fact, $a = \frac{(h-1)m}{\log h}$ and

$$\frac{b}{a} = \frac{m \log M - M \log m}{M - m} = \frac{\log M - h \log m}{h-1} = \frac{\log(h m^{1-h})}{h-1},$$

and this implies that

$$ae^{\frac{b-a}{a}} = \frac{(h-1)m}{\log h}(hm^{1-h})^{\frac{1}{h-1}}e^{-1} = \frac{(h-1)h^{\frac{1}{h-1}}}{e\log h} = S(h).$$

Hence, we obtain the second inequality of (6.5.10). □

We mention some basic properties of the Specht ratio $S(h)$ in [13, Theorem 2.16 and Theorem 2.17].

Lemma 6.5.5 *Let $h > 0$ and $\alpha \in \mathbb{R}$. Then*

(i) $S(1) = \lim_{h\to 1} S(h) = 1$;
(ii) $S(h) = S(h^{-1})$;
(iii) *A function $S(h)$ is strictly decreasing for $0 < h < 1$ and strictly increasing for $h > 1$;*
(iv) $\lim_{\alpha\to 0} S(h^\alpha)^{1/\alpha} = 1$;
(v) $\lim_{\alpha\to\infty} S(h^\alpha)^{1/\alpha} = h$ *for* $h > 1$ *and* $\lim_{\alpha\to\infty} S(h^\alpha)^{1/\alpha} = h^{-1}$ *for* $0 < h < 1$;
(vi) $\lim_{r\to 0} K(h^r, \frac{\alpha}{r}) = S(h^\alpha)$.

Tominaga [29] showed the following noncommutative matrix version of (6.5.8).

Theorem 6.5.6 *Let A and B be positive definite matrices such that $0 < m \le A, B \le M$ for some scalars $0 < m \le M$, and let $\alpha \in [0, 1]$. Put $h = \frac{M}{m}$. Then*

$$A \sharp_\alpha B \le (1-\alpha)A + \alpha B \le S(h)(A \sharp_\alpha B). \tag{6.5.11}$$

Proof Let $a \ne 1$. Put $f_a(t) = \frac{(1-t)a+t}{a^{1-t}}$. Then we have the constant $S(a) = \frac{(a-1)a^{\frac{1}{a-1}}}{e\log a}$ as the maximum of $f_a(t)$ for $t \in [0, 1]$ and the equation $f_a'(t) = 0$ has the following unique solution $t = t_0$:

$$t_0 = \frac{a}{a-1} - \frac{1}{\log a} \in [0, 1].$$

Thus, the maximum of $f_a(t)$ happens at $t = t_0$ and we have

$$\max_{0\le t\le 1} f_a(t) = f_a(t_0) = S(a).$$

Therefore, the inequality

$$S(a)\, a^{1-t} \ge (1-t)a + t \qquad \text{holds for all } t \in [0, 1].$$

Now, let C be a positive definite matrix such that $m \le C \le M$ for some scalars $0 < m < M$. Then it follows that

$$\max_{m\le a\le M} S(a)\, C^{1-\alpha} \ge (1-\alpha)C + \alpha I \qquad \text{for all } \alpha \in [0, 1].$$

Since $S(a)$ is decreasing for $0 < a < 1$ and increasing for $a > 1$, the maximum of $S(a)$ in $a \in [m, M]$ is given by $\max\{S(m), S(M)\}$, and hence

$$\max\{S(m), S(M)\}C^{1-\alpha} \geq (1-\alpha)C + \alpha I \qquad \text{for all } \alpha \in [0, 1].$$

Here we replace C with $B^{-1/2}AB^{-1/2}$. Then we have $h^{-1}I \leq B^{-1/2}AB^{-1/2} \leq hI$ and it follows from $S(h^{-1}) = S(h)$ in (ii) of Lemma 6.5.5 that

$$S(h)(B^{-1/2}AB^{-1/2})^{1-\alpha} \geq (1-\alpha)B^{-1/2}AB^{-1/2} + \alpha I.$$

Multiply both sides of this relation by $B^{1/2}$, and conclude that

$$S(h)B \sharp_{1-\alpha} A = S(h)B^{1/2}(B^{-1/2}AB^{-1/2})^{1-\alpha}B^{1/2} \geq (1-\alpha)A + \alpha B,$$

so we have the second inequality of the desired one (6.5.11) by $B \sharp_{1-\alpha} A = A \sharp_\alpha B$. For the first inequality, since $x^\alpha \leq \alpha x + 1 - \alpha$ for $x > 0$, it follows from the functional calculus that $(A^{-1/2}BA^{-1/2})^\alpha \leq \alpha A^{-1/2}BA^{-1/2} + (1-\alpha)I$, and hence $A \sharp_\alpha B \leq (1-\alpha)A + \alpha B$. □

The following lemma estimates the upper bound of the difference in the Jensen inequality for the logarithmic function.

Lemma 6.5.7 *Let $A_1, \ldots, A_k$ be positive definite matrices such that $m \leq A_i \leq M$ $(i = 1, \ldots, k)$ for some scalars $0 < m < M$. Let $x_1, \ldots, x_k$ be any finite number of vectors such that $\sum_{i=1}^k \|x_i\|^2 = 1$. Then*

$$\log\left(\sum_{i=1}^k \langle A_i x_i, x_i\rangle\right) - \sum_{i=1}^k \langle \log A_i x_i, x_i\rangle \leq \log S(h),$$

where $h = \frac{M}{m}$ and the Specht ratio $S(h)$ is defined by (6.5.9).

Proof Put $\bar{t} = \sum_{i=1}^k \langle A_i x_i, x_i\rangle$ and $\mu = \frac{\log M - \log m}{M-m}$. Then we have $m \leq \bar{t} \leq M$. By the concavity of $\log t$, we have

$$\begin{aligned}
&\sum_{i=1}^k \langle \log A_i x_i, x_i\rangle - \log\left(\sum_{i=1}^k \langle A_i x_i, x_i\rangle\right) \\
&\geq \sum_{i=1}^k \langle (\mu(A_i - mI) + \log m)x_i, x_i\rangle - \log\left(\sum_{i=1}^k \langle A_i x_i, x_i\rangle\right) \\
&= \mu(\bar{t} - m) + \log m - \log \bar{t} \\
&\geq \min_{m \leq t \leq M} \{\mu(t - m) + \log m - \log t\}.
\end{aligned}$$

Put $t = t_0 = \frac{M-m}{\log M - \log m}$ and obtain

$$\mu(t_0 - m) + \log m - \log t_0 = -\log S(h).$$

□

Since $\log t$ is matrix concave, we have

$$0 \le \log((1-\alpha)A + \alpha B) - ((1-\alpha)\log A + \alpha \log B)$$

for $A, B > 0$ and $\alpha \in [0, 1]$. Applying Lemma 6.5.7, we estimate the upper bound in which the Specht ratio appears.

Theorem 6.5.8 *Let A and B be positive definite matrices such that $m \le A, B \le M$ for some scalars $0 < m < M$ and $\alpha \in [0, 1]$. Then*

$$0 \le \log((1-\alpha)A + \alpha B) - ((1-\alpha)\log A + \alpha \log B) \le \log S(h),$$

where $h = \frac{M}{m}$.

Proof For a fixed $\alpha \in [0, 1]$ and a unit vector $x \in \mathbb{C}^n$, put $A_1 = A$ and $A_2 = B$, $x_1 = \sqrt{1-\alpha}\, x$ and $x_2 = \sqrt{\alpha}\, x$ in Lemma 6.5.7. Then we have

$$\begin{aligned}\log S(h) &\ge \log((1-\alpha)\langle Ax, x\rangle + \alpha\langle Bx, x\rangle) - ((1-\alpha)\langle \log Ax, x\rangle + \alpha\langle \log Bx, x\rangle) \\ &= \log\langle((1-\alpha)A + \alpha B)x, x\rangle - \langle((1-\alpha)\log A + \alpha \log B)x, x\rangle \\ &\ge \langle \log((1-\alpha)A + \alpha B)x, x\rangle - \langle((1-\alpha)\log A + \alpha \log B)x, x\rangle,\end{aligned}$$

where the last inequality is deduced by the concavity of $\log t$. □

In order to show a sharp reverse inequality to the Golden–Thompson trace inequality, we need the following theorem of [30], which is a reverse inequality of the Araki's inequality.

Theorem 6.5.9 *Let A and B be positive definite matrices in $\mathbb{M}_n$ such that $m \le B \le M$ for some scalars $0 < m < M$. Put $h = \frac{M}{m}$. For $p > 1$, there exist some unitaries U and V such that*

$$\frac{1}{K(h,p)} U(ABA)^p U^* \le A^p B^p A^p \le K(h,p) V(ABA)^p V^*. \tag{6.5.12}$$

The constant $K(h, p)$ and its inverse are optimal.

In particular, in the case of $p = 2$,

$$\frac{4Mm}{(M+m)^2} U(ABA)^2 U^* \le A^2 B^2 A^2 \le \frac{(M+m)^2}{4Mm} V(ABA)^2 V^*.$$

Proof By the minimax theorem (2.2.3), there exists a subspace $\mathscr{F}$ of codimension $k-1$ such that

$$\begin{aligned}\lambda_k((ABA)^p) &= \max_{y\in\mathscr{F},\|y\|=1}\langle y,(ABA)^p y\rangle\\ &= \max_{y\in\mathscr{F},\|y\|=1}\langle y,(ABA)y\rangle^p.\end{aligned}$$

On the other hand, still by the minimax theorem, for every subspace $\mathscr{E}$ of codimension $k-1$, we have

$$\begin{aligned}\lambda_k(A^pB^pA^p) &\le \max_{x\in\mathscr{E},\|x\|=1}\langle x,A^pB^pA^px\rangle = \max_{x\in\mathscr{E},\|x\|=1}\langle A^px,B^pA^px\rangle\\ &\le \max_{x\in\mathscr{E},\|x\|=1} K(h,p)\langle\frac{A^px}{\|A^px\|},B\frac{A^px}{\|A^px\|}\rangle^p\left\|A^px\right\|^2\\ &= \max_{x\in\mathscr{E},\|x\|=1} K(h,p)\langle A^px,BA^px\rangle^p\left\|A^px\right\|^{2-2p}\\ &= \max_{x\in\mathscr{E},\|x\|=1} K(h,p)\langle\frac{A^{p-1}x}{\left\|A^{p-1}x\right\|},ABA\frac{A^{p-1}x}{\left\|A^{p-1}x\right\|}\rangle^p\left\|A^{p-1}x\right\|^{2p}\left\|A^px\right\|^{2-2p}.\end{aligned}$$

Now, observe that

$$\begin{aligned}\left\|A^{p-1}x\right\|^{2p}\left\|A^px\right\|^{2-2p} &= \langle x,A^{2p-2}x\rangle^p\langle x,A^{2p}x\rangle^{1-p}\\ &= \langle x,(A^{2p})^{\frac{2p-2}{2p}}x\rangle^p\langle x,A^{2p}x\rangle^{1-p}\\ &\le \langle x,A^{2p}x\rangle^{p-1}\langle x,A^{2p}x\rangle^{1-p} = 1\end{aligned}$$

by the concavity of $t\mapsto t^{\frac{2p-2}{2p}}$. Therefore,

$$\lambda_k(A^pB^pA^p) \le \max_{x\in\mathscr{E},\|x\|=1} K(h,p)\langle\frac{A^{p-1}x}{\left\|A^{p-1}x\right\|},ABA\frac{A^{p-1}x}{\left\|A^{p-1}x\right\|}\rangle.$$

Since A is invertible, we choose $\mathscr{E}$ so that

$$\left\{\frac{A^{p-1}x}{\left\|A^{p-1}x\right\|} : x\in\mathscr{E},\ \|x\|=1\right\} = \{y\in\mathscr{F} : \|y\|=1\}.$$

Hence,

$$\begin{aligned}\lambda_k(A^pB^pA^p) &\le \max_{y\in\mathscr{F},\|y\|=1} K(h,p)\langle y,ABAy\rangle^p\\ &= K(h,p)\lambda_k((ABA)^p)\end{aligned}$$

and we obtain the right-hand side of (6.5.12).

To prove the left-hand side inequality of (6.5.12), it then suffices to take the inverse of the right-hand one. In fact, there exists a unitary matrix U such that

$$A^{-p}B^{-p}A^{-p} \le K(h,p)U(A^{-1}B^{-1}A^{-1})^pU^*,$$

and hence

$$\begin{aligned}A^pB^pA^p &\ge K(h,p)^{-1}(U(A^{-1}B^{-1}A^{-1})^pU^*)^{-1}\\ &= K(h,p)^{-1}U(ABA)^pU^*.\end{aligned}$$

It remains to check the optimality of the constants. For the right-hand side inequality it is obvious since (6.5.6) is sharp. The optimality of the left-hand side then follows since these two inequalities are equivalent. □

Theorem 6.5.10 *Let A and B be positive definite matrices such that $m \le B \le M$ for some scalars $0 < m < M$. Put $h = \frac{M}{m}$. For $0 < q < 1$, there exist some unitaries U and V such that*

$$K(h,q)U(ABA)^qU^* \le A^qB^qA^q \le \frac{1}{K(h,q)}V(ABA)^qV^*. \tag{6.5.13}$$

The constant $K(h,q)$ and its inverse are optimal.

Proof Since $\frac{1}{q} > 1$, it follows from Theorem 6.5.9 that

$$\frac{1}{K(h,\frac{1}{q})}U(ABA)^{\frac{1}{q}}U^* \le A^{\frac{1}{q}}B^{\frac{1}{q}}A^{\frac{1}{q}} \le K\left(h,\frac{1}{q}\right)V(ABA)^{\frac{1}{q}}V^*.$$

Replacing A and B with A^q, B^q, respectively, we have

$$\frac{1}{K(h^q,\frac{1}{q})}U(A^qB^qA^q)^{\frac{1}{q}}U^* \le ABA \le K\left(h^q,\frac{1}{q}\right)V(A^qB^qA^q)^{\frac{1}{q}}V^*,$$

and hence

$$\frac{1}{K(h^q,\frac{1}{q})^q}U(A^qB^qA^q)U^* \le (ABA)^q \le K\left(h^q,\frac{1}{q}\right)^qV(A^qB^qA^q)V^*.$$

By (6) of Lemma 6.5.3, we have the desired inequality (6.5.13). □

The following theorem can be found in [31].

Theorem 6.5.11 *Let A and B be Hermitian matrices in $\mathbb{M}_n$, and let h be the condition number of e^A. Then there exists a unitary V such that the following sharp inequality holds*

$$e^{A/2}e^{B}e^{A/2} \le S(h)Ve^{A+B}V^*. \tag{6.5.14}$$

In particular,

$$\operatorname{tr}(e^Ae^B) \le S(h)\operatorname{tr}(e^{A+B}).$$

Proof Note that $e^{A/2}e^{B}e^{A/2}$ is unitarily equivalent to $e^{B/2}e^{A}e^{B/2}$ as well as $e^{B/2}e^{A}e^{B/2} = (e^{B/2p})^p(e^{A/p})^p(e^{B/2p})^p$ for all $p > 1$. It follows from Theorem 6.5.9 that

$$e^{B/2}e^{A}e^{B/2} \le K(h^{\frac{1}{p}}, p)V_p(e^{B/2p}e^{A/p}e^{B/2p})^pV_p^*$$

for some unitaries $V_p (p > 1)$. By applying (vi) of Lemma 6.5.5 ($\lim_{p\to\infty} K(h^{\frac{1}{p}}, p) = S(h)$) and Theorem 6.1.2 (the Lie–Trotter formula), we get the desired inequality (6.5.14) for some unitary V.

In particular,

$$\operatorname{tr}(e^Ae^B) \le S(h)\operatorname{tr}(e^{A+B}).$$

□

Lemma 6.5.12 *Let A and B be positive definite matrices such that $0 < m \le A, B \le M$ for some scalars $0 < m < M$, and let $\alpha \in [0, 1]$. Put $h = \frac{M}{m}$. Then for each $0 < q \le p$, there exist unitary matrices U and V such that*

$$S(h^p)^{-\frac{1}{p}}V(A^p \sharp_\alpha B^p)^{\frac{1}{p}}V^* \le (A^q \sharp_\alpha B^q)^{\frac{1}{q}} \le S(h^p)^{\frac{1}{p}}U(A^p \sharp_\alpha B^p)^{\frac{1}{p}}U^*, \tag{6.5.15}$$

where $S(h)$ is defined as (6.5.9).

Proof By Theorem 6.5.6, we have

$$A \sharp_\alpha B \le (1-\alpha)A + \alpha B \le S(h)(A \sharp_\alpha B).$$

Since $0 < \frac{q}{p} \le 1$, it follows from the operator concavity of $t^{\frac{q}{p}}$ that

$$A^{\frac{q}{p}} \sharp_\alpha B^{\frac{q}{p}} \le (1-\alpha)A^{\frac{q}{p}} + \alpha B^{\frac{q}{p}} \le ((1-\alpha)A + \alpha B)^{\frac{q}{p}} \le S(h)^{\frac{q}{p}}(A \sharp_\alpha B)^{\frac{q}{p}}.$$

Replace A and B with A^p and B^p, respectively, and conclude that

$$A^q \sharp_\alpha B^q \le S(h^p)^{\frac{q}{p}}(A^p \sharp_\alpha B^p)^{\frac{q}{p}}. \tag{6.5.16}$$

In the case of $q > 1$, the Löwner–Heinz inequality asserts

$$(A^q \sharp_\alpha B^q)^{\frac{1}{q}} \le S(h^p)^{\frac{1}{p}}(A^p \sharp_\alpha B^p)^{\frac{1}{p}}.$$

In the case of $0 < q < 1$, by (6.5.16), we have

$$\begin{aligned}\lambda_k((A^q \natural_\alpha B^q)^{\frac{1}{q}}) &= \lambda_k(A^q \natural_\alpha B^q)^{\frac{1}{q}} \\ &= S(h^p)^{\frac{1}{p}} \lambda_k(A^p \natural_\alpha B^p)^{\frac{1}{p}} \\ &= S(h^p)^{\frac{1}{p}} \lambda_k((A^p \natural_\alpha B^p)^{\frac{1}{p}}),\end{aligned}$$

and hence we obtain the right-hand side of (6.5.15).

To prove the left-hand side inequality, we replace A and B with their inverses and apply $A^{-1} \natural_\alpha B^{-1} = (A \natural_\alpha B)^{-1}$. Then we have

$$(A^{-q} \natural_\alpha B^{-q})^{\frac{1}{q}} \leq S(h^{-p})^{\frac{1}{p}} V(A^{-p} \natural_\alpha B^{-p})^{\frac{1}{p}} V^*$$

for some unitary V. By taking the inverse of both sides and (ii) of Lemma 6.5.5, we obtain the desired inequality. □

As a corollary of Lemma 6.5.12, we have a reverse of Corollary 6.3.2.

Corollary 6.5.13 *Let A and B be positive definite matrices such that $0 < m \leq A, B \leq M$ for some scalars $0 < m < M$, and let $\alpha \in [0, 1]$. Put $h = \frac{M}{m}$. Then*

$$|||(A^q \natural_\alpha B^q)^{\frac{1}{q}}||| \leq S(h^p)^{\frac{1}{p}} |||(A^p \natural_\alpha B^p)^{\frac{1}{p}}||| \quad \textit{for all } 0 < q < p,$$

$$|||A^p \natural_\alpha B^p||| \leq S(h)^p \, |||(A \natural_\alpha B)^p||| \quad \textit{for all } 0 < p < 1$$

and

$$|||(A \natural_\alpha B)^p||| \leq S(h^p) \, |||A^p \natural_\alpha B^p||| \quad \textit{for all } p > 1.$$

We show reverses of the Golden–Thompson-type inequalities due to Ando, Hiai, and Petz [11, 15], see also [32, Lemma 3.1].

Theorem 6.5.14 *Let A and B be Hermitian matrices such that $m \leq A, B \leq M$ for some scalars $m < M$, and let $\alpha \in [0, 1]$. Then for each $p > 0$ there exist unitary matrices U and V such that*

$$\begin{aligned}&S(e^{p(M-m)})^{-\frac{1}{p}} V(e^{pA} \natural_\alpha e^{pB})^{\frac{1}{p}} V^* \\ &\leq e^{(1-\alpha)A+\alpha B} \leq S(e^{p(M-m)})^{\frac{1}{p}} U(e^{pA} \natural_\alpha e^{pB})^{\frac{1}{p}} U^*. \qquad (6.5.17)\end{aligned}$$

Proof By replacing A and B with e^A and e^B in Lemma 6.5.12, respectively, it follows that for each $0 < q < p$ there exists a unitary matrix $U_{p,q}$ such that

$$(e^{qA} \natural_\alpha e^{qB})^{\frac{1}{q}} \leq S(e^{p(M-m)})^{\frac{1}{p}} U_{p,q}(e^{pA} \natural_\alpha e^{pB})^{\frac{1}{p}} U_{p,q}^*.$$

By Theorem 6.3.11, we have

$$e^{(1-\alpha)A+\alpha B} = \lim_{q\to 0}(e^{qA} \natural_\alpha e^{qB})^{\frac{1}{q}}.$$

It follows that, for each $p > 0$, there exists a unitary matrix U such that

$$e^{(1-\alpha)A+\alpha B} \le S(e^{p(M-m)})^{\frac{1}{p}} U(e^{pA} \natural_\alpha e^{pB})^{\frac{1}{p}} U^*.$$

We also have the first inequality of (6.5.17) by a similar method as the proof of Lemma 6.5.12. □

In particular, we have the following theorem by (iv) of Lemma 6.5.5.

Theorem 6.5.15 *Let A and B be Hermitian matrices such that $m \le A, B \le M$ for some scalars $m < M$, and let $\alpha \in [0, 1]$. Then*

$$\left|\left|\left|(e^{pA} \natural_\alpha e^{pB})^{\frac{1}{p}}\right|\right|\right| \le \left|\left|\left|e^{(1-\alpha)A+\alpha B}\right|\right|\right| \le S(e^{p(M-m)})^{\frac{1}{p}} \left|\left|\left|(e^{pA} \natural_\alpha e^{pB})^{\frac{1}{p}}\right|\right|\right| \quad (6.5.18)$$

for all $p > 0$ and every unitarily invariant norm $|||\cdot|||$, and the right-hand side of (6.5.18) *converges to the middle hand side as $p \downarrow 0$. In particular,*

$$|||e^{2A} \natural e^{2B}||| \le |||e^{A+B}||| \le S(e^{2(M-m)})\, |||e^{2A} \natural e^{2B}||| \quad (6.5.19)$$

and

$$\operatorname{tr}(e^{2A} \natural e^{2B}) \le \operatorname{tr}(e^{A+B}) \le S(e^{2(M-m)}) \operatorname{tr}(e^{2A} \natural e^{2B}).$$

Next, we would like to present another estimate of the Golden–Thompson-type inequalities due to Ando, Hiai, and Petz. As a matter of fact, the upper bound $S(e^{p(M-m)})^{\frac{1}{p}}$ in (6.5.18) of Theorem 6.5.15 is constant for all values of $\alpha \in [0, 1]$. We demonstrate another order relation between $(A^q \natural_\alpha B^q)^{\frac{1}{q}}$ and $(A^p \natural_\alpha B^p)^{\frac{1}{p}}$ in terms of the generalized Kantorovich constant, which is discussed in [32, Lemma 4.1].

Lemma 6.5.16 *Let A and B be positive definite matrices such that $0 < m \le A, B \le M$ for some scalars $0 < m < M$, and let $\alpha \in [0, 1]$. Put $h = \frac{M}{m}$. Let $0 < q \le 1$. Then for each $0 < q \le p \le 1$, there exist unitary matrices U_1 and U_2 such that*

$$\begin{aligned} &K(h, p)^{\frac{\alpha}{p}} K(h^{2p}, \alpha)^{\frac{1}{p}} U_1 (A^p \natural_\alpha B^p)^{\frac{1}{p}} U_1^* \\ &\quad \le (A^q \natural_\alpha B^q)^{\frac{1}{q}} \le K(h, p)^{-\frac{\alpha}{p}} K(h^{2p}, \alpha)^{-\frac{1}{p}} U_2 (A^p \natural_\alpha B^p)^{\frac{1}{p}} U_2^* \end{aligned} \quad (6.5.20)$$

and for each $p \geq 1$*, there exist unitary matrices* V_1 *and* V_2 *such that*

$$
\begin{aligned}
&K(h^{2p}, \alpha)^{\frac{1}{p}} V_1 (A^p \natural_\alpha B^p)^{\frac{1}{p}} V_1^* \\
&\leq (A^q \natural_\alpha B^q)^{\frac{1}{q}} \leq K(h^{2p}, \alpha)^{-\frac{1}{p}} V_2 (A^p \natural_\alpha B^p)^{\frac{1}{p}} V_2^*,
\end{aligned}
\tag{6.5.21}
$$

where the generalized Kantorovich constant $K(h, p)$ *is defined as in* (6.5.5).

Proof For $0 < q < p \leq 1$ and every unit vector $x \in \mathbb{C}^n$,

$$
\begin{aligned}
\langle x, (A^q \natural_\alpha B^q)x\rangle^{\frac{1}{q}} &= \langle \frac{A^{\frac{q}{2}}x}{\left\|A^{\frac{q}{2}}x\right\|}, (A^{-\frac{q}{2}} B^q A^{-\frac{q}{2}})^{\alpha} \frac{A^{\frac{q}{2}}x}{\left\|A^{\frac{q}{2}}x\right\|}\rangle^{\frac{1}{q}} \left\|A^{\frac{q}{2}}x\right\|^{\frac{2}{q}} \\
&\leq \langle \frac{A^{\frac{q}{2}}x}{\left\|A^{\frac{q}{2}}x\right\|}, A^{-\frac{q}{2}} B^q A^{-\frac{q}{2}} \frac{A^{\frac{q}{2}}x}{\left\|A^{\frac{q}{2}}x\right\|}\rangle^{\frac{\alpha}{q}} \left\|A^{\frac{q}{2}}x\right\|^{\frac{2}{q}} \quad \text{(by } 0 < \alpha < 1\text{)} \\
&= \langle x, B^q x\rangle^{\frac{\alpha}{q}} \left\|A^{\frac{q}{2}}x\right\|^{\frac{2}{q}-\frac{2\alpha}{q}} \\
&\leq \langle x, Bx\rangle^{\alpha} \left\|A^{\frac{q}{2}}x\right\|^{\frac{2}{q}-\frac{2\alpha}{q}} \quad \text{(by } 0 < q < 1\text{)} \\
&= \langle x, Bx\rangle^{\frac{p\alpha}{p}} \left\|A^{\frac{q}{2}}x\right\|^{\frac{2}{q}-\frac{2\alpha}{q}} \\
&\leq (K(h, p)^{-1} \langle x, B^p x\rangle)^{\frac{\alpha}{p}} \left\|A^{\frac{q}{2}}x\right\|^{\frac{2}{q}-\frac{2\alpha}{q}} \quad \text{(by } 0 < p \leq 1 \text{ and (6.5.7))}
\end{aligned}
$$

$$
\begin{aligned}
&= K(h, p)^{-\frac{\alpha}{p}} \langle x, B^p x\rangle^{\frac{\alpha}{p}} \left\|A^{\frac{q}{2}}x\right\|^{\frac{2}{q}-\frac{2\alpha}{q}} \\
&= K(h, p)^{-\frac{\alpha}{p}} \langle \frac{A^{\frac{p}{2}}x}{\left\|A^{\frac{p}{2}}x\right\|}, A^{-\frac{p}{2}} B^p A^{-\frac{p}{2}} \frac{A^{\frac{p}{2}}x}{\left\|A^{\frac{q}{2}}x\right\|}\rangle^{\frac{\alpha}{p}} \left\|A^{\frac{p}{2}}x\right\|^{\frac{2\alpha}{p}} \left\|A^{\frac{q}{2}}x\right\|^{\frac{2}{q}-\frac{2\alpha}{q}} \\
&\leq K(h, p)^{-\frac{\alpha}{p}} K(h^{2p}, \alpha)^{-\frac{1}{p}} \langle x, A^p \natural_\alpha B^p x\rangle^{\frac{1}{p}} \left\|A^{\frac{p}{2}}x\right\|^{\frac{2\alpha}{p}-\frac{2}{p}} \left\|A^{\frac{q}{2}}x\right\|^{\frac{2}{q}-\frac{2\alpha}{q}} \quad \text{(by (6.5.7))} \\
&\leq K(h, p)^{-\frac{\alpha}{p}} K(h^{2p}, \alpha)^{-\frac{1}{p}} \langle A^p \natural_\alpha B^p x, x\rangle^{\frac{1}{p}}.
\end{aligned}
$$

The last inequality holds since it follows from $0 < q < p$ that

$$
\begin{aligned}
\left\|A^{\frac{p}{2}}x\right\|^{\frac{2\alpha}{p}-\frac{2}{p}} \left\|A^{\frac{q}{2}}x\right\|^{\frac{2}{q}-\frac{2\alpha}{q}} &= \langle A^p x, x\rangle^{\frac{\alpha-1}{p}} \langle A^q x, x\rangle^{\frac{1-\alpha}{q}} \\
&= \langle A^p x, x\rangle^{\frac{\alpha-1}{p}} \langle (A^p)^{\frac{q}{p}} x, x\rangle^{\frac{1-\alpha}{q}} \\
&\leq \langle A^p x, x\rangle^{\frac{\alpha-1}{p}} \langle A^p x, x\rangle^{\frac{1-\alpha}{p}} = 1.
\end{aligned}
$$

By the minimax theorem, there exists a subspace $\mathscr{F}$ of codimension $k-1$ such that

$$
\lambda_k((A^p \natural_\alpha B^p)^{\frac{1}{p}}) = \max_{y \in \mathscr{F}, \|x\|=1} \langle x, (A^p \natural_\alpha B^p)^{\frac{1}{p}} x\rangle = \max_{y \in \mathscr{F}, \|x\|=1} \langle x, (A^p \natural_\alpha B^p)x\rangle^{\frac{1}{p}}.
$$

Therefore, we have

$$\begin{aligned}\lambda_k((A^q \natural_\alpha B^q)^{\frac{1}{q}}) &\leq \max_{y\in\mathscr{F},\|x\|=1} \langle x, (A^q \natural_\alpha B^q)x\rangle^{\frac{1}{q}} \\ &\leq \max_{y\in\mathscr{F},\|x\|=1} K(h,p)^{-\frac{\alpha}{p}} K(h^{2p},\alpha)^{-\frac{1}{p}} \langle A^p \natural_\alpha B^p x, x\rangle^{\frac{1}{p}} \\ &= K(h,p)^{-\frac{\alpha}{p}} K(h^{2p},\alpha)^{-\frac{1}{p}} \lambda_k(A^p \natural_\alpha B^p)^{\frac{1}{p}}) \quad \text{by } \tfrac{1}{p} \geq 1.\end{aligned}$$

Hence, there exists a unitary matrix U_2 such that

$$(A^q \natural_\alpha B^q)^{\frac{1}{q}} \leq K(h,p)^{-\frac{\alpha}{p}} K(h^{2p},\alpha)^{-\frac{1}{p}} U_2 (A^p \natural_\alpha B^p)^{\frac{1}{p}} U_2^*.$$

By replacing A and B with their inverses, we have the first inequality of (6.5.20).

For $p \geq 1$, it follows that inequality (6.5.21) holds by a similar method. □

By Lemma 6.5.16, we have another reverse of the Golden–Thompson-type inequalities due to Ando, Hiai, and Petz; we refer readers to [32, Theorem 4.2].

Theorem 6.5.17 *Let A and B be Hermitian matrices such that $m \leq A, B \leq M$ for some scalars $m < M$, and let $\alpha \in [0,1]$. Then for every unitarily invariant norm* $|||\cdot|||$,

$$|||e^{(1-\alpha)A+\alpha B}||| \leq K(e^{M-m},p)^{-\frac{\alpha}{p}} K(e^{2p(M-m)},\alpha)^{-\frac{1}{p}} \left|\left|\left|(e^{pA} \natural_\alpha e^{pB})^{\frac{1}{p}}\right|\right|\right|$$

for all $0 < p \leq 1$ and

$$|||e^{(1-\alpha)A+\alpha B}||| \leq K(e^{2p(M-m)},\alpha)^{-\frac{1}{p}} \left|\left|\left|(e^{pA} \natural_\alpha e^{pB})^{\frac{1}{p}}\right|\right|\right|$$

for all $p \geq 1$. In particular,

$$|||e^{A+B}||| \leq \frac{e^{2M}+e^{2m}}{2e^M e^m} |||e^{2A} \sharp e^{2B}||| \tag{6.5.22}$$

and

$$\operatorname{tr} e^{A+B} \leq \frac{e^{2M}+e^{2m}}{2e^M e^m} \operatorname{tr}(e^{2A} \sharp e^{2B}).$$

Remark 6.5.18 (1) In Theorem 6.5.17, we have the constant

$$K(e^{M-m},p)^{-\frac{\alpha}{p}} K(e^{2p(M-m)},\alpha)^{-\frac{1}{p}} = 1$$

in the case of $(\alpha, p) = (0,1)$ and $(1,1)$.

(2) Compare the constants of (6.5.19) in Theorem 6.5.15 and of (6.5.22) in Theorem 6.5.17: If $\alpha = \frac{1}{2}$, then for each $p > 0$ it follows from Specht inequality (6.5.8) that

$$K(h^{2p}, \frac{1}{2})^{-\frac{1}{p}} = \left(\frac{h^{\frac{p}{2}} + h^{-\frac{p}{2}}}{2} \right)^{\frac{1}{p}} \le \left(S(h^p) \sqrt{h^{\frac{p}{2}} h^{-\frac{p}{2}}} \right)^{\frac{1}{p}} = S(h^p)^{\frac{1}{p}}.$$

Hence, for each $p \ge 1$

$$\begin{aligned} \left|\left|\left| e^{A+B} \right|\right|\right| &\le K(e^{4p(M-m)}, \frac{1}{2})^{-\frac{1}{p}} \left|\left|\left| (e^{2pA} \sharp e^{2pB})^{\frac{1}{p}} \right|\right|\right| \\ &\le S(e^{2p(M-m)})^{\frac{1}{p}} \left|\left|\left| (e^{2pA} \sharp e^{2pB})^{\frac{1}{p}} \right|\right|\right| . \end{aligned}$$

In particular, if we put $p = 1$, then

$$\left|\left|\left| e^{A+B} \right|\right|\right| \le \frac{e^{2M} + e^{2m}}{2e^M e^m} \left|\left|\left| e^{2A} \sharp e^{2B} \right|\right|\right| \le S(e^{2(M-m)}) \left|\left|\left| e^{2A} \sharp e^{2B} \right|\right|\right| .$$

Remark 6.5.19 Most of the results of Sect. 6.5 hold for operators on a Hilbert space.

6.6 Applications I (Hadamard Product Version)

A generalized Golden–Thompson inequality can be stated in the following form: For any pair A and B of Hermitian matrices and any unitarily invariant norm $|||\cdot|||$,

$$\left|\left|\left| e^{A+B} \right|\right|\right| \le \left|\left|\left| e^{A/2} e^B e^{A/2} \right|\right|\right| \le \left|\left|\left| e^A e^B \right|\right|\right| .$$

What can we say for the case of more than two Hermitian matrices? However, it is known that even for the trace norm, a modified generalization to three Hermitian matrices A, B and C is not valid. For instance,

$$\operatorname{tr}(e^{A+B+C}) \le \frac{1}{2} \left(\operatorname{tr}(e^A e^B e^C) + \operatorname{tr}(e^C e^B e^A) \right)$$

is not true in general.

In this section, we present upper and lower bounds of $\left|\left|\left| e^{A_1 + \cdots + A_k} \right|\right|\right|$ for a k-tuple $A_1, \ldots, A_k$ of Hermitian matrices in terms of norms of suitable Hadamard products, and their reverses.

The tensor product $\mathbb{M}_n \otimes \cdots \otimes \mathbb{M}_n$ of k copies of $\mathbb{M}_n$ is identified with $\mathbb{M}_{n^k}$ in a natural way. Then there exists a unique unital positive linear map from $\mathbb{M}_{n^k}$ to $\mathbb{M}_n$ that satisfies

$$\Phi(X_1 \otimes \cdots \otimes X_k) = X_1 \circ \cdots \circ X_k \qquad (X_i \in \mathbb{M}_n).$$

where $X_1 \circ \cdots \circ X_k$ is the Hadamard product of $X_1, \ldots, X_k$. It follows from the Choi–Davis–Jensen Theorem (Theorem 5.2.9 and Corollary 5.2.10) that

$$\Phi(\mathbf{X})^p \geq \Phi(\mathbf{X}^p) \qquad (\mathbf{X} > 0, 0 < p \leq 1),$$

and since $-\log t$ is matrix convex,

$$\log \Phi(\mathbf{X}) \geq \Phi(\log \mathbf{X}) \qquad (\mathbf{X} > 0). \tag{6.6.1}$$

Oppenheim [33] proved the following determinant inequality for the Hadamard product of positive definite matrices $A_1, \ldots, A_k$:

$$\det(A_1 \circ \cdots \circ A_k) \geq \det(A_1) \cdots \det(A_k).$$

Since for any positive definite A,

$$\log(\det(A)) = \operatorname{tr}(\log A) = \operatorname{tr}((\log A) \circ I),$$

it follows from the Oppenheim determinant inequality that

$$\begin{aligned} \operatorname{tr}(\log(A_1 \circ \cdots \circ A_k)) &= \log(\det(A_1 \circ \cdots \circ A_k)) \\ &\geq \log(\det(A_1) \cdots \det(A_k)) \\ &= \operatorname{tr}((\log A_1) \circ I) + \cdots + \operatorname{tr}((\log A_k) \circ I) \\ &= \operatorname{tr}((\sum_{i=1}^{k} \log A_i) \circ I). \end{aligned}$$

Ando [34] showed the following matricial generalization of Oppenheim's inequality.

Theorem 6.6.1 *Let $A_1, \ldots, A_k$ be any k-tuple of Hermitian matrices in $\mathbb{M}_n$. Then*

$$\log(e^{\alpha A_1} \circ \cdots \circ e^{\alpha A_k})^{1/\alpha} \geq \left(\sum_{i=1}^{k} A_i\right) \circ I \qquad (\alpha > 0) \tag{6.6.2}$$

and

$$\log(e^{-\alpha A_1} \circ \cdots \circ e^{-\alpha A_k})^{-1/\alpha} \leq \left(\sum_{i=1}^{k} A_i\right) \circ I \qquad (\alpha > 0). \tag{6.6.3}$$

In particular, if $A_1, \ldots, A_k$ are any k-tuple of positive definite matrices in $\mathbb{M}_n$, then

$$\log(A_1 \circ \cdots \circ A_k) \geq \left(\sum_{i=1}^{k} \log A_i\right) \circ I.$$

Proof Since

$$e^{A_1} \otimes \cdots \otimes e^{A_k} = \prod_{i=1}^{k} I \otimes \cdots \otimes I \otimes e^{A_i} \otimes I \otimes \cdots \otimes I$$

and $I \otimes \cdots \otimes I \otimes e^{A_i} \otimes I \otimes \cdots \otimes I$ $(i = 1, \ldots, k)$ are mutually commuting, we have

$$\begin{aligned} \log(e^{A_1} \otimes \cdots \otimes e^{A_k}) &= \sum_{i=1}^{k} \log(I \otimes \cdots \otimes I \otimes e^{A_i} \otimes I \otimes \cdots \otimes I) \\ &= \sum_{i=1}^{k} I \otimes \cdots \otimes I \otimes A_i \otimes I \otimes \cdots \otimes I. \end{aligned}$$

Then, since

$$\Phi(I \otimes \cdots \otimes I \otimes A_i \otimes I \otimes \cdots \otimes I) = A_i \circ I \qquad (i = 1, \ldots, k),$$

it follows from (6.6.1) that

$$\log(e^{\alpha A_1} \circ \cdots \circ e^{\alpha A_k})^{1/\alpha} \geq \left(\sum_{i=1}^{k} A_i\right) \circ I \qquad (\alpha > 0),$$

and correspondingly

$$\log(e^{-\alpha A_1} \circ \cdots \circ e^{-\alpha A_k})^{-1/\alpha} \leq \left(\sum_{i=1}^{k} A_i\right) \circ I \qquad (\alpha > 0).$$

□

The Hadamard product is defined in connection with a special choice of an orthonormal basis, while the notion of a unitarily invariant norm is independent of the chosen basis. A change of basis corresponds to selecting a unitary matrix, denoted as U. Therefore, for each unitary matrix U, we can consider a map $X \mapsto U^*XU$. This map is a $*$-isomorphism of the C^*-algebra $\mathbb{M}_n$ and preserves every unitarily invariant norm. Let $\mathscr{U}$ denote the set of unitary matrices in $\mathbb{M}_n$. Then Ando [34, Theorem 1] proved the following.

Theorem 6.6.2 *For any k-tuple $A_1, \ldots, A_k$ of Hermitian matrices and any unitarily invariant norm $|||\cdot|||$,*

$$\sup_{U \in \mathscr{U}} |||(e^{\alpha U^* A_1 U} \circ \cdots \circ e^{\alpha U^* A_k U})^{1/\alpha}||| \geq |||e^{A_1 + \cdots + A_k}||| \quad (\alpha > 0), \tag{6.6.4}$$

and

$$\left|\!\left|\!\left| e^{A_1+\cdots+A_k} \right|\!\right|\!\right| \geq \sup_{U\in\mathcal{U}} \left|\!\left|\!\left| (e^{-\alpha U^* A_1 U} \circ \cdots \circ e^{-\alpha U^* A_k U})^{-1/\alpha} \right|\!\right|\!\right| \quad (\alpha > 0). \qquad (6.6.5)$$

Proof Choose $V \in \mathcal{U}$, which diagonalizes $\sum_{i=1}^k A_i$. Then since

$$\sum_{i=1}^k V^* A_i V = V^* \left(\sum_{i=1}^k A_i \right) V = V^* \left(\sum_{i=1}^k A_i \right) V \circ I = \left(\sum_{i=1}^k V^* A_i V \right) \circ I,$$

it follows from (6.6.2) that

$$\log(e^{\alpha V^* A_1 V} \circ \cdots \circ e^{\alpha V^* A_k V})^{1/\alpha} \geq \sum_{i=1}^k V^* A_i V. \qquad (6.6.6)$$

It is well known that for Hermitian matrices X, Y the order relation $X \geq Y$ implies that for any nonnegative nondecreasing function $g(t)$,

$$\lambda_i(g(X)) = g(\lambda_i(X)) \geq g(\lambda_i(Y)) \geq \lambda_i(g(Y)) \, (\geq 0)$$

for $i = 1, \ldots, k$, and hence $|\!|\!| g(X) |\!|\!| \geq |\!|\!| g(Y) |\!|\!|$ for any unitarily invariant norm $|\!|\!| \cdot |\!|\!|$. Applying this principle with $g(t) = e^t$, we can derive from (6.6.6) that

$$\begin{aligned} \sup_{U\in\mathcal{U}} \left|\!\left|\!\left| (e^{\alpha U^* A_1 U} \circ \cdots \circ e^{\alpha U^* A_k U})^{1/\alpha} \right|\!\right|\!\right| &\geq \left|\!\left|\!\left| (e^{\alpha V^* A_1 V} \circ \cdots \circ e^{\alpha V^* A_k V})^{1/\alpha} \right|\!\right|\!\right| \\ &\geq \left|\!\left|\!\left| e^{V^* A_1 V + \cdots + V^* A_k V} \right|\!\right|\!\right| \\ &= \left|\!\left|\!\left| e^{A_1+\cdots+A_k} \right|\!\right|\!\right|, \end{aligned}$$

which proves (6.6.4).

According to the Schur theorem, the eigenvalues of a Hermitian matrix X form a sequence that majorizes the sequence of its diagonal entries, so that $|\!|\!| g(X) |\!|\!| \geq |\!|\!| g(X \circ I) |\!|\!|$ for any nonnegative convex function $g(t)$ and any unitarily invariant norm $|\!|\!| \cdot |\!|\!|$. Applying this principle with $g(t) = e^t$, we can derive from (6.6.3) that for any $U \in \mathcal{U}$

$$\begin{aligned} \left|\!\left|\!\left| e^{A_1+\cdots+A_k} \right|\!\right|\!\right| &= \left|\!\left|\!\left| e^{U^* A_1 U + \cdots + U^* A_k U} \right|\!\right|\!\right| \\ &\geq \left|\!\left|\!\left| e^{(\sum_{i=1}^k U^* A_i U)\circ I} \right|\!\right|\!\right| \\ &\geq \left|\!\left|\!\left| (e^{-\alpha U^* A_1 U} \circ \cdots \circ e^{-\alpha U^* A_k U})^{-1/\alpha} \right|\!\right|\!\right|, \end{aligned}$$

which proves (6.6.5). □

Next, we show a complementary inequality to the matrix generalization of Oppenheim's inequality in Theorem 6.6.1. Moreover, we demonstrate complementary inequalities to the Golden–Thompson-type inequalities on the Hadamard product in Theorem 6.6.2.

First, we establish a complementary inequality to the matrix generalization of Oppenheim's inequality. To prove this, we utilize the method established by Mond–Pečarić. Recall that the logarithmic mean $L(m, M)$ is defined for $0 < m \leq M$ as

$$L(m, M) = \frac{M - m}{\log M - \log m} \ (M > m) \quad \text{and} \quad L(m, m) = m.$$

It is evident that $m \leq L(m, M) \leq M$. The following theorem is attributed to [35, Theorem 1].

Theorem 6.6.3 *Let $A_1, \ldots, A_k$ be any k-tuple of positive definite matrices in $\mathbb{M}_n$ such that $m_i \leq A_i \leq M_i$ with $0 < m_i < M_i$ $(i = 1, \ldots, k)$. Let $m = m_1 \cdots m_k$ and $M = M_1 \cdots M_k$. Then*

$$\left(\sum_{i=1}^{k} \log A_i\right) \circ I \geq \alpha \log(A_1 \circ \cdots \circ A_k) + \beta I \tag{6.6.7}$$

for any $\alpha > 0$, where β is determined as the minimum of the function at $at + b - \alpha \log t$ on $[m, M]$ with

$$a \equiv \frac{1}{L(m, M)} = \frac{\log M - \log m}{M - m}$$

and

$$b \equiv \frac{-m}{L(m, M)} + \log m = \frac{M \log m - m \log M}{M - m}.$$

Proof By the definition

$$at + b \geq \alpha \log t + \beta \qquad \text{for all } t \in [m, M].$$

Since $0 < m \leq A_1 \circ \cdots \circ A_k \leq M$, applying this inequality to $A_1 \circ \cdots \circ A_k$, we have

$$a(A_1 \circ \cdots \circ A_k) + bI \geq \alpha \log(A_1 \circ \cdots \circ A_k) + \beta I.$$

Since $\log t$ is concave and the linear function $at + b$ interpolates $\log t$ at $t = m$ and $t = M$, we have $\log t \geq at + b$ for all $t \in [m, M]$, so that the inequality applied to $A_1 \otimes \cdots \otimes A_k$ and to $\Phi(\cdot)$ implies

$$\begin{aligned}\Phi(\log(A_1 \otimes \cdots \otimes A_k)) &\geq a\Phi(A_1 \otimes \cdots \otimes A_k) + b\Phi(I) \\ &= a(A_1 \circ \cdots \circ A_k) + bI,\end{aligned}$$

since $0 < m \le A_1 \otimes \cdots \otimes A_k \le M$. Also, it follows that

$$\begin{aligned}
&\Phi(\log(A_1 \otimes \cdots \otimes A_k)) \\
&= \Phi(\log(A_1 \otimes I \otimes \cdots \otimes I)) + \cdots + \Phi(\log(I \otimes \cdots I \otimes A_k)) \\
&= \Phi((\log A_1 \otimes I \otimes \cdots \otimes I) + \cdots + (I \otimes \cdots \otimes I \otimes \log A_k)) \\
&= (\log A_1) \circ I \circ \cdots \circ I + \cdots + I \circ \cdots \circ I \circ \log A_k \\
&= \left(\sum_{i=1}^{k} \log A_i \right) \circ I.
\end{aligned}$$

Therefore, we have

$$\begin{aligned}
\left(\sum_{i=1}^{k} \log A_i \right) \circ I &\ge a(A_1 \circ \cdots \circ A_k) + bI \\
&\ge \alpha \log(A_1 \circ \cdots \circ A_k) + \beta I.
\end{aligned}$$

□

Remark 6.6.4 $\beta = \beta(m, M, \alpha) = \min\{at + b - \alpha \log t : t \in [m, M]\}$ can be written explicitly as

$$\beta(m, M, \alpha) = \begin{cases} \alpha - \alpha \log(\alpha L(m, M)) \\ \quad + \frac{M \log m - m \log M}{M - m} & \text{if } m \le \alpha L(m, M) \le M, \\ (1 - \alpha) \log M & \text{if } M \le \alpha L(m, M), \\ (1 - \alpha) \log m & \text{if } \alpha L(m, M) \le m. \end{cases}$$

In fact, let $h(t) = at + b - \alpha \log t$. Then, by the definition, it follows that $\beta = \min_{m \le t \le M} h(t)$.

Now $h'(t_1) = 0$ if and only if $t_1 = \alpha L(m, M)$. If $m \le \alpha L(m, M) \le M$, then

$$\begin{aligned}
\beta = h(t_1) &= h(\alpha L(m, M)) \\
&= \alpha - \alpha \log(\alpha L(m, M)) + \frac{M \log m - m \log M}{M - m},
\end{aligned}$$

because $h''(t_1) = \alpha / t_1^2 > 0$. If $\alpha L(m, M) \ge M$, then $h(t)$ is nonincreasing for $t \le M$ so that $\beta = h(M) = (1 - \alpha) \log M$. If $\alpha L(m, M) \le m$, then $h(t)$ is nondecreasing for $t \ge m$, so that $\beta = h(m) = (1 - \alpha) \log m$.

As a corollary, we give an estimate of the lower bound of the difference of Oppenheim's inequality in [35, Corollary 2].

Corollary 6.6.5 *Let the hypothesis of Theorem 6.6.3 be satisfied. Then*

$$\left(\sum_{i=1}^{k}\log A_i\right)\circ I-\log(A_1\circ\cdots\circ A_k)\geq-\log S(h)I,$$

where the Specht ratio $S(h)$ is defined by (6.5.9) *and $h=\frac{M}{m}$.*

Proof If we put $\alpha=1$ in Remark 6.6.4, then it follows from $m\leq L(m,M)\leq M$ that

$$\beta=1+\frac{M\log m-m\log M}{M-m}-\log L(m,M)=-\log S(h),$$

and hence we have this corollary. □

We give complementary inequalities to the Golden–Thompson-type inequalities on the Hadamard product via the Specht ratio as shown in [35, Theorem 4].

Theorem 6.6.6 *Let $A_1,\ldots,A_k$ be any k-tuple of positive definite matrices in $\mathbb{M}_n$ such that $m_i\leq A_i\leq M_i$ with $0<m_i<M_i$ $(i=1,\ldots,k)$ and $|||\cdot|||$ any unitarily invariant norm. Then for any $\alpha>0$*

$$|||e^{A_1+\cdots+A_k}|||\geq e^{\beta_+(t)/t}\sup_{V\in\mathcal{U}}|||(e^{tV^*A_1V}\circ\cdots\circ e^{tV^*A_kV})^{\alpha/t}||| \tag{6.6.8}$$

for $t>0$, where $\beta_+(t)=\beta(e^{t(m_1+\cdots+m_k)},e^{t(M_1+\cdots+M_k)},\alpha)$ in Theorem 6.6.3, and

$$e^{-\beta_-(t)/t}\sup_{V\in\mathcal{U}}|||(e^{-tV^*A_1V}\circ\cdots\circ e^{-tV^*A_kV})^{-\alpha/t}|||\geq|||e^{A_1+\cdots+A_k}||| \tag{6.6.9}$$

for $t>0$, where $\beta_-(t)=\beta(e^{-t(m_1+\cdots+m_k)},e^{-t(M_1+\cdots+M_k)},\alpha)$ in Theorem 6.6.3.

Proof In Theorem 6.6.3, by replacing A_i with $e^{tV^*A_iV}$ $(i=1,\ldots,k)$ for $V\in\mathcal{U}$, we have

$$\left(\sum_{i=1}^{k}tV^*A_iV\right)\circ I\geq\alpha\log(e^{tV^*A_1V}\circ\cdots\circ e^{tV^*A_kV})+\beta_+(t)I,$$

where $\beta_+(t)=\beta(e^{t(m_1+\cdots+m_k)},e^{t(M_1+\cdots+M_k)},\alpha)$ in Theorem 6.6.3. If $X\geq Y$ for Hermitian matrices X,Y and g is a nonnegative nondecreasing function, then $|||g(X)|||\geq|||g(Y)|||$. Applying this principle to $g(s)=e^{s/t}$, we get

$$|||e^{V^*(A_1+\cdots+A_k)V\circ I}|||\geq e^{\beta_+(t)/t}|||(e^{tV^*A_1V}\circ\cdots\circ e^{tV^*A_kV})^{\alpha/t}|||.$$

Since the sequence of eigenvalues of a Hermitian matrix X majorizes the sequence of its diagonal entries, we have $|||g(X)||| \geq |||g(X \circ I)|||$ for any nonnegative convex function g. Therefore, it follows that

$$|||e^{A_1+\cdots+A_k}||| = |||e^{V^*(A_1+\cdots+A_k)V}||| \geq |||e^{V^*(A_1+\cdots+A_k)V\circ I}|||.$$

Hence, we have the desired inequality (6.6.8).

Next, since there exists $V \in \mathcal{U}$ such that $V^*(A_1 + \cdots + A_k)V$ is a diagonal matrix, we have $V^*(A_1 + \cdots + A_k)V = V^*(A_1 + \cdots + A_k)V \circ I$. Also, it follows that for $t > 0$

$$(V^*A_1V + \cdots + V^*A_kV) \circ I \leq -\frac{\alpha}{t}\log(e^{-tV^*A_1V} \circ \cdots \circ e^{-tV^*A_kV}) - \frac{\beta_-(t)}{t}I$$

and, consequently

$$\begin{aligned}|||e^{V^*A_1V+\cdots+V^*A_kV}||| &= |||e^{(V^*A_1V+\cdots+V^*A_kV)\circ I}|||\\ &\leq e^{-\beta_-(t)/t}\,|||(e^{-tV^*A_1V} \circ \cdots \circ e^{-tV^*A_kV})^{-\alpha/t}|||,\end{aligned}$$

which proves (6.6.9). □

If we put $\alpha = 1$ in Theorem 6.6.6, we have the following.

Theorem 6.6.7 *Let $A_1, \ldots, A_k$ be any k-tuple of positive definite matrices in $\mathbb{M}_n$ such that $m_i \leq A_i \leq M_i$ with $0 < m_i < M_i$ $(i = 1, \ldots, k)$ and $|||\cdot|||$ any unitarily invariant norm. Then,*

$$\begin{aligned}S(h^t)^{1/t} &\sup_{V\in\mathcal{U}} |||(e^{-tV^*A_1V} \circ \cdots \circ e^{-tV^*A_kV})^{-1/t}|||\\ &\geq |||e^{A_1+\cdots+A_k}|||\\ &\geq S(h^t)^{-1/t} \sup_{V\in\mathcal{U}} |||(e^{tV^*A_1V} \circ \cdots \circ e^{tV^*A_kV})^{1/t}|||\end{aligned}$$

for any $t > 0$, where $h = e^{M_1+\cdots+M_k}/e^{m_1+\cdots+m_k}$ and $S(h)$ is the Specht ratio.

Proof If $\alpha = 1$ in Theorem 6.6.6, then we have

$$\begin{aligned}\beta_+(t) &= \beta(e^{t(m_1+\cdots+m_k)}, e^{t(M_1+\cdots+M_k)}, 1)\\ &= \beta(e^{-t(M_1+\cdots+M_k)}, e^{-t(m_1+\cdots+m_k)}, 1)\\ &= \beta_-(t).\end{aligned}$$

Hence, it follows that $e^{-\beta_+(t)} = e^{-\beta_-(t)} = S(h^t)$, and this implies Theorem 6.6.7. □

6.7 Applications II (Multivariable Version)

By Theorem 6.4.2, we have the complemented Golden–Thompson trace inequality: Let A and B be Hermitian matrices and $\alpha \in [0, 1]$. Then

$$|||(e^{pA} \natural_\alpha e^{pB})^{1/p}||| \leq |||e^{(1-\alpha)A+\alpha B}|||$$

for all $p > 0$. In this section, we aim to consider its multivariable version. For this, we present the multivariable version of matrix α-geometric mean $\natural_\alpha$ for $\alpha \in [0, 1]$: In 2014, Lawson, Lim and Pálfia [36, 37] established the formulation of the geometric mean for k (≥ 3) positive definite matrices, which is an interesting extension of the matrix α-geometric mean in the Kubo–Ando theory. They showed that there exists a unique positive definite solution to the Karcher equation

$$\sum_{i=1}^{k} \omega_i \log X^{-\frac{1}{2}} A_i X^{-\frac{1}{2}} = 0 \tag{6.7.1}$$

for given k positive definite matrices $A_1, \ldots, A_k$, where $\omega = (\omega_1, \ldots, \omega_k)$ is a weight vector, that is, $\omega_1, \ldots, \omega_k \geq 0$ and $\sum_{i=1}^k \omega_i = 1$. We say that the solution X of (6.7.1) is the *Karcher mean* for k positive definite matrices $A_1, \ldots, A_k$ and denote it by $G_K(\omega; A_1, \ldots, A_k)$. In the case of $k = 2$, the Karcher mean $G_K((1-\alpha, \alpha); A, B)$ coincides with the matrix α-geometric mean $A \natural_\alpha B$. Compared to equation (6.3.4) of Lemma 6.3.4, Karcher equation (6.7.1) is a k-variable version of (6.3.4). Here, let $\mathbb{P}_n$ be the set of all $n \times n$ positive definite matrices in $\mathbb{M}_n$. The Thompson metric on $\mathbb{P}_n$ is defined by $d(A, B) = \|\log(A^{-1/2} B A^{-1/2})\|$. It is known that d is a complete metric on $\mathbb{P}_n$ and $d(A, B) = \max\{\log M(B/A), \log M(A/B)\}$, where $M(B/A) = \inf\{\alpha > 0 : B \leq \alpha A\} = \lambda_1(A^{-1/2} B A^{-1/2})$, the largest eigenvalue of $A^{-1/2} B A^{-1/2}$.

Lemma 6.7.1 *The Thompson metric on $\mathbb{P}_n$ has the following properties:*

(i) $d(A, B) = d(A^{-1}, B^{-1}) = d(MAM^*, MBM^*)$ *for any invertible* $M \in \mathbb{M}_n$*;*
(ii) $d(A \natural B, A) = d(A \natural B, B) = \frac{1}{2} d(A, B)$*;*
(iii) $d(A \natural_\alpha B, C \natural_\alpha D) \leq (1-\alpha) d(A, C) + \alpha d(B, D)$ *for* $\alpha \in [0, 1]$.

We list some properties of the Karcher mean, which we need later.

Theorem 6.7.2 *The Karcher mean satisfies the following properties:*

(P1) Consistency with scalars: $G_K(\omega; A_1, \ldots, A_k) = A_1^{\omega_1} \cdots A_k^{\omega_n}$ *if the* A_i*'s commute;*
(P2) Joint homogeneity: $G_K(\omega; a_1 A_1, \ldots, a_k A_k) = a_1^{\omega_1} \cdots a_k^{\omega_k} G_K(\omega; A_1, \ldots, A_k)$*;*
(P3) Permutation invariance: $G_K(\omega_\sigma; A_{\sigma(1)}, \ldots, A_{\sigma(k)}) = G_K(\omega; A_1, \ldots, A_k)$ *where* $\omega_\sigma = (\omega_{\sigma(1)}, \ldots, \omega_{\sigma(k)})$ *and* σ *is any permutation;*

(P4) *Monotonicity: If* $B_i \le A_i$ *for all* $i = 1, \dots, k$, *then*

$$G_K(\omega; B_1, \dots, B_k) \le G_K(\omega; A_1, \dots, A_k);$$

(P5) $d(G_{\mathrm{K}}(\omega; A_1, \dots, A_k), G_{\mathrm{K}}(\omega; B_1, \dots, B_k)) \le \max_{1 \le i \le k} d(A_i, B_i)$;

(P6) *Congruence invariance:* $T^* G_{\mathrm{K}}(\omega; A_1, \dots, A_k) T = G_{\mathrm{K}}(\omega; T^* A_1 T, \dots, T^* A_k T)$ *for every invertible matrix* T;

(P7) *Self-duality:* $G_{\mathrm{K}}(\omega; A_1^{-1}, \dots, A_k^{-1})^{-1} = G_{\mathrm{K}}(\omega; A_1, \dots, A_k)$;

(P8) *Information monotonicity:* $\Phi(G_{\mathrm{K}}(\omega; A_1, \dots, A_k)) \le G_{\mathrm{K}}(\omega; \Phi(A_1), \dots, \Phi(A_k))$
for any unital positive linear map Φ;

(P9) *Determinant identity:* $\det(G_{\mathrm{K}}(\omega; A_1, \dots, A_k)) = \prod_{i=1}^k \det(A_i)^{\omega_i}$;

(P10) *Arithmetic–geometric–harmonic weighted mean inequality:*

$$\left(\sum_{i=1}^k \omega_i A_i^{-1}\right)^{-1} \le G_{\mathrm{K}}(\omega; A_1, \dots, A_k) \le \sum_{i=1}^k \omega_i A_i.$$

We refer the reader to [38–41] for more information on the Karcher mean.

The following theorem due to Yamazaki [41, Theorem 1] is a k-variable version of Lemma 6.3.7.

Theorem 6.7.3 *Let* $A_1, \dots, A_k$ *be positive definite matrices, and* $\omega = (\omega_1, \dots, \omega_k)$ *a weight vector. Then*

$$\omega_1 \log A_1 + \cdots + \omega_k \log A_k \le 0 \quad \textit{implies} \quad G_{\mathrm{K}}(\omega; A_1, \dots, A_k) \le I.$$

Proof If $\omega_1 \log A_1 + \cdots + \omega_k \log A_k \le 0$, then there exists a positive definite matrix $A \in \mathbb{P}_n$ such that $A \ge I$ and

$$\frac{\omega_1}{2} \log A_1 + \cdots + \frac{\omega_k}{2} \log A_k + \frac{1}{2} \log A = 0.$$

Then $\omega_0 = (\frac{\omega_1}{2}, \dots, \frac{\omega_k}{2}, \frac{1}{2})$ is a weight vector and by the definition of the Karcher mean, we have

$$G_{\mathrm{K}}(\omega_0; A_1, \dots, A_k, A) = I.$$

Define a sequence $\{G_\ell\}_{\ell=0}^\infty \subset \mathbb{P}_n$ by

$$G_{\ell+1} = G_{\mathrm{K}}(\omega_1; A_1, \dots . A_k, G_\ell) \quad \text{and} \quad G_0 = G_{\mathrm{K}}(\omega_1; A_1, \dots, A_k, I)$$

and obtain

$$I \ge G_0 \ge G_1 \ge \cdots \ge G_\ell \ge \cdots \ge 0.$$

In fact, by (P4), $A \geq I$ implies that

$$I = G_K(\omega_1; A_1, \ldots, A_k, A) \geq G_K(\omega_1; A_1, \ldots, A_k, I) = G_0$$

and $G_0 = G_K(\omega_1; A_1, \ldots, A_k, I) \geq G_K(\omega_1; A_1, \ldots, A_k, G_0) = G_1$.

Therefore, the sequence $\{G_\ell\}_{\ell=1}^{\infty}$ converges to a positive semidefinite matrix.

Let $X = G_K(\omega; A_1, \ldots, A_k)$. We shall show that G_ℓ converges to X. Note that

$$0 = \sum_{i=1}^{k} \omega_i \log X^{-1/2} A_i X^{-1/2} = \sum_{i=1}^{k} \frac{\omega_i}{2} \log X^{-1/2} A_i X^{-1/2} + \frac{1}{2} \log X^{-1/2} X X^{-1/2},$$

and hence

$$G_K(\omega; A_1, \ldots, A_k, X) = X.$$

Then, by the nonexpansive property (P5) of the Karcher mean, we have

$$\begin{aligned} d(X, G_\ell) &= d(G_K(\omega_0; A_1, \ldots, A_k, X), G_K(\omega_0; A_1, \ldots, A_k, G_{\ell-1})) \\ &\leq \frac{1}{2} d(X, G_{\ell-1}) \\ &\leq \cdots \\ &\leq \left(\frac{1}{2}\right)^{\ell} d(X, G_0) \to 0 \quad \text{as } \ell \to \infty, \end{aligned}$$

and hence $G_\ell \to X$ as $\ell \to \infty$. Since $\{G_\ell\}_{\ell=0}^{\infty}$ is a contractive and decreasing sequence, we have

$$G_K(\omega; A_1, \ldots, A_k) = X \leq I.$$

□

The following theorem due to Yamazaki [41, Theorem 3] is an extension of the Ando–Hiai inequality (Theorem 6.3.1), because $G_K((1-\alpha, \alpha); A, B) = A \natural_\alpha B$.

Theorem 6.7.4 *Let $A_1, \ldots, A_k$ be positive definite matrices and ω a weight vector. Then $G_K(\omega; A_1, \ldots, A_k) \leq I$ implies that $G_K(\omega; A_1^p, \ldots, A_k^p) \leq I$ for all $p \geq 1$.*

Proof Let $\omega = (\omega_1, \ldots, \omega_k)$ and $X = G_K(\omega; A_1, \ldots, A_k) \leq I$. Then for $p \in [1, 2]$, we have

$$\begin{aligned} 0 &= p(\omega_1 \log X^{1/2} A_1^{-1} X^{1/2} + \cdots + \omega_k \log X^{1/2} A_k^{-1} X^{1/2}) \\ &= \omega_1 \log(X^{1/2} A_1^{-1} X^{1/2})^p + \cdots + \omega_k \log(X^{1/2} A_k^{-1} X^{1/2})^p \\ &\leq \omega_1 \log(X^{1/2} A_1^{-p} X^{1/2}) + \cdots + \omega_k \log(X^{1/2} A_k^{-p} X^{1/2}), \end{aligned}$$

where the last inequality holds since $\log t$ is operator monotone, $X \leq I$, and t^p is operator convex for $p \in [1, 2]$. It is equivalent to

$$\omega_1 \log X^{-1/2} A_1^p X^{-1/2} + \cdots + \omega_k \log X^{-1/2} A_k^p X^{-1/2} \leq 0$$

and it follows from Theorem 6.7.3 that

$$G_{\mathrm{K}}(\omega; X^{-1/2} A_1^p X^{-1/2}, \ldots, X^{-1/2} A_k^p X^{-1/2}) \leq I.$$

Hence, we have

$$G_{\mathrm{K}}(\omega; A_1^p, \ldots, A_k^p) \leq X = G_{\mathrm{K}}(\omega; A_1, \ldots, A_k) \leq I$$

for $p \in [1, 2]$ by (P6). Repeat this procedure for $G_{\mathrm{K}}(\omega; A_1^p, \ldots, A_k^p) \leq I$, and complete the proof. □

By Theorem 6.7.4 and the determinant identity (P9), we arrive at the following Ando–Hiai inequality for the Karcher mean due to Hiai and Petz [42].

Theorem 6.7.5 *Let $A_1, \ldots, A_k$ be positive definite matrices and ω a weight vector. Then*

$$G_{\mathrm{K}}(\omega; A_1^p, \ldots, A_k^p) \prec_{\log} G_{\mathrm{K}}(\omega; A_1, \ldots, A_k)^p \quad \text{for all } p \geq 1$$

or equivalently

$$G_{\mathrm{K}}(\omega; A_1^p, \ldots, A_k^p)^{1/p} \prec_{\log} G_{\mathrm{K}}(\omega; A_1^q, \ldots, A_k^q)^{1/q} \quad \text{for all } 0 < q \leq p.$$

The following lemma is a Karcher mean version of the Lie–Trotter formula in Hiai and Petz [42], also see [43].

Lemma 6.7.6 *Let $A_1, \ldots, A_k$ be positive definite matrices and ω be a weight vector. Then*

$$\lim_{p \to 0} G_{\mathrm{K}}(\omega; A_1^p, \ldots, A_k^p)^{1/p} = e^{\sum_{i=1}^k \omega_i \log A_i}.$$

Proof We can assume that there exist some scalars $0 < m \leq M$ such that $m \leq A_i \leq M$ for all $i = 1, \ldots, k$. Put $h = \frac{M}{m}$. It follows that

$$0 \leq \log(\sum_{i=1}^k \omega_i A_i) - \sum_{i=1}^k \omega_i \log A_i \leq \log S(h),$$

where $S(h)$ is the Specht ratio defined by (6.5.9). Replace A_i with A_i^p for $i = 1, \ldots, k$ and $p > 0$, respectively, and write down

$$0 \leq \log(\sum_{i=1}^k \omega_i A_i^p) - \sum_{i=1}^k \omega_i \log A_i^p \leq \log S(h^p);$$

so,

$$0 \le \log(\sum_{i=1}^{k} \omega_i A_i^p)^{1/p} - \sum_{i=1}^{k} \omega_i \log A_i \le \log S(h^p)^{1/p}.$$

Since $S(h^p)^{1/p} \to 1$ as $p \to 0$ by (iv) of Lemma 6.5.5, it follows that

$$\lim_{p\to 0}(\sum_{i=1}^{k} \omega_i A_i^p)^{1/p} = e^{\sum_{i=1}^{k} \omega_i \log A_i}.$$

On the other hand, since

$$0 \le \log(\sum_{i=1}^{k} \omega_i A_i^{-1}) - \sum_{i=1}^{k} \omega_i \log A_i^{-1} \le \log S(h^{-1}),$$

it follows from (ii) of Lemma 6.5.5 that $S(h) = S(h^{-1})$, and hence

$$0 \ge \log(\sum_{i=1}^{k} \omega_i A_i^{-1})^{-1} - \sum_{i=1}^{k} \omega_i \log A_i \ge -\log S(h),$$

which yields

$$0 \ge \log(\sum_{i=1}^{k} \omega_i A_i^{-p})^{-1/p} - \sum_{i=1}^{k} \omega_i \log A_i \ge -\log S(h^p)^{1/p}$$

for $p > 0$. Hence, we have

$$\lim_{p\to 0}(\sum_{i=1}^{k} \omega_i A_i^{-p})^{-1/p} = e^{\sum_{i=1}^{k} \omega_i \log A_i}.$$

Since, by (P10) of Theorem 6.7.2,

$$\log(\sum_{i=1}^{k} \omega_i A_i^{-p})^{-1/p} \le \log(G_K(\omega; A_1^p, \dots, A_k^p)^{1/p} \le \log(\sum_{i=1}^{k} \omega_i A_i^p)^{1/p}$$

for $p > 0$, we achieve the requested result. □

By Theorem 6.7.5 and Lemma 6.7.6, we have the following theorem, which is a k-variable version of the complemented Golden–Thompson inequality (Theorem 6.3.13).

Theorem 6.7.7 *Let $A_1, \ldots, A_k$ be positive definite matrices and ω a weight vector. Then*

$$G_K(\omega; e^{pA_1}, \ldots, e^{pA_k})^{1/p} \prec_{\log} e^{\sum_{i=1}^k \omega_i A_i} \quad \text{for all } p > 0.$$

For the weight vector $\tilde{\omega} = (1/k, \ldots .1/k)$,

$$|||G_K(\tilde{\omega}; e^{kpA_1}, \ldots, e^{kpA_k})^{1/p}||| \leq |||e^{A_1+\cdots+A_k}||| \quad \text{for all } p > 0.$$

In particular,

$$\mathrm{tr}(G_K(\tilde{\omega}; e^{kA_1}, \ldots, e^{kA_k})) \leq \mathrm{tr}(e^{A_1+\cdots+A_k}).$$

If $k = 2$, then $\mathrm{tr}(e^{2A_1} \sharp e^{2A_2}) \leq \mathrm{tr}(e^{A_1+A_2})$.

We prove some reverse inequalities to the complemented Golden–Thompson ones for the Karcher means. For this, we show a matrix version of Theorem 6.5.4.

Lemma 6.7.8 *Let $A_1, \ldots, A_k$ be positive definite matrices such that $m \leq A_i \leq M$ for some scalars $0 < m < M$ and $i = 1, \ldots, k$, and ω a weight vector. Put $h = \frac{M}{m}$. Then*

$$\sum_{i=1}^k \omega_i A_i \leq S(h)\, e^{\sum_{i=1}^k \omega_i \log A_i} \tag{6.7.2}$$

where the Specht ratio $S(h)$ is defined by (6.5.9).

Proof Put $A = \mathrm{diag}(A_1, \ldots, A_k)$, $y = (\sqrt{\omega_1}\, x, \ldots, \sqrt{\omega_k}\, x)^t$ for every unit vector $x \in \mathbb{C}^n$. By Theorem 6.5.4, since $m \leq A \leq M$, we have

$$\langle Ay, y\rangle \leq S(h)\, e^{\langle \log Ay, y\rangle}.$$

Hence, it follows from the Jensen inequality that

$$\begin{aligned}
\langle (\sum_{i=1}^k \omega_i A_i)x, x\rangle &= \langle Ay, y\rangle \\
&\leq S(h)\, e^{\langle \log Ay, y\rangle} \\
&= S(h)\, e^{\langle \sum_{i=1}^k \omega_i \log A_i x, x\rangle} \\
&\leq S(h)\, \langle e^{\sum_{i=1}^k \omega_i \log A_i} x, x\rangle \qquad \text{(by Theorem 6.5.4)}
\end{aligned}$$

for every unit vector $x \in \mathbb{C}^n$ and we obtain (6.7.2):

$$\sum_{i=1}^k \omega_i A_i \leq S(h)\, e^{\sum_{i=1}^k \omega_i \log A_i}.$$

□

Although there is no order relation between $e^A \sharp_\alpha e^B$ and $e^{(1-\alpha)A+\alpha B}$, the complemented Golden–Thompson inequality asserts that

$$|||e^A \sharp_\alpha e^B||| \le |||e^{(1-\alpha)A+\alpha B}|||$$

for every unitarily invariant norm $|||\cdot|||$. Here, we present the estimate under the matrix order in terms of the Specht ratio.

Theorem 6.7.9 *Let $A_1, \ldots, A_k$ be Hermitian matrices such that $m \le A_i \le M$ for some scalars $m < M$ and $i = 1, \ldots, k$, and ω a weight vector. Then*

$$S(e^{M-m})^{-1} e^{\sum_{i=1}^k \omega_i A_i} \le G_{\mathrm{K}}(\omega; e^{A_1}, \ldots, e^{A_k}) \le S(e^{M-m})\, e^{\sum_{i=1}^k \omega_i A_i}, \tag{6.7.3}$$

where the Specht ratio $S(h)$ is defined by (6.5.9).

Proof Since $e^{A_1}, \ldots, e^{A_k}$ are positive definite matrices such that $e^m \le e^{A_i} \le e^M$ for $i = 1, \ldots, k$, it follows from Lemma 6.7.8 that

$$G_{\mathrm{K}}(\omega; e^{A_1}, \ldots, e^{A_k}) \le \sum_{i=1}^k \omega_i e^{A_i} \le S(e^{M-m})\, e^{\sum_{i=1}^k \omega_i A_i}.$$

Since $e^{-M} \le e^{-A_i} \le e^{-m}$ for $i = 1, \ldots, k$, the generalized condition number of e^{-A_i} for $i = 1, \ldots, k$ is $e^{-m}/e^{-M} = e^{M-m}$. Hence, we have

$$G_{\mathrm{K}}(\omega; e^{-A_1}, \ldots, e^{-A_k}) \le S(e^{M-m}) e^{\sum_{i=1}^k -\omega_i A_i}$$

and by taking the inverse of both sides

$$G_{\mathrm{K}}(\omega; e^{-A_1}, \ldots, e^{-A_k})^{-1} \ge S(e^{M-m})^{-1} (e^{\sum_{i=1}^k -\omega_i A_i})^{-1}.$$

Therefore, it follows from the self-duality (P7) in Theorem 6.7.2 that

$$G_{\mathrm{K}}(\omega; e^{A_1}, \ldots, e^{A_k}) \ge S(e^{M-m})^{-1} e^{\sum_{i=1}^k \omega_i A_i},$$

and so we have the desired inequality (6.7.3). □

The following theorem is a k-variable version of Theorem 6.5.15.

Theorem 6.7.10 *Let $A_1, \ldots, A_k$ be Hermitian matrices such that $m \le A_i \le M$ for some scalars $m < M$ and $i = 1, \ldots, k$, and ω a weight vector. Then*

$$\begin{aligned} \left|\left|\left| G_{\mathrm{K}}(\omega; e^{pA_1}, \ldots, e^{pA_k})^{\frac{1}{p}} \right|\right|\right| &\le \left|\left|\left| e^{\sum_{i=1}^k \omega_i A_i} \right|\right|\right| \\ &\le S(e^{p(M-m)})^{\frac{1}{p}} \left|\left|\left| G_{\mathrm{K}}(\omega; e^{pA_1}, \ldots, e^{pA_k})^{\frac{1}{p}} \right|\right|\right| \end{aligned} \tag{6.7.4}$$

for all $p > 0$ and every unitarily invariant norm $|||\cdot|||$, and the right-hand side of (6.7.4) converges to the middle hand side as $p \downarrow 0$.

Proof The first inequality follows from Theorem 6.7.7.

For $p > 0$, since $e^{pm} \le e^{pA_i} \le e^{pM}$ for $i = 1, \dots, k$, the generalized condition number of e^{pA_i} is $e^{p(M-m)}$. Thus, it follows from Theorem 6.7.7 that

$$e^{\sum_{i=1}^k \omega_i pA_i} \le S(e^{p(M-m)})\, G_K(\omega; e^{pA_1}, \dots, e^{pA_k}).$$

Then there exists a unitary matrix V such that

$$(e^{\sum_{i=1}^k \omega_i pA_i})^{\frac{1}{p}} \le S(e^{p(M-m)})^{\frac{1}{p}} V^* G_K(\omega; e^{pA_1}, \dots, e^{pA_k})^{\frac{1}{p}} V$$

for all $p > 0$, and we have the second inequality in (6.7.4).

The last assertion follows from the Lie–Trotter formula for the Karcher mean (Lemma 6.7.6) and (iv) of Lemma 6.5.5. □

Finally, we would like to comment on a reverse of the arithmetic–geometric mean inequality. The Karcher mean satisfies the arithmetic–geometric–harmonic mean inequality: For a k-tuple of positive definite matrices $A_1, \dots, A_k$ such that $m \le A_i \le M$ for some scalars $0 < m \le M$ and $i = 1, \dots, k$, and $h = \frac{M}{m}$

$$\left(\sum_{i=1}^k \omega_i A_i^{-1}\right)^{-1} \le G_K(\omega; A_1, \dots, A_k) \le \sum_{i=1}^k \omega_i A_i. \tag{6.7.5}$$

By Theorem 6.5.6, as a reverse inequality of (6.7.5), we would expect the following:

$$\sum_{i=1}^k \omega_i A_i \le S(h)\, G_K(\omega; A_1, \dots, A_k), \tag{6.7.6}$$

where $S(h)$ is the Specht ratio. Unfortunately, we do not know whether inequality (6.7.6) holds or not. By the Kantorovich inequality (6.5.2), we have

$$\sum_{i=1}^k \omega_i A_i \le \frac{(M+m)^2}{4Mm} G_K(\omega; A_1, \dots, A_k). \tag{6.7.7}$$

Since it follows from Lemma 2.1 in [40] that

$$S(h) \le \frac{(M+m)^2}{4Mm} \le S(h)^2 \le S(h^2),$$

the estimate of (6.7.7) is much rougher than that of (6.7.6). However, we do not know whether the refinement of (6.7.7) holds or not. Thus, we have two open problems:

(1) Does the Specht inequality (6.7.6) for the Karcher mean hold?
(2) If the Specht inequality (6.7.6) for the Karcher mean is not true, can (6.7.7) be refined?

Remark 6.7.11 Lawson and Lim established the notion of the Karcher mean for positive invertible operators on a Hilbert space [36]. Therefore, most of the results in Sect. 6.7 hold for operators on a Hilbert space.

6.8 Exercises and Problems

Exercise 6.8.1 Show that for Hermitian matrices A and B, $AB = BA$ if and only if $e^{A+B} = e^A e^B$.

Exercise 6.8.2 Prove (2.2.5): If A and B are positive semidefinite matrices in $\mathbb{M}_n$, show that the following implications hold

$$A \prec_{w\log} B \Longrightarrow A \prec_w B \Longrightarrow |||A||| \le |||B|||$$

for any unitarily invariant norm.

Exercise 6.8.3 Let $A \in \mathbb{M}_n$ be a positive semidefinite matrix with eigenvalues $\lambda_1(A) \ge \cdots \ge \lambda_n(A)$. Prove that the largest eigenvalue of $C_k(A)$ is the product of the first k largest eigenvalues of A; that is,

$$\lambda_1(C_k(A)) = \prod_{i=1}^{k} \lambda_i(A).$$

Exercise 6.8.4 Let A and B be positive operators on a Hilbert space. Show that

$$\left\|(A^{1/2}BA^{1/2})^r\right\| \le \left\|A^{r/2}B^r A^{r/2}\right\| \qquad \text{for all } r \ge 1$$

and

$$\left\|(A^{1/2}BA^{1/2})^r\right\| \ge \left\|A^{r/2}B^r A^{r/2}\right\| \qquad \text{for all } 0 < r \le 1.$$

Exercise 6.8.5 Let A and B be self-adjoint operators on a Hilbert space. Show that

$$\left\|e^{A/2}e^B e^{A/2}\right\| \le \left\|e^A e^B\right\|.$$

Exercise 6.8.6 Let A and B be self-adjoint operators on a Hilbert space. Show that

$$\left\|e^{A+B}\right\| \le \left\|(e^{pA/2}e^{pB}e^{pA/2})^{1/p}\right\|$$

for all $p > 0$.

Exercise 6.8.7 If A and B are positive invertible operators on a Hilbert space and $\alpha, \beta \in [0, 1]$, show that the following properties hold

(1) $A \leq A_1$ and $B \leq B_1$ imply $A \natural_\alpha B \leq A_1 \natural_\alpha B_1$.
(2) $A \natural_\alpha B = B \natural_{1-\alpha} A$.
(3) $A \natural_{\alpha\beta} B = A \natural_\alpha (A \natural_\beta B)$.

Exercise 6.8.8 Let A and B be self-adjoint operators on a Hilbert space. Show that for each $\alpha \in [0, 1]$

$$\left\|(e^{pA} \natural_\alpha e^{pB})^{1/p}\right\| \leq \left\|e^{(1-\alpha)A+\alpha B}\right\| \qquad \text{for all } p > 0.$$

Exercise 6.8.9 Let $K(m, M, \alpha)$ be the generalized Kantorovich constant for positive real numbers $0 < m < M$ and $\alpha \in \mathbb{R}\backslash\{0, 1\}$. Prove that

$$K(m, M, \alpha) = \max\left\{\frac{a(t-m)+m^\alpha}{t^\alpha} : t \in [m, M]\right\}$$

where $a = \frac{M^\alpha - m^\alpha}{M-m}$. Moreover, show that

$$\lim_{\alpha \to 1} K(m, M, \alpha) = \lim_{\alpha \to 0} K(m, M, \alpha) = 1.$$

Exercise 6.8.10 Let $h = \frac{M}{m}$ for positive real numbers $0 < m < M$, and $S(h)$ be the Specht ratio. If we put $\beta = 1 - \log\left(\frac{M-m}{\log M - \log m}\right) + \frac{M \log m - m \log M}{M-m}$, show that

$$e^{-\beta} = S(h).$$

Exercise 6.8.11 Prove (P10) of Theorem 6.7.2: The arithmetic–geometric–harmonic mean inequality for the Karcher mean

$$\left(\sum_{i=1}^{k} \omega_i A_i^{-1}\right)^{-1} \leq G_{\mathrm{K}}(\omega; A_1, \ldots, A_k) \leq \sum_{i=1}^{k} \omega_i A_i$$

holds.

Exercise 6.8.12 Let $A_1, \ldots, A_k$ be positive definite matrices and ω a weight vector. Show that

$$G_{\mathrm{K}}(\omega; A_1^p, \ldots, A_k^p) \leq \|G_{\mathrm{K}}(\omega; A_1, \ldots, A_k)\|^{p-1} G_{\mathrm{K}}(\omega; A_1, \ldots, A_k)$$

for all $p \geq 1$.

Exercise 6.8.13 Let $A_1, \ldots, A_k$ be positive definite matrices such that $m \leq A_i \leq M$ for some scalars $0 < m < M$, and ω a weight vector. Show that

$$\sum_{i=1}^{k} \omega_i A_i \leq \frac{(M+m)^2}{4Mm} G_{\mathrm{K}}(\omega; A_1, \ldots, A_k).$$

6.9 Notes, Hints, and References

Here, we present some hints and references for the exercises and problems.

Apply the power series expansion for the exponential function $e^{A+B} = I + (A + B) + \frac{1}{2}(A+B)^2 + \cdots$ to solve Exercise 6.8.1. For Exercise 6.8.2, refer to [6, Theorem IV.2.2]. Exercise 6.8.3. can be found in [9, Compound matrices]. The inequalities in Exercise 6.8.4 are known as the *Araki–Cordes inequalities*. To prove the first one, it is sufficient to verify that $A^{r/2} B^r A^{r/2} \leq I$ implies $(A^{1/2} B A^{1/2})^r \leq I$ for all $r \geq 1$, and then use Theorem 1.4.3.

To prove the inequality in Exercise 6.8.5, utilize $r(X) \leq \|X\|$ and $r(Y) = \|Y\|$ for normal Y. The Lie–Trotter formula (Theorem 6.1.2) and Remark 6.2.5 can be used to solve Exercise 6.8.6. For (2) of Exercise 6.8.7, first show that for $p \in \mathbb{R}$, $(X^* A^2 X)^p = X^* A (A X X^* A)^{p-1} A X$ for $A > 0$ and invertible X. To solve Exercise 6.8.8, employ Remark 6.3.3 to show $\left\|(e^{pA} \sharp_\alpha e^{pB})^{1/p}\right\| \leq \left\|(e^{qA} \sharp_\alpha e^{qB})^{1/q}\right\|$ for $0 < q < p$.

We refer the reader to [44, Theorem 2.54] for the last part of Exercise 6.8.9. A solution to Exercise 6.8.10 requires the computation of $e^{-\beta}$. Since the Karcher mean is the solution of (6.7.1), use the inequality $\log x \leq x - 1$ in Exercise 6.8.11. To solve Exercise 6.8.12, combine the proofs of Theorem 6.7.4 and Theorem 6.3.8. Finally, for Exercise 6.8.13, first show the inequality $\sum_{i=1}^{k} \omega_i A_i \leq \frac{(M+m)^2}{4Mm} \left(\sum_{i=1}^{n} \omega_i A_i^{-1}\right)^{-1}$, and then utilize (P10) of Theorem 6.7.2.

References

1. W. So, Equality cases in matrix exponential inequality. SIAM J. Matrix Anal. Appl. **13**, 1154–1158 (1992)
2. F. Hiai, Equality cases in matrix norm inequalities of Golden–Thompson type. Linear Multilinear Algebra **36**, 239–249 (1994)
3. S. Golden, Lower bounds for the Helmholtz function. Phys. Rev. (2) **137**, 1127–1128 (1965)
4. K. Symanzik, Proof and refinements of an inequality of Feynman. J. Math. Phys. **6**, 1155–1156 (1965)
5. C.J. Thompson, Inequality with applications in statistical mechanics. J. Math. Phys. **6**, 1812–1813 (1965)
6. R. Bhatia, *Matrix Analysis* (Springer, New York, 1997)
7. F. Hiai, D. Petz, *Introdduction to Matrix Analysis and Applications*. (Springer, 2014)
8. A.W. Marshall, I. Olkin, *Inequalities: Theory of majorization and its applications, Mathematics in Science and Engineering*, Vol. 143. (Academic, 1979)

9. F. Zhang, *Matrix Theory, Basic Results and Techniques*, Universitext, 2nd edn. (Springer, New York, 2011)
10. H. Araki, On an inequality of Lieb and Thirring. Lett. Math. Phys. **19**, 167–170 (1990)
11. T. Ando, F. Hiai, Log-majorization and complementary Golden-Thompson type inequalities. Linear Algebra Appl. **197**(198), 113–131 (1994)
12. M. Fujii, T. Furuta, E. Kamei, Furuta's inequality and its application to Ando's theorem. Linear Algebra Appl. **179**, 161–169 (1993)
13. M. Fujii, J. Mićić Hot, J. Pečarić, Y. Seo, *Recent Developments of Mond–Pečarić method in operator inequalities*, Monographs in Inequalities 4, Element, Zagreb (2012)
14. M. Uchiyama, Some exponential operator inequalities. Math. Inequal. Appl. **2**, 469–471 (1999)
15. F. Hiai, D. Petz, The Golden-Thompson trace inequality is complemented. Linear Algebra Appl. **181**, 153–185 (1993)
16. R. Bhatia, P. Grover, Norm inequalities related to the matrix geometric mean. Linear Algebra Appl. **437**, 726–733 (2012)
17. F. Kubo, T. Ando, Means of positive linear operators. Math. Ann. **246**, 205–224 (1980)
18. J.I. Fujii, Y. Seo, Tsallis relative operator entropy with negative parameters. Adv. Oper. Theory **1**, 219–236 (2016)
19. M. Kian, Y. Seo, Norm inequalities related to the matrix geometric mean of negative power. Sci. Math. Jpn. **82**, 191–200 (2019)
20. F. Hiai, Matrix analysis: matrix monotone functions, matrix means, and majorization. Inter. Inf. Sci. **16**, 39–248 (2010)
21. J.I. Fujii, An external version of the Jensen operator inequality. Sci. Math. Jpn. **73**, 125–128 (2011)
22. A.M. Turing, Rounding off-errors in matrix processes. Quart. J. Mech. Appl. Math. **1**, 287–308 (1948)
23. L.V. Kantorovich, *Functional analysis and applied mathematics*. Uspehi Mat. Nauk. **3**, 89-185 (1948). Translated from the Russian, National Bureau of Standards, Report 1509, March 7 (1952)
24. W. Greub, W. Rheinboldt, On a generalization of an inequality of L.V. Kantorovich. Proc. Am. Math. Soc. **10**, 407–415 (1959)
25. T. Furuta, Operator inequalities associated with Hölder-McCarthy and Kantorovich inequalities. J. Inequal. Appl. **2**, 137–148 (1998)
26. J.I. Fujii, Y. Seo, T. Yamazaki, Norm inequalities for matrix geometric means of positive definite matrices. Linear Multilinear Algebra **64**, 512–526 (2016)
27. W. Specht, Zur Theorie der elementaren Mittel. Math. Z. **74**, 91–98 (1960)
28. J.I. Fujii, Y. Seo, Determinant for positive operators. Sci. Math. **1**, 153–156 (1998)
29. M. Tominaga, Specht's ratio in the Young inequality. Sci. Math. Jpn. **55**, 585–588 (2002)
30. J.-C. Bourin, Reverse inequality to Araki's inequality comparison of $A^p Z^p A^p$ and $(AZA)^p$. Math. Inequal. App. **8**, 373–378 (2005)
31. J.-C. Bourin, Y. Seo, Reverse inequality to Golden-Thompson type inequalities: comparison of e^{A+B} and $e^A e^B$. Linear Algebra Appl. **426**, 312–316 (2007)
32. Y. Seo, Reverses of the Golden-Thompson type inequalities due to Ando-Hiai-Petz. Banach J. Math. Anal. **2**, 140–149 (2008)
33. A. Oppenheim, Inequalities connected with definite Hermitian forms. J. London Math. Soc. **5**, 114–119 (1930)
34. T. Ando, Hadamard products and Golden-Thompson inequalities. Linear Algebra Appl. **241**(243), 105–112 (1996)
35. Y. Seo, Specht's ratio and complementary inequalities to Golden-Thompson type inequalities on the Hadamard product. Linear Algebra Appl. **320**, 15–22 (2000)
36. J. Lawson, Y. Lim, Karcher means and Karcher equations of positive definite operators. Trans. Am. Math. Soc. Ser. B **1**, 1–22 (2014)
37. Y. Lim, M. Pálfia, Matrix power means and the Karcher mean. J. Funct. Anal. **262**, 1498–1514 (2012)

38. R. Bhatia, R.I. Karandikar, Monotonicity of the matrix geometric mean. Math. Ann. **353**, 1453–1467 (2012)
39. D.A. Bini, B. Meini, F. Poloni, An effective matrix geometric mean satisfying the Ando–Li–Mathias properties. Math. Comp. **79**, 437–452 (2010)
40. J.I. Fujii, Y. Seo, On the Ando–Li–Mathias mean and the Karcher mean of positive definite matrices. Linear Multilinear Algebra **63**, 636–649 (2015)
41. T. Yamazaki, Riemannian mean and matrix inequalities related to the Ando–Hiai inequality and chaotic order. Oper. Matrices **6**, 577–588 (2012)
42. F. Hiai, D. Petz, Riemannian metrics on positive definite matrices related to means II. Linear Algebra Appl. **436**, 2117–2136 (2012)
43. J.I. Fujii, M. Fujii, Y. Seo, The Golden–Thompson–Segal type inequalities related to the weighted geometric mean due to Lawson–Lim. J. Math. Inequal. **3**, 511–518 (2009)
44. T. Furuta, J. Mićić Hot, J. Pečarić, Y. Seo, *Mond–Pečarić Method in Operator Inequalities*, Monographs in Inequalities 1, Element, Zagreb (2005)

Chapter 7
Quantum Relative Entropy

In this chapter, we demonstrate matrix trace inequalities on quantum relative entropies. Some of the results presented here also apply to the infinite-dimensional case.

Firstly, we examine the fundamental properties of the Umegaki relative entropy and the FK (Fujii–Kamei) relative entropy. These serve as quantum generalizations of the classical Kullback–Leibler divergence between probability distributions. Next, we introduce two quantum Tsallis relative entropies. These entropies extend the Umegaki relative entropy and the FK relative entropy by introducing a single parameter. We then provide estimates for the relationships between these two quantum Tsallis relative entropies.

Finally, we explore the von Neumann entropy and the Umegaki relative entropy, which are referred to as the "entanglement entropy" and "entanglement measure", respectively. These measures have recently gained attention and have proved to be significant in physics.

7.1 Quantum Entropy and Quantum Relative Entropy

In 1932, von Neumann [1] showed that quantum mechanics can be described within the framework of Hilbert spaces. In quantum mechanics, density matrices play an essential role in the finite-dimensional Hilbert space $\mathscr{H} = \mathbb{C}^n$. A Hermitian matrix $D \in \mathbb{M}_n$ is called a *density matrix* if it is positive semidefinite and the sum of its eigenvalues is equal to 1. Utilizing the rank-one operators presented in (1.1.5), a density matrix D can be expressed in the following spectral decomposition:

$$D = \sum_{k=1}^{n} \lambda_k (e_k \otimes e_k^*)$$

A. M. Bikchentaev et al., *Trace Inequalities*, Forum for Interdisciplinary Mathematics,
https://doi.org/10.1007/978-981-97-6520-1_7

for an orthonormal basis (e_k), and $\lambda_k \geq 0$ for $k = 1, \ldots, n$ with $\sum_{k=1}^n \lambda_k = 1$. When the physical system is in a state described by the density D, the expected value of a (Hermitian) observable $A \in \mathbb{M}_n$ is

$$\operatorname{tr}(AD) = \operatorname{tr}\left(\sum_{k=1}^n \lambda_k (Ae_k \otimes e_k^*)\right) = \sum_{j=1}^n \left\langle \sum_{k=1}^n \lambda_k (Ae_k \otimes e_k^*) e_j, e_j \right\rangle = \sum_{j=1}^n \lambda_j \left\langle Ae_j, e_j \right\rangle.$$

Thus, von Neumann defined a quantity for the density matrix D that expresses its uncertainty or entropy, as

$$S(D) = -\operatorname{tr}(D \log D) = \operatorname{tr}(\eta(D)) = \sum_{i=1}^n \langle \eta(D)\, e_i, e_i \rangle = -\sum_{i=1}^n \lambda_i \log \lambda_i, \quad (7.1)$$

which is called the *von Neumann entropy* of D, where the entropy function η is defined by $\eta(x) = -x \log x$ $(x > 0)$ and $\eta(0) = -0 \cdot \log 0 = 0$.

In 2006, Ryu and Takayanagi [2] pointed out that the Hawking–Bekenstein entropy and von Neumann entropy are connected through the AdS/CFT correspondence (Anti de Sitter/conformal field theory correspondence). This connection led to the term "entanglement entropy", which is used to describe it, as it represents the level of quantum entanglement. For further information, refer to [3].

The fundamental properties of the von Neumann entropy are listed below.

Theorem 7.1.1 *Let D, D_1, and D_2 be density matrices in $\mathbb{M}_n$. Then the von Neumann entropy has the following properties:*

(i) **Positivity**: $0 \leq S(D) \leq \log n$;
(ii) **Unitary invariance**: $S(UDU^*) = S(D)$ *for any unitary* U;
(iii) **Concavity**: *If* $D = (1-\lambda)D_1 + \lambda D_2$ *for some* $\lambda \in [0, 1]$, *then*

$$S(D) \geq (1-\lambda)S(D_1) + \lambda S(D_2);$$

(iv) **Additivity**: $S(D_1 \otimes D_2) = S(D_1) + S(D_2)$.

Proof Let $D = \sum_{i=1}^n \lambda_i (e_i \otimes e_i^*)$ be the spectral decomposition of D, where (e_i) is an orthonormal basis and $\lambda_i \geq 0$ for $i = 1, \ldots, n$ with $\sum_{i=1}^n \lambda_i = 1$ (see Sect. 1.3).

For (i), since $\lambda_i \geq 0$ for $i = 1, \ldots, n$ with $\sum_{i=1}^n \lambda_i = 1$, we have $-\lambda_i \log \lambda_i \geq 0$ for $i = 1, \ldots, n$, and hence $S(D) \geq 0$. It follows from the concavity of $\log x$ that

$$S(D) - \log n = \sum_{i=1}^n \lambda_i \log \frac{1}{\lambda_i n} \leq \log \left(\sum_{i=1}^n \frac{1}{n}\right) = \log 1 = 0,$$

where $\lambda_i > 0$ for i with $\sum_{i=1}^n \lambda_i = 1$.

For (ii), for any unitary U, it follows that

$$S(UDU^*) = -\operatorname{tr}(UDU^* \log(UDU^*)) = -\operatorname{tr}(UDU^* U (\log D) U^*)$$

$$= -\operatorname{tr}(D\log D) = \operatorname{tr}(\eta(D)) = S(D).$$

For (iii), the matrix concavity of $\eta(x)$ implies concavity of $S(D)$. However, we present an elementary proof for the convenience of readers: Since $\lambda_i = \langle De_i, e_i\rangle$ for $i = 1, \dots, n$ and the entropy function $\eta(x)$ is concave, it follows that

$$\begin{aligned} S(D) &= \sum_{i=1}^{n} \eta(\langle De_i, e_i\rangle) = \sum_{i=1}^{n} \eta((1-\lambda)\langle D_1 e_i, e_i\rangle + \lambda\langle D_2 e_i, e_i\rangle) \\ &\geq \sum_{i=1}^{n} ((1-\lambda)\eta(\langle D_1 e_i, e_i\rangle) + \lambda\eta(\langle D_2 e_i, e_i\rangle)) \geq (1-\lambda)S(D_1) + \lambda S(D_2). \end{aligned}$$

The last inequality follows from the concavity of $\eta(t)$: For the spectral decomposition $D_1 = \sum_{j=1}^{n} \mu_j (f_j \otimes f_j^*)$, we have

$$\begin{aligned} S(D_1) &= \sum_{i=1}^{n} \langle \eta(D_1)e_i, e_i\rangle = \sum_{i,j=1}^{n} |\langle e_i, f_j\rangle|^2 \eta(\mu_j) \\ &\leq \sum_{i=1}^{n} \eta\left(\sum_{j=1}^{n} |\langle e_i, f_j\rangle|^2 \mu_j\right) = \sum_{i=1}^{n} \eta(\langle D_1 e_i, e_i\rangle) \end{aligned}$$

and similarly, $S(D_2) \leq \sum_{i=1}^{n} \eta(\langle D_2 e_i, e_i\rangle)$.

For (iv), it follows that

$$\begin{aligned} S(D_1 \otimes D_2) &= -\operatorname{tr}((D_1 \otimes D_2)\log(D_1 \otimes D_2)) \\ &= -\operatorname{tr}((D_1 \otimes D_2)\log(D_1 \otimes I)(I \otimes D_2)) \\ &= -\operatorname{tr}((D_1 \otimes D_2)(\log(D_1 \otimes I) + \log(I \otimes D_2))) \\ &= -\operatorname{tr}((D_1 \otimes D_2)(\log D_1 \otimes I + I \otimes \log D_2)) \\ &= -\operatorname{tr}(D_1 \log D_1 \otimes D_2 + D_1 \otimes D_2 \log D_2) \\ &= S(D_1) + S(D_2). \end{aligned}$$

□

Nakamura and Umegaki [4] extended the concept of von Neumann entropy by showing that the entropy function $\eta(t)$ is a matrix concave function on $[0, \infty)$ and the operator entropy is given by $\eta(D) = -D\log D$ for a positive semidefinite matrix D.

Shannon [5] defined the so-called Shannon entropy in information theory. Let X be a random variable that takes values in the discrete alphabet χ and has probability distribution $P_X(x) = Pr\{X = x\}$ $(x \in \chi)$. In this case, the entropy of the discrete random variable X is defined as

$$H(X) = H(P_X) = E_{P_X}\{-\log P_X(X)\} = -\sum_{x\in\chi} P_X(x)\log P_X(x).$$

In simple terms, entropy measures the randomness and uncertainty of a random variable, and it represents the average amount of information that can be obtained by observing the random variable.

Five years after Shannon's discovery of information entropy, Kullback and Leibler [6] introduced a new information quantity. Suppose that two random variables X and Y on a discrete alphabet χ have the probability distribution functions $P(x)$ and $Q(x)$, respectively. The divergence between the two random variables X and Y, or the Kullback–Leibler divergence, is a quantity that expresses the degree of separation between the two probability distributions:

$$D(X||Y) = D(P||Q) = E_P \log\frac{P(X)}{Q(X)} = \sum_{x\in\chi} P(x)\log\frac{P(x)}{Q(x)}.$$

Although the Kullback–Leibler divergence is not symmetric with respect to X and Y, it effectively captures the differences between two distributions, leading to a geometric "distance" structure in the space of distributions. In particular, we know that $D(P||Q)$ is positive except when the distributions P and Q coincide.

For any pair of density matrices A and B, the quantum relative entropy of A with respect to B is defined as follows

$$S_U(A||B) := \begin{cases} \mathrm{tr}(A(\log A - \log B)) & \text{if } \mathrm{supp}A \le \mathrm{supp}B \\ +\infty & \text{otherwise,} \end{cases} \tag{7.2}$$

where $\mathrm{supp}A$ is the support projection of A, which is the orthogonal projection onto $\overline{\mathrm{ran}(A)}$. This definition was first introduced by Umegaki as a quantum generalization of the classical Kullback–Leibler divergence between probability distributions. It is referred to as *Umegaki relative entropy*.

In recent years, there has been active discussion of entanglement entropy in physics. This involves considering an "entanglement measure" with desirable properties concerning relative entropy. The Umegaki relative entropy plays a key role in inducing important properties of entanglement entropy. For further information, see [2, 3]. The Umegaki relative entropy has the following fundamental properties.

Theorem 7.1.2 *Let A, A_1, A_2 and B, B_1, B_2 be density matrices in $\mathbb{M}_n$. Then the Umegaki relative entropy satisfies the following properties:*

(i) **Non-negativity**: *$S_U(A||B) \ge 0$, and $S_U(A||B) = 0$ if and only if $A = B$;*
(ii) **Additivity**: *$S_U(A_1 \otimes A_2||B_1 \otimes B_2) = S_U(A_1||B_1) + S_U(A_2||B_2)$;*
(iii) **Joint convexity**: *For any $\lambda \in [0, 1]$,*

$$S_U((1-\lambda)A_1 + \lambda A_2||(1-\lambda)B_1 + \lambda B_2) \le (1-\lambda)S_U(A_1||B_1) + \lambda S_U(A_2||B_2);$$

(iv) **Monotonicity**: *For any trace-preserving completely positive linear map* Φ, *it holds that*

$$S_U(\Phi(A)||\Phi(B)) \leq S_U(A||B).$$

Proof If we let $\alpha \to 0$ in Theorem 7.1.4, then we arrive at Theorem 7.1.2. □

Remark 7.1.3 The concept of generalized channels in information theory was first introduced by Echigo and Nakamura [7] as a positive linear map on von Neumann algebras. Subsequently, the completely positive linear maps were formulated in the quantum information theory.

From the need for a non-additive (nonextensive) extension of Bolzmann–Gibbs statistical mechanics, Tsallis [8] considered an extended entropy. This extended entropy, denoted as $S_\alpha(X)$, is defined as

$$S_\alpha(X) = -\sum_x P_X(x)^{1-\alpha} \ln_\alpha P_X(x)$$

with one-parameter α as an extension of Shannon entropy, where α-logarithm is defined by $\ln_\alpha(x) = (x^\alpha - 1)/\alpha$ for any nonnegative real numbers α and x, and $P_X(x) = P_r\{X = x\}$ represents the probability distribution of the given random variable X. In fact, if $\alpha \to 0$, the Tsallis entropy converges to the Shannon entropy $-\sum_x P_X(x) \log P_X(x)$, since α-logarithm uniformly converges to the natural logarithm as $\alpha \to 0$. The Tsallis entropy plays an essential role in nonextensive statistics.

In 2003, Abe [9, 10] defined the quantum Tsallis relative entropy of A with respect to B as follows: For two density matrices A and B and for $\alpha \in [-1, 1]\backslash\{0\}$,

$$\mathrm{D}_\alpha(A||B) \equiv \begin{cases} \mathrm{tr}(A^{1-\alpha}(\ln_\alpha A - \ln_\alpha B)) = \frac{1}{\alpha}(1 - \mathrm{tr}(A^{1-\alpha}B^\alpha)) & \text{if } \mathrm{supp}A \leq \mathrm{supp}B \\ +\infty & \text{otherwise,} \end{cases}$$

where $0^{1-\alpha} \cdot 0^\alpha = 0^1 = 0$ for $\alpha < 0$.

Moreover, the Tsallis entropy of a density matrix A is defined by

$$S_\alpha(A) = -\mathrm{D}_\alpha(A|I) = -\mathrm{tr}(A^{1-\alpha} \ln_\alpha A).$$

If $\alpha \to 0$, then the Tsallis relative entropy and the Tsallis entropy converge to the Umegaki relative entropy $S_U(A||B)$ and the von Neumann entropy $S(A)$, respectively. Therefore, they can be seen as one-parameter extensions of the Umegaki relative entropy and the von Neumann entropy.

Furthermore, this entropy satisfies the following properties.

Theorem 7.1.4 *Let* A, A_1, A_2 *and* B, B_1, B_2 *be density matrices in* $\mathbb{M}_n$. *Then the Tsallis relative entropy satisfies the following properties:*

(i) **Non-negativity**: *For any* $\alpha \in [-1, 1]\backslash\{0\}$, $D_\alpha(A||B) \geq 0$, *and for* $0 < \alpha < 1$, $D_\alpha(A||B) = 0$ *if and only if* $A = B$;

(ii) **Pseudoadditivity**: *For any* $\alpha \in [-1, 1]\backslash\{0\}$,

$$D_\alpha(A_1 \otimes A_2||B_1 \otimes B_2) = D_\alpha(A_1||B_1) + D_\alpha(A_2||B_2) - \alpha D_\alpha(A_1||B_1)D_\alpha(A_2||B_2);$$

(iii) **Joint convexity**: *For any* $\alpha \in [0, 1]$,

$$D_\alpha((1-\lambda)A_1 + \lambda A_2||(1-\lambda)B_1 + \lambda B_2) \leq (1-\lambda)D_\alpha(A_1||B_1) + \lambda D_\alpha(A_2||B_2)$$

for all $\lambda \in [0, 1]$;

(iv) **Monotonicity**: *For any trace-preserving completely positive linear map* Φ *and any* $\alpha \in [0, 1]$,

$$D_\alpha(\Phi(A)||\Phi(B)) \leq D_\alpha(A||B).$$

Proof By restricting to supp$B$, we can assume that B is positive definite.

For (i), let $A = \sum_{i=1}^n \lambda_i(e_i \otimes e_i^*)$ and $B = \sum_{j=1}^n \mu_j(f_j \otimes f_j^*)$ be the spectral decompositions of A and B, respectively. For $-1 < \alpha < 1, \alpha \neq 0$, it follows from Theorem 2.1.8 that

$$\begin{aligned} D_\alpha(A||B) &= \frac{1}{\alpha}\left(1 - \operatorname{tr}(A^{1-\alpha}B^\alpha)\right) = \frac{1}{\alpha}\left(1 - \sum_{i,j=1}^n |\langle f_j, e_i\rangle|^2 \lambda_i^{1-\alpha}\mu_j^\alpha\right) \\ &\geq \frac{1}{\alpha}\left(1 - \sum_{i,j=1}^n |\langle f_j, e_i\rangle|^2\left((1-\alpha)\lambda_i + \alpha\mu_j\right)\right) \\ &= \frac{1}{\alpha}(1 - (1-\alpha)\operatorname{tr} A - \alpha \operatorname{tr} B) = 0. \end{aligned}$$

Suppose that $D_\alpha(A||B) = 0$ for $0 < \alpha < 1$. Therefore,

$$\sum_{i,j=1}^n |\langle f_j, e_i\rangle|^2 \lambda_i^{1-\alpha}\mu_j^\alpha = \sum_{i,j=1}^n |\langle f_j, e_i\rangle|^2\left((1-\alpha)\lambda_i + \alpha\mu_j\right).$$

If $\lambda_i \neq \mu_j$ for $i \neq j$, then we have $\lambda_i^{1-\alpha}\mu_j^\alpha < (1-\alpha)\lambda_i + \alpha\mu_j$ and so $\langle f_j, e_i\rangle = 0$. If $\lambda_i \neq \mu_i$, then we have $\langle f_i, e_i\rangle = 0$ and this implies $\langle f_i, e_k\rangle = 0$ for all $k = 1, \ldots, n$, which is a contradiction. Hence, we have $\lambda_i = \mu_i$ for all $i = 1, \ldots, n$. Therefore,

$$Ae_i = \lambda_i e_i = \mu_i e_i = \mu_i \sum_{j=1}^n \langle f_j, e_i\rangle f_j$$

$$
\begin{aligned}
&= \mu_i \langle f_i, e_i \rangle f_i = \sum_{j=1}^{n} \langle f_j, e_i \rangle \mu_j f_j \\
&= \sum_{j=1}^{n} \langle f_j, e_i \rangle B f_j = B \left(\sum_{j=1}^{n} \langle f_j, e_i \rangle f_j \right) = B e_i
\end{aligned}
$$

for $i = 1, \ldots, n$ and so $A = B$. If $A = B$, then we easily have $D_\alpha(A||B) = 0$.

For (ii), it follows that

$$
\begin{aligned}
D_\alpha(A_1 \otimes A_2 || B_1 \otimes B_2) &= \frac{1}{\alpha}(1 - \mathrm{tr}(A_1^{1-\alpha} \otimes A_2^{1-\alpha} \cdot B_1^\alpha \otimes B_2^\alpha)) \\
&= \frac{1}{\alpha}(1 - \mathrm{tr}(A_1^{1-\alpha} B_1^\alpha \otimes A_2^{1-\alpha} B_2^\alpha)) \\
&= \frac{1}{\alpha}(1 - \mathrm{tr}(A_1^{1-\alpha} B_1^\alpha)\,\mathrm{tr}(A_2^{1-\alpha} B_2^\alpha)) \\
&= D_\alpha(A_1||B_1) + D_\alpha(A_2||B_2) - \alpha D_\alpha(A_1||B_1) D_\alpha(A_2||B_2).
\end{aligned}
$$

For (iii), since $(A, B) \mapsto \mathrm{tr}(A^{1-\alpha} B^\alpha)$ is jointly concave by the Lieb concavity theorem [11], it follows that $D_\alpha(A||B)$ is jointly convex for $\alpha \in [0, 1]$.

For (iv), since Φ is a completely positive linear map, it follows from [12] that $\Phi(A) = \sum_{i=1}^{m} V_i A V_i^*$, where $V_i \in \mathbb{M}_n$ $(i = 1, \ldots, m)$ and $\sum_{i=1}^{m} V_i^* V_i = I$, since Φ is trace-preserving. Let (e_i) be an orthonormal basis for $\mathbb{C}^m$ and let x be a unit vector in $\mathbb{C}^m$. Put

$$
W = \sum_{i=1}^{m} V_i \otimes (e_i \otimes x^*).
$$

Then, we have $W^* W = I \otimes P_x$, where $P_x = x \otimes x^*$ is a projection. There exists a unitary matrix U such that the polar decomposition $W = U|W| = U(I \otimes P_x)$ holds. This implies

$$
U(A \otimes P_x)U^* = W(A \otimes I)W^* = \sum_{i=1}^{m} V_i A V_i^* \otimes (e_i \otimes e_i^*).
$$

Since there exists a family of unitary matrices U_j $(j = 1, \ldots, m)$ such that

$$
\sum_{j=1}^{m} U_j (e_i \otimes e_i^*) U_j^* = I \qquad \text{for } i = 1, \ldots, m,
$$

we have

$$\begin{aligned}
&\sum_{i=1}^{m} \frac{1}{m}(I \otimes U_j)(U(A \otimes P_x)U^*)(I \otimes U_j)^* \\
&= \sum_{j=1}^{m} \frac{1}{m}(I \otimes U_j)\left(\sum_{i=1}^{m} V_i A V_i^* \otimes (e_i \otimes e_i^*)\right)(I \otimes U_j)^* \\
&= \sum_{j=1}^{m} \frac{1}{m} \sum_{i=1}^{m} \left(V_i A V_i^* \otimes U_j(e_i \otimes e_i^*)U_j^*\right) \\
&= \sum_{i=1}^{m} \left(V_i A V_i^* \otimes \frac{1}{m} \sum_{j=1}^{m} U_j(e_i \otimes e_i^*)U_j^*\right) \\
&= \sum_{j=1}^{m} \left(V_i A V_i^*) \otimes \frac{1}{m} I\right) \\
&= \left(\sum_{i=1}^{m} V_i A V_i^*\right) \otimes \frac{1}{m} I \\
&= \Phi(A) \otimes \frac{1}{m} I.
\end{aligned}$$

Notice that from (ii), we have $D_\alpha(A \otimes Q||B \otimes Q) = D_\alpha(A||B)$ for any density matrices A, B, Q. Hence, it follows from the joint convexity of D_α that

$$\begin{aligned}
D_\alpha(\Phi(A)||\Phi(B)) &= D_\alpha(\Phi(A) \otimes \frac{1}{m} I||\Phi(B) \otimes \frac{1}{m} I) \\
&\le \sum_{j=1}^{m} p_j D_\alpha((I \otimes U_j)(U(A \otimes P_x)U^*)(I \otimes U_j)^*) \\
&= D_\alpha(U(A \otimes P_x)U^*||U(B \otimes P_x)U^*) \\
&= D_\alpha(A \otimes P_x||B \otimes P_x) \\
&= D_\alpha(A||B),
\end{aligned}$$

as desired. □

The following theorem is due to Ruskai and Stillinger [13], which establishes a relation between the quantum Tsallis relative entropy and the Umegaki relative entropy.

Theorem 7.1.5 *Let A and B be density matrices. Then*

$$\mathrm{D}_\alpha(A||B) \le S_U(A||B) \le \mathrm{D}_{-\alpha}(A||B) \tag{7.3}$$

for all $0 < \alpha \le 1$.

Proof By restricting to supp$B$, we can assume that B is positive definite. Let $A = \sum_{i=1}^{n} \lambda_i (e_i \otimes e_i^*)$ and $B = \sum_{j=1}^{n} \mu_j (f_j \otimes f_j^*)$ be the spectral decompositions of A and B, respectively. Then

$$\begin{aligned} S_U(A||B) &= \mathrm{tr}(A \log A - A \log B) \\ &= \sum_{j=1}^{n} \left\langle \left(\left(\sum_{i=1}^{n} \lambda_i \log \lambda_i (e_i \otimes e_i^*) - \left(\sum_{i=1}^{n} \lambda_i (e_i \otimes e_i^*) (\sum_{k=1}^{n} \log \mu_k (f_k \otimes f_k^* \right) \right) \right) f_j, f_j \right\rangle \\ &= \sum_{i,j=1}^{n} \langle f_j, e_i \rangle \langle (\lambda_i \log \lambda_i - \lambda_i \log \mu_j) e_i, f_j \rangle \\ &= \sum_{i,j}^{n} |\langle e_i, f_j \rangle|^2 (\lambda_i \log \lambda_i - \lambda_i \log \mu_j) \end{aligned}$$

and

$$D_\alpha(A||B) = \sum_{i,j=1}^{n} |\langle e_i, f_j \rangle|^2 \frac{\lambda_i \left(1 - \left(\frac{\lambda_i}{\mu_j}\right)^{-\alpha}\right)}{\alpha}.$$

Since $1 - x^{-1} \leq \log x$ for $x > 0$, we have

$$\frac{1 - x^{-\alpha}}{\alpha} \leq \log x$$

for $0 < \alpha \leq 1$, and hence

$$\frac{\lambda_i \left(1 - \left(\frac{\lambda_i}{\mu_j}\right)^{-\alpha}\right)}{\alpha} \leq \lambda_i \log \lambda_i - \lambda_i \log \mu_j$$

for all $i, j = 1, \ldots, n$. Therefore, we have $\mathrm{D}_\alpha(A||B) \leq S_U(A||B)$. Similarly, we have the second inequality in (7.3). □

Theorem 7.1.6 *The Tsallis relative entropy $D_\alpha(A||B)$ is monotonically nonincreasing on $\alpha \in [-1, 1] \backslash \{0\}$.*

Proof It follows from the fact that $F(x) = \frac{1-a^x}{x}$, for $a > 0$, is monotonically nonincreasing on $[-1, 1]$. □

7.2 Another Quantum Tsallis Relative Entropies

In this section, we introduce another formulation of the quantum Tsallis relative entropy of order $\alpha \in [-1, 1] \backslash \{0\}$. To do this, we first need to establish some preliminaries. Fujii and Kamei [14] introduced the relative operator entropy which is a relative version of the operator entropy defined by Nakamura and Umegaki [4]: For positive semidefinite matrices A and B, the relative operator entropy is defined by

$$S(A|B) = B^{1/2}\eta(B^{-1/2}AB^{-1/2})B^{1/2},$$

if B is invertible, where η is the entropy function $\eta(t) = -t\log t$. In addition, if A is invertible, then

$$S(A|B) = A^{1/2}(\log A^{-1/2}BA^{-1/2})A^{1/2}.$$

In fact, since $X\eta(X^*X) = \eta(XX^*)X$, applying it for $X = A^{1/2}B^{-1/2}$, we have

$$\begin{aligned} S(A|B) &= -A^{1/2}(\log A^{1/2}B^{-1}A^{1/2})A^{1/2} \\ &= -A^{1/2}\log(XX^*)XB^{1/2} = -A^{1/2}X\log(X^*X)B^{1/2} \\ &= -B^{1/2}X^*X\log(X^*X)B^{1/2} = B^{1/2}\eta(X^*X)B^{1/2} \\ &= B^{1/2}\eta(B^{-1/2}AB^{-1/2})B^{1/2}. \end{aligned}$$

Since $S(A|B)$ has the right-term monotone decreasing property of $S(A|B+\varepsilon)$ as $\varepsilon \downarrow 0$, we can define $S(A|B)$ for non-invertible A and B as

$$S(A|B) = \lim_{\varepsilon\downarrow 0} S(A|B+\varepsilon)$$

if the limit exists in $\mathbb{M}_n$. However, in general, $S(A|B)$ may not exist. It should be noted that according to [15], $S(A|B)$ exists if and only if $\mathrm{supp}A \le \mathrm{supp}B$. Also, since $S(A|I) = -A\log A = \eta(A)$, the relative operator entropy is a relative version of the operator entropy introduced by Nakamura and Umegaki.

If $f : [0,\infty) \to \mathbb{R}$ is a matrix convex function, then the perspective function g associated with f is defined by

$$g(B, A) = A^{\frac{1}{2}} f(A^{-\frac{1}{2}}BA^{-\frac{1}{2}})A^{\frac{1}{2}}$$

for any Hermitian matrix B and any positive definite matrix A; see [16]. We can introduce a more general case: Let $\widetilde{B} = (B_1, \ldots, B_n)$ and $\widetilde{A} = (A_1, \ldots, A_n)$ be n-tuples of Hermitian and positive definite matrices, respectively. Then the noncommutative f-divergence functional Θ (see [17]) is defined by

$$\Theta(\widetilde{B}, \widetilde{A}) = \sum_{i=1}^{n} A_i^{\frac{1}{2}} f(A_i^{-\frac{1}{2}} B_i A_i^{-\frac{1}{2}})A_i^{\frac{1}{2}}.$$

Next, recall that $X\natural_q Y$ is defined by $X^{\frac{1}{2}}\left(X^{-\frac{1}{2}}YX^{-\frac{1}{2}}\right)^q X^{\frac{1}{2}}$ for any real number q and any positive definite matricess X and Y. For $p \in [0, 1]$, the operator $X\natural_p Y$ coincides with the well-known p-power mean of X and Y.

Furuta [18] defined a parametric extension of the entropy by

$$S_p(A|B) = A^{\frac{1}{2}}\left(A^{-\frac{1}{2}}BA^{-\frac{1}{2}}\right)^p \log\left(A^{-\frac{1}{2}}BA^{-\frac{1}{2}}\right)A^{\frac{1}{2}},$$

where $p \in [0, 1]$ and A and B are positive definite matrices. He also proved some entropy inequalities as follows: if $\{A_1, \ldots, A_n\}$ and $\{B_1, \ldots, B_n\}$ are two sequences of positive definite matrices such that $\sum_{j=1}^n A_j \natural_p B_j \le I$, then

$$\begin{aligned} &\log\left[\sum_{j=1}^n (A_j \natural_{p+1} B_j) + t_0\left(I - \sum_{j=1}^n A_j \natural_p B_j\right)\right] - (\log t_0)\left(I - \sum_{j=1}^n A_j \natural_p B_j\right) \\ &\ge \sum_{j=1}^n S_p(A_j|B_j) && (7.4) \\ &\ge -\log\left[\sum_{j=1}^n (A_j \natural_{p-1} B_j) + t_0\left(I - \sum_{j=1}^n A_j \natural_p B_j\right)\right] && (7.5) \\ &\quad + (\log t_0)\left(I - \sum_{j=1}^n A_j \natural_p B_j\right) \end{aligned}$$

for a fixed real number $t_0 > 0$; see [19] for other extensions. Note that $S_1(A|B) = BA^{-1}S(A|B)$.

By virtue of the relative operator entropy, we define the quantum relative entropy as

$$S_{FK}(A||B) = \begin{cases} -\operatorname{tr}(S(A|B)) & \text{if } \operatorname{supp}A \le \operatorname{supp}B \\ +\infty & \text{otherwise.} \end{cases} \tag{7.6}$$

The quantum quantity $\operatorname{tr}(A(\log A^{1/2}B^{-1}A^{1/2}))$ is first proposed by Belavkin and Staszewski [20] in the framework of C*-algebras. Since we treat $S_{FK}(A||B)$ as the negative of the trace of the relative operator entropy $S(A|B)$, we call it the *FK relative entropy* or the BS relative entropy in [20].

We demonstrate that the relative operator entropy possesses numerous desirable properties.

Theorem 7.2.1 *Let A and B be positive semidefinite matrices in $\mathbb{M}_n$. Under the existence of relative operator entropies, the following properties hold:*

(i) **Right monotonicity**: *$B \le C$ implies that $S(A|B) \le S(A|C)$;*
(ii) **Homogenity**: *$S(\alpha A|\alpha B) = \alpha S(A|B)$ for $\alpha > 0$;*
(iii) **Upper bounded**: *$S(A|B) \le B - A$;*
(iv) **Subadditivity**: *$S(A + B|C + D) \ge S(A|C) + S(B|D)$;*
(v) **Joint concavity**: *If $A = (1 - \lambda)A_1 + \lambda A_2$ and $B = (1 - \lambda)B_1 + \lambda B_2$ for some $\lambda \in [0, 1]$, then*

$$S(A|B) \geq (1-\lambda)S(A_1|B_1) + \lambda S(A_2|B_2);$$

(vi) **Information monotonicity**: *If* Φ *is a unital positive linear map, then*

$$\Phi(S(A|B)) \leq S(\Phi(A)|\Phi(B)).$$

Proof For (i), since $B \leq C$, it follows that $B + \varepsilon \leq C + \varepsilon$ for all $\varepsilon > 0$, and thus it follows from the existence of $S(A|B)$ that for each $\varepsilon > 0$,

$$\begin{aligned} S(A|B) &\leq S(A|B+\varepsilon) = -A^{1/2}\left(\log A^{1/2}(B+\varepsilon)^{-1}A^{1/2}\right)A^{1/2} \\ &\leq -A^{1/2}\left(\log A^{1/2}(C+\varepsilon)^{-1}A^{1/2}\right)A^{1/2} \\ &= S(A|C+\varepsilon). \end{aligned}$$

As $\varepsilon \downarrow 0$, it follows that $S(A|C)$ exists and $S(A|B) \leq S(A|C)$.

For (ii) and (iii), they easily follow from the definition of the relative operator entropy.

For (iv), for each $\varepsilon > 0$, put $X_\varepsilon = (C+\varepsilon)^{1/2}(C+D+2\varepsilon)^{-1/2}$ and $Y_\varepsilon = (D+\varepsilon)^{1/2}(C+D+2\varepsilon)^{-1/2}$. Since $X_\varepsilon^* X_\varepsilon + Y_\varepsilon^* Y_\varepsilon = I$ and $\eta(t)$ is matrix concave, it follows from Theorem 5.1.5 and the existence of $S(A|C)$ and $S(B|D)$ that

$$\begin{aligned} &S(A+B|C+D+2\varepsilon) \\ &= (C+D+2\varepsilon)^{1/2}\eta\left[(C+D+2\varepsilon)^{-1/2}(A+B)(C+D+2\varepsilon)^{-1/2}\right](C+D+2\varepsilon)^{1/2} \\ &= (C+D+2\varepsilon)^{1/2}\eta\left[X_\varepsilon^*(C+\varepsilon)^{-1/2}A(C+\varepsilon)^{-1/2}X_\varepsilon + \right. \\ &\qquad\qquad \left. Y_\varepsilon^*(D+\varepsilon)^{-1/2}B(D+\varepsilon)^{-1/2}Y_\varepsilon\right](C+D+2\varepsilon)^{1/2} \\ &\geq (C+D+2\varepsilon)^{1/2}\left[X_\varepsilon^*\eta(C^{-1/2}AC^{-1/2})X_\varepsilon + Y_\varepsilon^*\eta(D^{-1/2}BD^{-1/2})Y_\varepsilon\right](C+D+2\varepsilon)^{1/2} \\ &= (C+\varepsilon)^{1/2}\eta((C+\varepsilon)^{-1/2}A(C+\varepsilon)^{-1/2})(C+\varepsilon)^{1/2} + \\ &\qquad\qquad (D+\varepsilon)^{1/2}\eta((D+\varepsilon)^{-1/2}B(D+\varepsilon)^{-1/2})(D+\varepsilon)^{1/2} \\ &= S(A|C+\varepsilon) + S(B|D+\varepsilon) \\ &\geq S(A|C) + S(B|D). \end{aligned}$$

As $\varepsilon \downarrow 0$, it follows that $S(A+B|C+D)$ exists and $S(A+B|C+D) \geq S(A|C) + S(B|D)$, so $S(A|B)$ is subadditive.

For (v), it follows from (ii) and (iv).

For (vi), by restricting to supp$B$, we can assume that B is positive definite, and so is $\Phi(B)$ because Φ is unital. Define a unital positive linear map by

$$\Phi_B(X) = \Phi(B)^{-1/2}\Phi(B^{1/2}XB^{1/2})\Phi(B)^{-1/2}.$$

Since the entropy function $\eta(t)$ is matrix concave, it follows from Theorem 5.2.9 (Choi–Davis–Jensen theorem) that

$$\begin{aligned}\Phi(S(A|B)) &= \Phi(B^{1/2}\eta(B^{-1/2}AB^{-1/2})B^{1/2})B^{1/2}) \\ &= \Phi(B)^{1/2}\Phi_B(\eta(B^{-1/2}AB^{-1/2}))\Phi(B)^{1/2} \\ &\leq \Phi(B)^{1/2}\eta(\Phi_B(B^{-1/2}AB^{-1/2}))\Phi(B)^{1/2} \\ &= S(\Phi(A)|\Phi(B)),\end{aligned}$$

and hence $S(A|B)$ has information monotonicity. □

By Theorem 7.2.1, we have the following properties of the FK relative entropy.

Theorem 7.2.2 *Let A, A_1, A_2 and B, B_1, B_2 be density matrices in $\mathbb{M}_n$. Under the existence of relative operator entropies, the FK relative entropy satisfies the following properties:*

(i) **Positivity**: *$S_{FK}(A||B) \geq 0$, and $S_{FK}(A||B) = 0$ if and only if $A = B$;*
(ii) **Additivity**: *$S_{FK}(A_1 \otimes A_2||B_1 \otimes B_2) = S_{FK}(A_1||B_1) + S_{FK}(A_2||B_2)$;*
(iii) **Joint convexity**: *For any $\lambda \in [0, 1]$,*

$$S_{FK}((1-\lambda)A_1 + \lambda A_2||(1-\lambda)B_1 + \lambda B_2) \leq (1-\lambda)S_{FK}(A_1||B_1) + \lambda S_{FK}(A_2||B_2);$$

(iv) **Monotonicity**: *For any trace-preserving positive linear map Φ,*

$$S_{FK}(\Phi(A)||\Phi(B)) \leq S_{FK}(A||B).$$

Proof If we let $\alpha \to 0$ in Theorem 7.2.3, then we reach Theorem 7.2.2. □

Moreover, Yanagi, Kuriyama, and Furuichi [21] have been advancing research on the Tsallis relative operator entropy as an operator generalization of the Tsallis relative entropy. This operator entropy is considered a parametric extension of the relative operator entropy introduced by Fujii–Kamei: For positive definite matrices A and B, the Tsallis relative operator entropy is defined by

$$T_\alpha(A|B) = \frac{A \natural_\alpha B - A}{\alpha} \qquad \text{for } \alpha \in [-1, 1]\backslash\{0\},$$

where the binary operation $\natural_\alpha$ is defined by

$$A \natural_\alpha B = A^{1/2}(A^{-1/2}BA^{-1/2})^\alpha A^{1/2} \qquad \text{for } \alpha \notin [0, 1].$$

For non-invertible case, in the case of $\alpha \in (0, 1]$, it follows from the monotonicity of $\sharp_\alpha$ that $A \sharp_\alpha B$ is defined as $\lim_{\varepsilon\downarrow 0}(A+\varepsilon) \sharp_\alpha (B+\varepsilon)$. In the case of $\alpha \in [-1, 0)$, it follows that $A \natural_\alpha (B+\varepsilon)$ is monotone increasing on $\varepsilon \downarrow 0$ by (ii) in Lemma 6.3.15. In fact, for $\delta > 0$ and $\varepsilon > \varepsilon' > 0$,

$$(A+\delta) \natural_\alpha (B+\varepsilon) \leq (A+\delta) \natural_\alpha (B+\varepsilon')$$

by (ii) in Lemma 6.3.15. Since $B+\varepsilon$ and $B+\varepsilon'$ are invertible and

$$(A+\delta) \natural_\alpha (B+\varepsilon) = (A+\delta)^{1/2}\left((A+\delta)^{1/2}(B+\varepsilon)^{-1}(A+\delta)^{1/2}\right)^{-\alpha}(A+\delta)^{1/2},$$

it follows from $\delta \to 0$ that

$$A \natural_\alpha (B+\varepsilon) \le A \natural_\alpha (B+\varepsilon').$$

Hence, $A \natural_\alpha B$ for $\alpha \in [-1, 0)$ is defined as the limit if it exists:

$$A \natural_\alpha B = \lim_{\varepsilon \downarrow 0} A \natural_\alpha (B+\varepsilon).$$

Note that it follows from [22] that $A \natural_\alpha B$ for $\alpha \in [-1, 0)$ exists if and only if $\mathrm{supp} A \le \mathrm{supp} B$.

Now, we introduce another quantum Tsallis relative entropy of order $\alpha \in [-1, 1]\backslash\{0\}$ defined by

$$NT_\alpha(A||B) = \begin{cases} -\operatorname{tr}(T_\alpha(A|B)) = -\operatorname{tr}\left(\frac{A \natural_\alpha B - A}{\alpha}\right) & \text{if } \mathrm{supp} A \le \mathrm{supp} B \\ +\infty & \text{otherwise} \end{cases}$$

for density matrices A and B. If A and B commute, then we have $D_\alpha(A||B) = NT_\alpha(A||B)$ for all $\alpha \in [-1, 1]\backslash\{0\}$. Generally, the two quantum Tsallis relative entropies are different. In Theorem 7.3.2 later, we show a relation between the two quantum Tsallis relative entropies.

We have the following properties of the quantum relative entropy NT_α of order $\alpha \in [-1, 1]\backslash\{0\}$, see also [22].

Theorem 7.2.3 *Let A and B be density matrices in $\mathbb{M}_n$. Under the existence of Tsallis relative operator entropies, the following properties of the quantum Tsallis relative entropy NT_α of order $\alpha \in [-1, 1]\backslash\{0\}$ hold:*

(i) **Positivity**: *$NT_\alpha(A||B) \ge 0$, and $NT_\alpha(A||B) = 0$ if and only if $A = B$;*
(ii) **Pseudoadditivity**:

$$NT_\alpha(A_1 \otimes A_2||B_1 \otimes B_2) = NT_\alpha(A_1||B_1) + NT_\alpha(A_2||B_2) + \alpha NT_\alpha(A_1||B_1)NT_\alpha(A_2||B_2);$$

(iii) **Joint convexity**: *If $A = (1-\lambda)A_1 + \lambda A_2$ and $B = (1-\lambda)B_1 + \lambda B_2$ for some $\lambda \in [0, 1]$, then*

$$NT_\alpha(A||B) \le (1-\lambda)NT_\alpha(A_1||B_1) + \lambda NT_\alpha(A_2||B_2);$$

(iv) **Monotonicity**: *For any trace-preserving positive linear map Φ*

$$NT_\alpha(\Phi(A)||\Phi(B)) \le NT_\alpha(A||B).$$

Proof For (i), if $\alpha \in (0, 1]$, then it follows that $A \sharp_\alpha B \le (1-\alpha)A + \alpha B$, and hence

$$\begin{aligned} NT_\alpha(A||B) &= -\operatorname{tr}\left(\frac{A \sharp_\alpha B - A}{\alpha}\right) \\ &\ge -\operatorname{tr}\left(\frac{(1-\alpha)A + \alpha B - A}{\alpha}\right) \\ &= -\operatorname{tr}(B - A) = 0 \qquad (\text{by tr } A = \text{tr } B = 1). \end{aligned}$$

Suppose that $NT_\alpha(A||B) = 0$ for $\alpha \in (0, 1]$. Since $\operatorname{tr}((1-\alpha)A + \alpha B - A \sharp_\alpha B) \ge 0$, we have $A \sharp_\alpha B = (1-\alpha)A + \alpha B$ and hence $A = B$.

If $\alpha \in [-1, 0)$, then it follows that $A \natural_\alpha B \ge (1-\alpha)A + \alpha B$, and hence

$$NT_\alpha(A||B) \ge -\operatorname{tr}(B - A) = 0.$$

Suppose that $NT_\alpha(A||B) = 0$ for $\alpha \in [-1, 0)$. Since $\operatorname{tr}(A \natural_\alpha B - (1-\alpha)A - \alpha B) \ge 0$, we have $A \natural_\alpha B = (1-\alpha)A + \alpha B$, and hence $A = B$.

For (ii), since $(A_1 \otimes A_2) \natural_\alpha (B_1 \otimes B_2) = (A_1 \natural_\alpha B_1) \otimes (A_2 \natural_\alpha B_2)$ for $\alpha \in [-1, 1]\backslash\{0\}$, it follows that

$$\begin{aligned} &(A_1 \natural_\alpha B_1) \otimes (A_2 \natural_\alpha B_2) - A_1 \otimes A_2 \\ &= (A_1 \natural_\alpha B_1 - A_1) \otimes (A_2 \natural_\alpha B_2 - A_2) + A_1 \otimes (A_2 \natural_\alpha B_2 - A_2) + (A_1 \natural_\alpha B_1 - A_1) \otimes A_2, \end{aligned}$$

and hence

$$NT_\alpha(A_1 \otimes A_2||B_1 \otimes B_2) = NT_\alpha(A_1||B_1) + NT_\alpha(A_2||B_2) + \alpha NT_\alpha(A_1||B_1) NT_\alpha(A_2||B_2).$$

For (iii), since $\sharp_\alpha$ is jointly concave for $\alpha \in [0, 1]$ and $\natural_\alpha$ is jointly convex for $\alpha \in [-1, 0)$ by (iii) in Lemma 6.3.15, we have (iii).

For (iv), since $\Phi(A \sharp_\alpha B) \le \Phi(A) \sharp_\alpha \Phi(B)$ for $\alpha \in [0, 1]$ by the Ando theorem 5.3.4, and $\Phi(A \natural_\alpha B) \ge \Phi(A) \natural_\alpha \Phi(B)$ for $\alpha \in [-1, 0]$ by (vi) in Lemma 6.3.15, we have the monotonicity of NT_α for $\alpha \in [-1, 1]\backslash\{0\}$. □

We show the following relation between the quantum Tsallis relative entropy NT_α of order $\alpha \in [-1, 1]\backslash\{0\}$ and the FK relative entropy.

Theorem 7.2.4 *Let A and B be density matrices. Then*

$$\mathrm{NT}_\alpha(A||B) \le S_{FK}(A||B) \le \mathrm{NT}_{-\alpha}(A||B) \tag{7.7}$$

for all $0 < \alpha \le 1$.

Proof We can assume that A and B are positive definite.
Since

$$\frac{x^{-\alpha}-1}{-\alpha} \leq \log x \leq \frac{x^{\alpha}-1}{\alpha}$$

for $x > 0$ and $\alpha \in (0, 1]$, it follows that

$$\frac{(A^{-1/2}BA^{-1/2})^{-\alpha}-I}{-\alpha} \leq \log(A^{-1/2}BA^{-1/2}) \leq \frac{(A^{-1/2}BA^{-1/2})^{\alpha}-I}{\alpha},$$

and hence

$$\frac{A \natural_{-\alpha} B - A}{-\alpha} \leq S(A|B) \leq \frac{A \sharp_{\alpha} B - A}{\alpha}.$$

Therefore, we have the desired inequality (7.7). □

Theorem 7.2.5 *The Tsallis relative entropy $NT_\alpha(A||B)$ is monotonically nonincreasing on $\alpha \in [-1, 1]$.*

Proof It follows from the fact that $F(x) = \frac{a^x-1}{x}$, for $a > 0$, is monotonically increasing on $[-1, 1]$. □

7.3 Comparison of Two Quantum Tsallis Relative Entropies

For two quantum Tsallis relative entropies, it follows from Theorem 7.1.5 and Theorem 7.2.4 that

$$D_\alpha(A||B) \leq S_U(A||B) \leq D_{-\alpha}(A||B) \qquad \text{for } 0 < \alpha \leq 1$$

and

$$NT_\alpha(A||B) \leq S_{FK}(A||B) \leq NT_{-\alpha}(A||B) \qquad \text{for } 0 < \alpha \leq 1. \tag{7.8}$$

Here, it is natural to investigate the relation between $D_\alpha(A||B)$ and $NT_\alpha(A||B)$ in the case when $\alpha \in [-1, 1]\backslash\{0\}$. For this, we need the following Bebiano, Lemos, and Providência result (the so-called BLP inequality) in [23, Theorem 2.1].

Theorem 7.3.1 *Let A and B be positive semidefinite matrices in $\mathbb{M}_n$. Then*

$$\left|\left|\left| A^{\frac{1+t}{2}} B^t A^{\frac{1+t}{2}} \right|\right|\right| \leq \left|\left|\left| A^{1/2}(A^{s/2}B^sA^{s/2})^{\frac{t}{s}}A^{1/2} \right|\right|\right| \tag{7.9}$$

for all $s \geq t \geq 0$ and any unitarily invariant norm $|||\cdot|||$.

Proof We can assume that A and B are positive definite by continuity. It is sufficient to show that

$$A^{1/2}(A^{s/2}B^sA^{s/2})^{\frac{t}{s}}A^{1/2} \le I \qquad \text{implies} \qquad A^{\frac{1+t}{2}}B^tA^{\frac{1+t}{2}} \le I.$$

By assumption, we have $(A^{s/2}B^sA^{s/2})^{\frac{t}{s}} \le A^{-1}$. Putting $p = s/t(\ge 1), r = s$ and $q = s/t(\ge 1)$ in the Furuta inequality (Theorem 1.4.4), it follows that

$$\left(A^{-s/2}\left((A^{s/2}B^sA^{s/2})^{t/s}\right)^{s/t}A^{-s/2}\right)^{t/s} \le \left(A^{-s/2}B^{-s/t}A^{-s/2}\right)^{t/s}$$

and this implies

$$B^t \le A^{-1-t},$$

and hence we get $A^{\frac{1+t}{2}}B^tA^{\frac{1+t}{2}} \le I$. □

The next result reads as follows.

Theorem 7.3.2 *Let A and B be density matrices in $\mathbb{M}_n$. Then*

$$D_\alpha(A||B) \le NT_\alpha(A||B) \qquad \textit{for all } \alpha \in [-1, 1]\backslash\{0\}. \tag{7.10}$$

Proof By restricting to supp$B$, we can assume that B is positive definite. Suppose that $\alpha \in [-1, 0)$. Since the trace norm $\|\cdot\|_1$ is unitarily invariant norm, putting $s = 1$ and $t = -\alpha$ in (7.9), we get

$$\begin{aligned} \mathrm{tr}(A \natural_\alpha B) &= \mathrm{tr}(A^{1/2}(A^{1/2}B^{-1}A^{1/2})^{-\alpha}A^{1/2}) \\ &\ge \mathrm{tr}(A^{\frac{1-\alpha}{2}}(B^{-1})^{-\alpha}A^{\frac{1-\alpha}{2}}) \\ &= \mathrm{tr}(A^{1-\alpha}B^\alpha), \end{aligned}$$

from which we get the desired inequality (7.10):

$$D_\alpha(A||B) = -\,\mathrm{tr}\left(\frac{A^{1-\alpha}B^\alpha - A}{\alpha}\right) \le -\,\mathrm{tr}\left(\frac{A \natural_\alpha B - A}{\alpha}\right) = -\,\mathrm{tr}(T_\alpha(A||B)) = NT_\alpha(A||B).$$

Suppose that $\alpha \in (0, 1]$. Then

$$\begin{aligned} \mathrm{tr}(A \sharp_\alpha B) &\le \mathrm{tr}(e^{(1-\alpha)\log A + \alpha \log B}) && \text{(by Theorem 6.3.13)} \\ &= \mathrm{tr}(e^{(1-\alpha)\log A}e^{\alpha \log B}) && \text{(by the Golden–Thompson inequality (6.1.1))} \\ &= \mathrm{tr}(A^{1-\alpha}B^\alpha), \end{aligned}$$

which yields the result. □

If we let $\alpha \to 0$ in Theorem 7.3.2, then we arrive at the following result due to Hiai and Petz [24] (Fig. 7.1).

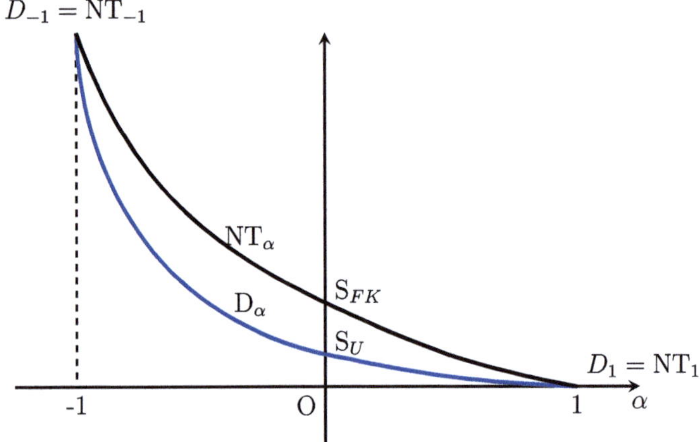

Fig. 7.1 The relation between D_α and NT_α

Theorem 7.3.3 *Let A and B be density matrices in* $\mathbb{M}_n$. *Then*

$$S_U(A||B) \leq S_{FK}(A||B).$$

Remark 7.3.4 If we put $\alpha = 1$ in Theorem 7.3.2, then $D_1(A||B) = NT_1(A||B) = 0$ and if we put $\alpha = -1$, then $D_{-1}(A||B) = NT_{-1}(A||B) = \mathrm{tr}(A^2B^{-1} - A)$.

Finding bounds for various types of entropies and relative entropies is of special interest due to its applications. We recommend readers to have a look at [25–28] and the references therein.

Umegaki firstly defined the relative entropy in the context of semifinite von Neumann algebras. Araki [29] later extended this definition to general von Neumann algebras.

The concepts of quantum information theory and entanglement have become increasingly important in quantum field theory and string theory. Witten [30] explored the entanglement properties of quantum field theory, utilizing Araki's notion of relative entropy in the context of general von Neumann algebras. The term von Neumann algebra is defined in the next chapter.

7.4 Exercises and Problems

Exercise 7.4.1 Let A and B be density matrices. Show the second inequality of (7.3) in Theorem 7.1.5.

Exercise 7.4.2 Let A and B be positive semidefinite matrices in $\mathbb{M}_n$. Show that the Arithmetic–Geometric–Harmonic mean inequality

$$\left((1-\alpha)A^{-1} + \alpha B^{-1}\right)^{-1} \leq A \,\sharp_\alpha\, B \leq (1-\alpha)A + \alpha B \quad (\alpha \in [0, 1])$$

Exercise 7.4.3 Let A and B be positive semidefinite matrices in $\mathbb{M}_n$. Show that

$$A \natural_\alpha B \geq (1-\alpha)A + \alpha B$$

holds for $\alpha \in [-1, 0)$.

Exercise 7.4.4 Let A, A_1, A_2 and B, B_1, B_2 be positive semidefinite matrices in $\mathbb{M}_n$, and $A = (1-\lambda)A_1 + \lambda A_2$ and $B = (1-\lambda)B_1 + \lambda B_2$ for some $\lambda \in [0, 1]$. Show that

$$(1-\lambda)(A_1 \sharp_\alpha B_1) + \lambda(A_2 \sharp_\alpha B_2) \leq A \sharp_\alpha B$$

holds for $\alpha \in [0, 1]$.

Exercise 7.4.5 Let A and B be positive definite matrices in $\mathbb{M}_n$. Show that $(A, B) \mapsto \operatorname{tr}(A^{1-\alpha}B^\alpha)$ is jointly concave for $\alpha \in [0, 1]$.

Exercise 7.4.6 Let A and B be positive semidefinite matrices in $\mathbb{M}_n$, and Φ a unital positive linear map. Show that

$$\Phi(A \sharp_\alpha B) \leq \Phi(A) \sharp_\alpha \Phi(B)$$

holds for $\alpha \in [0, 1]$.

Exercise 7.4.7 Complete the proof of Theorem 7.3.1.

Problem 7.4.8 Let A and B be positive semidefinite matrices in $\mathbb{M}_n$. Then show that $A \natural_\alpha B$ for $\alpha \in [-1, 0)$ exists if and only if $\operatorname{supp} A \leq \operatorname{supp} B$.

7.5 Notes, Hints, and References

We give some hints and references for the exercises and problems.

For Exercise 7.4.1, use $\log x \leq \frac{x^\alpha - 1}{\alpha} = \frac{1-x^\alpha}{-\alpha}$ for $\alpha \in (0, 1]$ and $x > 0$, and carefully observe that the proof of the first inequality of (7.3) holds for $\alpha \in [0, 1]$. To solve Exercise 7.4.2, first use $\left(\alpha x^{-1} + 1 - \alpha\right)^{-1} \leq x^\alpha \leq \alpha x + 1 - \alpha$ for $\alpha \in (0, 1]$ and $x > 0$ to show it for positive definite case. Then, replace A and B by $A + \varepsilon$ and $B + \varepsilon$ for all $\varepsilon > 0$, respectively, for the positive semidefinite case, and finally, let $\varepsilon \to 0$. Apply the fact that $x^\alpha \geq \alpha x + 1 - \alpha$ for $\alpha \in [-1, 0)$ and $x > 0$ to solve Exercise 7.4.3.

For Exercise 7.4.4, note that $\sharp_\alpha$ for $\alpha \in (0, 1]$ is positively homogeneous. Employ Theorem 5.1.5 to show $A_1 \sharp_\alpha B_1 + A_2 \sharp_\alpha B_2 \leq (A_1 + A_2) \sharp_\alpha (B_1 + B_2)$ for $\alpha \in (0, 1]$. Also, carefully observe the proofs of Lemma 6.3.15. For Exercise 7.4.5, see [11, Theorem 1] and [31, Theorem IX.6.1].

If A is positive definite, then it follows from the positivity of Φ that $\Phi(A)$ is positive definite. Consider the map $\Psi(X) = \Phi(A)^{-\frac{1}{2}}\Phi(A^{\frac{1}{2}}XA^{\frac{1}{2}})\Phi(A)^{-\frac{1}{2}}$, and use

Corollary 5.2.10 (by examining the proof of Lemma 6.3.15) to solve Exercise 7.4.6. A careful investigation of the proof of Theorem 6.3.16 helps to solve Exercise 7.4.7. For Exercise 7.4.8, we refer to [22] and notice that $\overline{\mathrm{ran}(A)} = \mathrm{ran}(A) = \mathrm{ran}(A^{1/2})$ for a positive semidefinite A.

References

1. J. von Neumann, *Die mathematischen grundlagen der quantenmechanik* (Springer, Berlin, 1932)
2. S. Ryu, T. Takayanagi, Aspects of holographic entanglement entropy. JHEP **08**, 045 (2006). hep-th/0605073
3. M. Rangamani, T. Takayanagi, *Holographic Entanglement Entropy*, Lecture Notes in Physics, vol. 931 (Springer, Cham, 2017)
4. M. Nakamura, H. Umegaki, A note on the entropy for operator algebras. Proc. Japan. Acad. **37**, 149–154 (1961)
5. C.R. Shannon, A mathematical theory of communication. Bull Syst. Tech. J. **27**, 379–423 and 623–656 (1948)
6. S. Kullback, R. Leibler, On information and sufficiency. Ann. Math. Stat. **22**, 79–86 (1951)
7. M. Echigo, M. Nakamura, A remark on the concept of channels. Proc. Japan Acad. **38**, 307–309 (1962)
8. C. Tsallis, Possible generalization of Bolzmann–Gibbs statistics. J. Stat. Phys. **52**, 479–487 (1988)
9. S. Abe, Nonadditive generalization of the quantum Kullback–Leibler divergence for measuring the degree of purification. Phys. Rev. A **68**, 032302 (2003)
10. S. Abe, Monotone decrease of the quantum nonadditive divergence by projective measurements. Phys. Lett., A **312**, 336–338 (2003)
11. E.H. Lieb, Convex trace functions and the Wigner–Yanase–Dyson conjecture. Adv. Math. **11**, 267–288 (1973)
12. M.D. Choi, Completely positive linear maps on complex matrices. Linear Albegra Appl. **10**, 285–290 (1975)
13. M.B. Ruskai, F.H. Stillinger, Convexity inequalities for estimating free energy and relative entropy. J. Phys. A **23**, 2421–2437 (1990)
14. J.I. Fujii, E. Kamei, Relative operator entropy in noncommutative information theory. Math. Japon. **34**, 341–348 (1989)
15. J.I. Fujii, Y. Seo, The relative operator entropy and the Karcher mean. Linear Algebra Appl. **542**, 4–34 (2018)
16. E.G. Effros, A matrix convexity approach to some celebrated quantum inequalities. Proc. Natl. Acad. Sci. USA **106**, 1006–1008 (2009)
17. M.S. Moslehian, M. Kian, Non-commutative f-divergence functional. Math. Nachr. **286**, 1514–1529 (2013)
18. T. Furuta, Parametric extensions of Shannon inequality and its reverse one in Hilbert space operators. Linear Algebra Appl. **381**, 219–235 (2004)
19. M.S. Moslehian, F. Mirzapour, A. Morassaei, Operator entropy inequalities. Colloq. Math. **130**, 159–168 (2013)
20. V.P. Belavkin, P. Staszewski, C^*-algebraic generalization of relative entropy and entropy. Ann. de ℓ'I. H. P., Sect. A **37**, 51–58 (1982)
21. K. Yanagi, K. Kuriyama, S. Furuichi, Generalized Shannon inequalities based on Tsallis relative operator entropy. Linear Algebra Appl. **394**, 109–118 (2005)
22. J.I. Fujii, Y. Seo, Tsallis relative operator entropy with negative parameters. Adv. Oper. Theory **1**, 219–236 (2016)

23. N. Bebiano, R. Lemos, J. da Providência, Inequalities for quantum relative entropy. Linear Algebra Appl. **401**, 159–172 (2005)
24. F. Hiai, D. Petz, The Golden–Thompson trace inequality is complemented. Linear Algebra Appl. **181**, 153–185 (1993)
25. K. Yanagi, Refinements of bounds for entropy and relative entropy. Linear Nonlinear Anal. **8**, 197–215 (2022)
26. K.M.R. Audenaert, J. Eisert, Continuity bounds on the quantum relative entropy. J. Math. Phys. **46**, 102104, 21 (2005)
27. M. Raïssouli, M.S. Moslehian, S. Furuichi, Relative entropy and Tsallis entropy of two accretive operators. C. R. Math. Acad. Sci. Paris **355**, 687–693 (2017)
28. A. Winter, Tight uniform continuity bounds for quantum entropies: conditional entropy, relative entropy distance and energy constraints. Comm. Math. Phys. **347**, 291–313 (2016)
29. H. Araki, Relative entropy of states of von Neumann algebras. Publ. RIMS, Kyoto Univ. **11**, 809–833 (1976)
30. E. Witten, Notes on some entanglement properties of quantum field theory. https://arxiv.org/pdf/1803.04993.pdf
31. R. Bhatia, *Matrix Analysis* (Springer, New York, 1997)

Chapter 8
Trace Inequalities for von Neumann Algebras

Traces and weights on C^*-algebras are fundamental tools in operator theory and its applications. For a C^*-algebra $\mathscr{A}$, we denote its subset of projections by $\mathscr{A}_{\mathrm{pr}}$. A projection $P \in \mathscr{A}_{\mathrm{pr}}$ is said to be a *minimal projection* (or *atom*) if $P\mathscr{A}P = \mathbb{C}P$ indicating that P majorizes no other nonzero projection in $\mathscr{A}$.

A linear functional φ on $\mathscr{A}$ is called a *Hermitian functional* if $\varphi(A^*) = \overline{\varphi(A)}$ for all $A \in \mathscr{A}$ and it is called a *positive functional* if $\varphi(\mathscr{A}_+) \subset \mathbb{R}_+$.

Two projections P and Q are said to be *Murray–von Neumann equivalent* if there exists a partial isometry U such that $U^*U = P$ and $UU^* = Q$. In this case, we write $P \sim Q$. If $P \leq Q$, then we say that P is a *subprojection* of Q. If P is equivalent to a subprojection of Q, we write $P \prec Q$.

Definition 8.0.1 We say that $P \in \mathscr{A}_{\mathrm{pr}}$ is

- an *abelian projection* if the algebra $P\mathscr{A}P$ is commutative;
- a *finite projection* if $P \sim Q \leq P$ implies $P = Q$;
- an *infinite projection* if it is not finite;
- a *purely infinite projection* if there is no nonzero finite projection $Q \leq P$ in $\mathscr{A}$;
- a *properly infinite projection* if $P \sim P_1$, $P \sim P_2$ with $P_1, P_2 \leq P$ and $P_1P_2 = 0$.

It is clear that an abelian projection is finite, and a minimal projection is abelian. Any projection equivalent to a finite (respectively, abelian, infinite, properly infinite) projection is also finite (respectively, abelian, infinite, properly infinite). It should be noted that 0 is properly infinite (it is also finite).

A unital C^*-algebra $\mathscr{A}$ is said to be *finite* (respectively, *infinite*, *properly infinite*, *purely infinite*) depending on the property of the identity I.

We say that a map $\varphi : \mathscr{A}_+ \to [0, \infty]$ is a *weight* on a C^*-algebra $\mathscr{A}$, if $\varphi(X + Y) = \varphi(X) + \varphi(Y)$, $\varphi(\lambda X) = \lambda\varphi(X)$ for all $X, Y \in \mathscr{A}_+$, $\lambda \geq 0$ (here, $0 \cdot (\infty)$ is considered as 0). Introduce the set

$$\mathscr{A}_{\varphi+} = \{X \in \mathscr{A}_+ \colon \ \varphi(X) < \infty\},$$

A. M. Bikchentaev et al., *Trace Inequalities*, Forum for Interdisciplinary Mathematics,
https://doi.org/10.1007/978-981-97-6520-1_8

and let $\mathscr{A}_\varphi$ be the linear span of $\mathscr{A}_{\varphi+}$ for a weight φ. We can always extend the restriction $\varphi|_{\mathscr{A}_{\varphi+}}$ to a functional on $\mathscr{A}_\varphi$ by linearity. This extension is denoted by the same letter φ. Such an extension tells us that finite weights (that is, $\varphi(X) < \infty$ for all $X \in \mathscr{A}_+$) are essentially the same as positive functionals on $\mathscr{A}$.

A weight φ is said to be

- *faithful*, if $\varphi(X) = 0 \ \ (X \in \mathscr{A}_+)$ implies that $X = 0$;
- a *trace*, if $\varphi(Z^*Z) = \varphi(ZZ^*)$ for every $Z \in \mathscr{A}$.

If φ is a trace, then $\varphi(XY) = \varphi(YX)$ for all $X \in \mathscr{A}_\varphi$ and $Y \in \mathscr{A}$. A trace φ on a C^*-algebra $\mathscr{A}$ is said to be *semifinite* if $\varphi(A) = \sup\{\varphi(B) : \ B \in \mathscr{A}_+, \ B \leq A, \ \varphi(B) < \infty\}$ for every $A \in \mathscr{A}_+$.

8.1 Traces on Von Neumann Algebra

Recall that the weak (operator) topology on $\mathbb{B}(\mathscr{H})$ is defined by the family of seminorms $A \mapsto |\langle A\xi, \eta\rangle|, \xi, \eta \in \mathscr{H}$; the strong (operator) topology on $\mathbb{B}(\mathscr{H})$ is defined by the family of seminorms $A \mapsto \|A\xi\|, \xi \in \mathscr{H}$.

The σ-weak (operator) topology on $\mathbb{B}(\mathscr{H})$ is defined by the family of seminorms

$$A \mapsto \left|\sum_{n=1}^{\infty}\langle A\xi_n, \eta_n\rangle\right|, \quad \sum_{n=1}^{\infty}\|\xi_n\|^2 < \infty, \quad \sum_{n=1}^{\infty}\|\eta_n\|^2 < \infty \quad ((\xi_n)_{n=1}^{\infty}, (\eta_n)_{n=1}^{\infty} \text{ in } \mathscr{H}).$$

For any set $\mathcal{X} \subset \mathbb{B}(\mathscr{H})$, we construct the commutant

$$\mathcal{X}^c = \{Y \in \mathbb{B}(\mathscr{H}) : \ XY = YX \ \text{ for all } \ X \in \mathcal{X}\}.$$

Definition 8.1.1 A von Neumann algebra $\mathscr{A}$ acting on a Hilbert space $\mathscr{H}$ is a $*$-subalgebra of the algebra $\mathbb{B}(\mathscr{H})$ such that $\mathscr{A} = \mathscr{A}^{cc}$.

Every von Neumann algebra is a unital C^*-algebra. Let $\mathscr{A}$ be a von Neumann algebra. Let $\mathscr{A}_*$ be the space of all σ-weakly continuous linear functionals on $\mathscr{A}$. Then the dual of $\mathscr{A}_*$ is $\mathscr{A}$ and the weak*-topology on $\mathscr{A}$ is nothing more than the σ-weak topology. Let $\mathcal{Z}(\mathscr{A}) := \mathscr{A} \cap \mathscr{A}^c$ be *the center* of $\mathscr{A}$. A *factor* is a von Neumann algebra $\mathscr{A}$ with trivial center $\mathcal{Z}(\mathscr{A}) = \mathbb{C}I$.

A weight φ on a von Neumann algebra $\mathscr{A}$ is called

- *a normal weight*, if $\varphi(\sup X_i) = \sup \varphi(X_i)$ for every bounded increasing net $\{X_i\}$ in $\mathscr{A}_+$.

A weight φ on $\mathscr{A}$ is normal if and only if it is σ-weakly lower semicontinuous. In other words, for each $A \in \mathscr{A}_+$, it holds that $\varphi(A) \leq \liminf_i \varphi(A_i)$ for every net $(A_i) \subset \mathscr{A}_+$, which σ-weakly converges to A.

Example 8.1.2 (i) The algebra $\mathbb{B}(\mathscr{H})$ is a von Neumann algebra. In particular, the matrix algebra $\mathbb{M}_n$ equipped with the operator norm is a von Neumann algebra. Consider the restriction of the canonical trace tr to the set $\mathcal{C}_{1+}$ of all positive trace class operators and define the extension of this restriction to the cone $\mathbb{B}(\mathscr{H})_+$ by considering $\operatorname{tr} X = \infty$ whenever $X \in \mathbb{B}(\mathscr{H})_+ \setminus \mathcal{C}_1$. Then, tr is a faithful normal semifinite trace on $\mathbb{B}(\mathscr{H})$. Since $\operatorname{tr}(\cdot) = \sum_{i \in J} \langle (\cdot)\xi_i, \xi_i \rangle$, where $(\xi_i)_{i \in J}$ is any orthonormal basis in $\mathscr{H}$ (see also Remark 2.3.2), the trace tr is normal and faithful. Put $P_a := \sum_{i \in a} \langle (\cdot)\xi_i, \xi_i \rangle$, where $a \subset J$ is a finite set. Then $P_a \in \mathbb{B}(\mathscr{H})_{\mathrm{pr}}$, and $P_a \nearrow I$. Therefore, $X^{1/2} P_a X^{1/2} \nearrow X$ for any $X \in \mathbb{B}(\mathscr{H})_+$; moreover, $X^{1/2} P_a X^{1/2} \in \mathcal{C}_{1+}$. Also, $\operatorname{tr} X = \sup_a \operatorname{tr}(X^{1/2} P_a X^{1/2})$ and $X^{1/2} P_a X^{1/2} \leq X$, meaning that tr is semifinite.

(ii) Let (Ω, Σ, ν) be a locally finite measure space. Then the $*$-algebra $\mathscr{A} = L_\infty(\Omega, \nu)$ acts naturally as a von Neumann algebra of operators on the Hilbert space $L_2(\Omega, \nu)$ by multiplicators $(A\phi)(\omega) := A(\omega)\phi(\omega)$, $\omega \in \Omega$, and is commutative. In fact, $\mathscr{A} = \mathscr{A}^c$ and $\mathscr{A}$ is a maximal abelian subalgebra in $\mathscr{B}(L_2(\Omega, \nu))$. The trace $\varphi(\cdot) = \int_\Omega \cdot \, d\nu$, given by integration with respect to ν, is faithful, normal and semifinite. Indeed, every commutative von Neumann algebra looks like an L_∞ algebra acting by multiplication on L_2 of a locally finite measure space, with multiplicity.

(iii) Suppose that τ is a normalized ($\tau(I) = 1$) normal faithful finite trace on $\mathscr{A}$. Let $d \in \mathbb{N}$ and $\mathbb{M}_d \otimes \mathscr{A} \cong \mathbb{M}_d(\mathscr{A})$ be the algebra of all $d \times d$ matrices with entries in $\mathscr{A}$. Then the trace $\overline{\tau}$ is defined by

$$\overline{\tau}(X) = \tau\left(\overline{\operatorname{tr}} X\right) := \frac{1}{d} \sum_{i=1}^{d} \tau(X_{ii}).$$

It is a unital normal faithful finite trace on $\mathbb{M}_d(\mathscr{A})$, where $X = \left(X_{ij}\right)_{d \times d}$ and $\overline{\operatorname{tr}} = \frac{1}{d} \operatorname{tr}$.

If $\mathscr{A}$ is a von Neumann algebra, then the set $\mathscr{A}_{\mathrm{pr}}$ is a complete lattice. Every unitary in $\mathscr{A}^c$ leaves each $P_i\mathscr{H}$ invariant, so it leaves the intersection $\bigcap_{i \in J} P_i\mathscr{H}$ invariant as well. Therefore, it commutes with P, that is, $P \in \mathscr{A}$. Obviously, P is the greatest lower bound of $(P_i)_{i \in J}$.

Proposition 8.1.3 *(i) If $(P_\lambda)_{\lambda \in \Lambda}$ is a family of projections in $\mathcal{C}_1$, then*

$$\operatorname{tr}\left(\vee_{\lambda \in \Lambda} P_\lambda\right) \leq \sum_{\lambda \in \Lambda} \operatorname{tr} P_\lambda;$$

(ii) If $(P_n)_{n=1}^\infty$ is a decreasing sequence of projections and $P_1 \in \mathcal{C}_1$, then

$$\operatorname{tr}(\wedge_n P_n) = \lim_{n \to \infty} \operatorname{tr} P_n.$$

Proof (i) It suffices to show that $\operatorname{tr}(P \vee Q) \leq \operatorname{tr} P + \operatorname{tr} Q$. It follows from Exercise 8.8.2 that $\operatorname{tr}(P \vee Q) - \operatorname{tr} Q = \operatorname{tr} P - \operatorname{tr}(P \wedge Q)$. Hence,

$$\operatorname{tr}(P \vee Q) = \operatorname{tr} P + \operatorname{tr} Q - \operatorname{tr}(P \wedge Q) \leq \operatorname{tr} P + \operatorname{tr} Q.$$

(ii) $(P_1 - P_n)$ is a norm-bounded increasing sequence of projections. Since $\sup_n(P_1 - P_n) = P_1 - \wedge_n P_n$ and the trace is a normal positive linear functional (see Example 8.1.2), we have

$$\begin{aligned}\operatorname{tr} P_1 - \operatorname{tr}(\wedge_n P_n) &= \operatorname{tr}(P_1 - \wedge_n P_n) \\ &= \operatorname{tr}(\sup_n(P_1 - P_n)) \\ &= \sup_n \operatorname{tr}(P_1 - P_n) \\ &= \operatorname{tr} P_1 - \lim_{n\to\infty} \operatorname{tr} P_n,\end{aligned}$$

whence

$$\operatorname{tr}(\wedge_n P_n) = \lim_{n\to\infty} \operatorname{tr} P_n.$$

□

There exists a unique projection $P_f \in \mathcal{Z}(\mathscr{A})$ such that P_f is finite and $I - P_f$ is properly infinite. $\mathscr{A} P_f$ is called the *finite part* of $\mathscr{A}$ and $\mathscr{A}(I - P_f)$ is said to be the *properly infinite part*; we have a direct sum

$$\mathscr{A} = \mathscr{A}_{P_f} \oplus \mathscr{A}_{I-P_f}.$$

It is important to note that $\mathscr{A}_{P_f}$ does not always contain all finite projections of $\mathscr{A}$ in general. For example, if $\mathscr{A} = \mathbb{B}(\mathscr{H})$, then $P_f = 0$.

Definition 8.1.4 A von Neumann algebra $\mathscr{A}$ is said to be of *type I* if every projection in $\mathcal{Z}(\mathscr{A})$ majorizes a nonzero abelian projection in $\mathscr{A}$. If there is no nonzero finite projection in $\mathscr{A}$, then it is said to be of *type III*. If $\mathscr{A}$ has no nonzero abelian projection and if each nonzero projection in $\mathcal{Z}(\mathscr{A})$ majorizes a nonzero finite projection in $\mathscr{A}$, then it is said to be of *type II*. If $\mathscr{A}$ is finite and of type II, then it is said to be of *type* II_1. If $\mathscr{A}$ is of type II and has no nonzero central finite projection, then it is said to be of *type* II_∞.

Every von Neumann algebra $\mathscr{A}$ is uniquely decomposable into the direct sum of those of type I, type II_1, type II_∞, and type III [1, Chap. V, Theorem 1.19]. Thus,

$$\mathscr{A} = \mathscr{A}_{P_I} \oplus \mathscr{A}_{P_{II_1}} \oplus \mathscr{A}_{P_{II_\infty}} \oplus \mathscr{A}_{P_{III}},$$

where projections P_I, P_{II_1}, P_{II_∞}, and P_{III} in $\mathcal{Z}(\mathscr{A})$ are such that $P_I + P_{II_1} + P_{II_\infty} + P_{III} = I_{\mathscr{A}}$. If $P_{III} = 0$, then $\mathscr{A}$ is said to be a *semifinite von Neumann algebra*. $\mathscr{A}$

is semifinite if and only if $\mathscr{A}$ admits a faithful semifinite normal trace ([1, Chap. V, Theorem 2.15]). The commutant $\mathscr{A}^c$ of a von Neumann algebra $\mathscr{A}$ is semifinite if and only if $\mathscr{A}$ is semifinite ([1, Chap. V, Corollary 2.23]). On a semifinite factor, any semifinite normal traces are proportional ([1, Chap. V, Corollary 2.32]).

Definition 8.1.5 A von Neumann algebra $\mathscr{A}$ is said to be

- a *continuous von Neumann algebra*, if it contains no nonzero abelian projections;
- of *type* I_n if there are abelian projections such that $P_i \sim P_j$, $i, j = 1, \dots, n$, $P_1 + \cdots + P_n = I_{\mathscr{A}}$;
- of *type* I_∞, if it is properly infinite of type I.

Thus, $\mathscr{A}$ is of type II (respectively, II_∞), if it is continuous and semifinite (respectively, properly infinite of type II). A von Neumann algebra $\mathscr{A}$ is of type I (respectively, type II, type III) if and only if $\mathscr{A}^c$ is of type I (respectively, type II, type III) ([1, Chap. V, Corollary 2.24]).

Consider a von Neumann algebra $\mathscr{A}$. Let $U \in \mathscr{A}$ be a unitary operator. There exists an automorphism α of $\mathscr{A}$, defined by the formula $\alpha(A) = U^*AU$ for all $A \in \mathscr{A}$. It follows from [2, Theorem 1.4] that the automorphism α can be represented as a finite product of involutions $U_S : A \mapsto SAS$, where S is a symmetry in $\mathscr{A}$. In other words, $U^*AU = S_1 \cdots S_m A S_m \cdots S_1$ with unitaries $S_1, \dots, S_m \in \mathscr{A}_{\mathrm{sa}}$. Moreover, if $\mathscr{A}$ does not have any finite type I direct summand, then the unitary operator U itself can be represented as a finite product of symmetries from $\mathscr{A}_{\mathrm{sa}}$ [2, Theorem 1.6]. Applying Upmeier's results, one can see that [3, Theorem 1.4.2] shows us that a weight φ on a von Neumann algebra $\mathscr{A}$ is a trace if and only if $\varphi(SAS) = \varphi(A)$ for any positive operator $A \in \mathscr{A}_+$ and any symmetry $S \in \mathscr{A}_{\mathrm{sa}}$.

For a positive linear functional φ on $\mathscr{A}$ the following conditions are equivalent:

(i) φ is tracial;
(ii) $\varphi(AB) = \varphi(BA)$ for all $A, B \in \mathscr{A}$;
(iii) $\varphi(UAU^*) = \varphi(A)$ for all $A \in \mathscr{A}_+$ and a unitary $U \in \mathscr{A}$.

A positive normal linear functional φ on a von Neumann algebra $\mathscr{A}$ is tracial if and only if $\varphi(P) = \varphi(Q)$ for all $P, Q \in \mathscr{A}_{\mathrm{pr}}$ with $PQ = 0$ and $P \sim Q$ (see [4], [5, Lemma 2]).

Consider a tracial positive normal linear functional φ on a von Neumann algebra $\mathscr{A}$, and positive numbers p and q such that $1/p + 1/q = 1$, then the following holds

- Hölder's inequality [1, Chap. IX, Theorem 2.13], [6, Theorem 5]:

$$\varphi(|XY|) \leq \varphi(X^p)^{1/p}\varphi(Y^q)^{1/q} \quad \text{for all} \quad X, Y \in \mathscr{A}_+;$$

- Cauchy–Schwarz inequality [7, Theorem 4.21]:

$$\varphi(|XY|^{1/2}) \leq \varphi(X)^{1/2}\varphi(Y)^{1/2} \quad \text{for all} \quad X, Y \in \mathscr{A}_+;$$

- Golden–Thompson inequality [8, Theorem 4]:

$$\varphi(e^{X+Y}) \le \varphi(e^{X/2}e^{Y}e^{X/2}) \text{ for all } X, Y \in \mathscr{A}_{\text{sa}}; \tag{8.1.1}$$

- Peierls–Bogoliubov inequality [8, Theorem 7]:

$$\varphi(e^X)\exp\frac{\varphi(e^{X/2}Ye^{X/2})}{\varphi(e^X)} \le \varphi(e^{X+Y}) \text{ for all } X, Y \in \mathscr{A}_+;$$

- Young's inequality [9]:

$$\varphi(|XY|) \le \frac{\varphi(X^p)}{p} + \frac{\varphi(Y^q)}{q} \text{ for all } X, Y \in \mathscr{A}_+;$$

- Araki–Lieb–Thirring inequality [10, 11] (see (3.2.27)):

$$\varphi((X^{1/2}YX^{1/2})^{rp}) \le \varphi((X^{r/2}Y^rX^{r/2})^p) \text{ for all } r \ge 1, \ p > 0 \text{ and } X, Y \in \mathscr{A}_+.$$

In addition, the inequality is reversed if $0 < r \le 1$.

The Araki–Lieb–Thirring inequality resembles the Golden–Thompson inequality (see [12, Sect. 8]). For all $X, Y \in \mathscr{A}_+$, the inequalities

$$0 \le \varphi(XY) \le \varphi(|XY|) \le \varphi(X^p)^{1/p}\varphi(Y^q)^{1/q} \le \frac{\varphi(X^p)}{p} + \frac{\varphi(Y^q)}{q}$$

hold; see [8, Theorem 5] and [9].

Let $A \in \mathbb{M}_n$ be positive semidefinite, and let $B \in \mathbb{M}_n$ be Hermitian. If we take $\varphi = \text{tr}$, $r = 1/2$, $p = 1$, $Y = B^2$, and $X = |A|^2$ in the Araki–Lieb–Thirring inequality, then

$$\text{tr}(|B|A) = \text{tr}\left(A^{1/2}\,(|B|^2)^{1/2}\,A^{1/2}\right) \le \text{tr}\left(\left(A\,|B|^2\,A\right)^{1/2}\right) = \text{tr}(|BA|).$$

It can be said that any of the given trace inequalities is sharp in the following sense: only the trace of all positive linear functionals satisfies the inequality. It is known that if we limit ourselves only to projections of a von Neumann algebra $\mathscr{A}$, then each of Hölder, Cauchy–Schwarz, Golden–Thompson, Peierls–Bogoliubov, Young, Araki–Lieb–Thirring inequalities characterizes the tracial functionals among all positive normal functionals; see [4, 6, 13–19]. As a consequence, we obtain certain criteria for the commutativity of von Neumann algebras and C^*-algebras.

8.2 Trace Inequalities for Quantum Statistical Mechanics

For simplicity, we restrict ourselves to finite traces. In this section, we present some Golden–Thompson type inequalities for a tracial functional φ on a von Neumann algebra $\mathscr{A}$. The proof of the following Trotter formula can be found in [20, Theorem 8.31, p. 295].

The *graph of an operator* A is the set $G(A) = \{(\xi, A\xi) \mid \xi \in \mathfrak{D}(A)\} \subset \mathscr{H} \times \mathscr{H}$. An operator B is called an *extension* of A if $G(A) \subseteq G(B)$. This is written as $A \subseteq B$. A densely defined operator A is called *symmetric* if $\langle A\xi, \eta\rangle = \langle \xi, A\eta\rangle$ for all $\xi, \eta \in \mathfrak{D}(A)$. By definition, the *adjoint operator* A^* acts on the subspace $\mathfrak{D}(A^*) \subseteq \mathscr{H}$ consisting of the elements η for which there is a $\zeta \in \mathscr{H}$ such that $\langle A\xi, \eta\rangle = \langle \xi, \zeta\rangle$, for every $\xi \in \mathfrak{D}(A)$. Setting $A^*\eta = \zeta$, defines the linear operator A^*. An unbounded densely defined operator A is called a *self-adjoint unbounded operator* if $G(A) = G(A^*)$. A symmetric operator A is said to be *essentially self-adjoint* if the closure of A is self-adjoint. A symmetric operator A is always *closable*, that is, the closure of the graph of A is the graph of an operator. An operator B is *closed* if and only if $G(B)$ is a closed subset of the Cartesian product $\mathscr{H} \times \mathscr{H}$.

Lemma 8.2.1 *It holds that*

$$e^{A+B} = \text{sot}-\lim_{n\to\infty} (e^{A/n}e^{B/n})^n$$

if both statements

(a) A and B are self-adjoint operators and bounded above,
(b) A + B is essentially self-adjoint

hold.

In fact, $e^{A+B} = \text{sot}-\lim_{n\to\infty} (e^{A/n}e^{B/n})^{n/2}(e^{B/n}e^{A/n})^{n/2}$ for A and B as in Lemma 8.2.1. Apply the Trotter formula to $e^{(A+B)/2}$. Then for the subsequence $m = 2n$

$$e^{(A+B)/2} = \text{sot}-\lim_{m\to\infty} (e^{A/m}e^{B/m})^{m/2}. \tag{8.2.1}$$

Since A is bounded above, e^A is a bounded, self-adjoint operator and $\|e^{A/m}\| = \|e^A\|^{1/m}$. Hence, $\|e^{A/m}e^{B/m}\|^{m/2} \le (\|e^A\|\,\|e^B\|)^{1/2}$, and the sequence in (8.2.1) is bounded independently of m. Then, reversing the roles of A and B in Lemma 8.2.1, and using the fact that multiplication is continuous on bounded sets in the strong operator topology, gives the desired result.

Theorem 8.2.2 *For all $X, Y \in \mathscr{A}_+$ and $p \in \mathbb{N} \cup \{0\}$,*

$$\varphi((XY)^{2^{p+1}}) \le \varphi((XY)^{2^p}(YX)^{2^p}) \tag{8.2.2}$$

$$\le \varphi((X^2Y^2)^{2^p}) \tag{8.2.3}$$

$$\le \varphi(X^{2^{p+1}}Y^{2^{p+1}}). \tag{8.2.4}$$

Proof Since $0 \le \varphi(Y^{1/2}(XY)^{2^{p+1}-1}XY^{1/2}) = \varphi((XY)^{2^{p+1}})$ all terms are real and positive. Note that, for any positive linear functional ϕ on $\mathscr{A}$, the Cauchy–Schwarz inequality holds

$$|\phi(A^*B)|^2 \leq \phi(A^*A)\phi(B^*B) \;\text{ for all }\; A, B \in \mathscr{A}. \tag{8.2.5}$$

Assume that $\phi(A^*B) \neq 0$. For $\lambda \in \mathbb{R}$ and $C = \dfrac{\phi(B^*A)}{|\phi(A^*B)|} + \lambda B$, we know that

$$0 \leq \phi(C^*C) = \lambda^2\phi(B^*B) + 2\lambda|\phi(A^*B)| + \phi(A^*A) = f(\lambda)$$

and via the positivity of the trinomial $f(\lambda)$ we obtain (8.2.5).

The theorem readily follows from (8.2.5) for $p = 0$, that is,

$$\varphi((XY)^2) = \varphi((XY)(XY)) \leq \varphi((XY)(YX)) = \varphi(X^2Y^2).$$

Obviously, relation (8.2.2) is a special case of (8.2.5) and relation (8.2.4) follows from the repeated application of inequality (8.2.3). Therefore, it suffices to give an inductive proof of (8.2.3) under the assumption that (8.2.4) is true for all p. To prove this, we introduce the notation

$$\alpha_m = (XY)^{2^m}(YX)^{2^m}, \quad \beta_m = (YX)^{2^m}(XY)^{2^m}$$

and note that

$$\varphi((\alpha_m)^r) = \varphi((\beta_m)^r) = \varphi((\alpha_{m-1}\beta_{m-1})^r). \tag{8.2.6}$$

Now

$$\varphi((XY)^{2^{p+1}}) = \varphi((XY)^{2^p}(XY)^{2^p}) \leq \varphi((XY)^{2^p}(YX)^{2^p}) = \varphi(\alpha_p)$$

and $\varphi((X^2Y^2)^{2^p}) = \varphi((\alpha_0)^{2^p})$. Hence, (8.2.3) is satisfied if

$$\varphi(\alpha_p) \leq \varphi((\alpha_0)^{2^p}). \tag{8.2.7}$$

Then we prove (8.2.3) by proving (8.2.7) under the inductive assuption that (8.2.3) and (8.2.4) hold for all $k < p$. Relation (8.2.6) for $k < p$ implies that

$$\varphi((\alpha_{p-k})^{2^k}) = \varphi((\alpha_{p-k-1}\beta_{p-k-1})^{2^k}) \leq \varphi((\alpha_{p-k-1})^{2^k}(\beta_{p-k-1})^{2^k}) \leq \varphi((\alpha_{p-(k+1)})^{2^{k+1}}),$$

where the first inequality follows from (8.2.4) and the second from (8.2.5) and (8.2.6). This shows us that $\varphi((\alpha_{p-k})^{2^k})$ increases with k so that (8.2.7) is satisfied as required. □

In particular,

$$\varphi((XY)^{2^p}(YX)^{2^p}) \leq \varphi(X^{2^p}Y^{2^{p+1}}X^{2^p}). \tag{8.2.8}$$

Theorem 8.2.3 *The inequality* $\varphi(e^{A+B}) \leq \varphi(e^{A/2}e^Be^{A/2})$ *holds if*

(a) A and B are self-adjoint operators and bounded above,

(b) $A + B$ is essentially self-adjoint.

Proof Let $X_p = (e^{A/2^{p+1}} e^{B/2^{p+1}})^{2^p} (e^{B/2^{p+1}} e^{A/2^{p+1}})^{2^p}$. It then follows from the Trotter formula (see the text after Lemma 8.2.1) that $X_p \to e^{A+B}$ strongly as $p \to \infty$. By (8.2.8) we obtain $\varphi(X_p) \le \varphi(X_0)$ for all p. If $\{A_n\} \subset \mathscr{A}$ is a sequence of positive operators converging weakly to $A \in \mathscr{A}$, then $\varphi(A) \le \underline{\lim}_{n\to\infty} \varphi(A_n)$ (by Fatou's lemma; see Theorem 8.3.22). Hence,

$$\varphi(e^{A+B}) \le \underline{\lim}_{p\to\infty} \varphi(X_p) \le \varphi(X_0) = \varphi(e^{A/2} e^B e^{A/2}).$$

□

Theorems 8.2.2 and 8.2.3 are proved in [8] and generalize results of Golden [21] and Thompson [22]. The following two theorems also are proved in [8].

Theorem 8.2.4 *The function $f(x) = \log \varphi(e^{A+xB})$ is convex on $\mathbb{R}$ if*

(a) B is a bounded and self-adjoint operator;
(b) A is a self-adjoint and bounded above operator.

Proof For all $0 \le t \le 1$ and $x, y \in \mathbb{R}$, we obtain

$$\begin{aligned} f(tx + (1-t)y) &= \log \varphi(e^{t(A+xB)+(1-t)(A+yB)}) \\ &\le \log \varphi(e^{t(A+xB)} e^{(1-t)(A+yB)}) \\ &\le \log (\varphi(e^{A+xB}))^t (\varphi(e^{A+yB}))^{1-t} \\ &= tf(x) + (1-t)f(y), \end{aligned}$$

where the first inequality follows by Theorem 8.2.3 and the second by the Hölder inequality (see Theorem 8.4.5(i) below). □

Theorem 8.2.5 (Peierls–Bogoliubov Inequality). *The inequality*

$$\log \frac{\varphi(e^{A+B})}{\varphi(e^A)} \ge \frac{\varphi(e^A B)}{\varphi(e^A)}$$

is true if A and B meet the conditions of Theorem 8.2.4.

Proof If f is a convex function and $a < b$, then

$$f'(a) \le \frac{f(b) - f(a)}{b - a}.$$

Let $f(x)$ be as in Theorem 8.2.4 and $a = 0$, $b = 1$. To compute $f'(0)$, let $g(x) = \log \varphi(e^A e^{xB})$ and note that $f(x) \le g(x)$ for all x, $f(0) = g(0)$, and

$$g'(x) = \frac{\varphi(e^A B e^{xB})}{\varphi(e^A e^{xB})}.$$

Therefore,

$$f'(0) = g'(0) = \frac{\varphi(e^A B)}{\varphi(e^A)} \le f(1) - f(0) = \log \frac{\varphi(e^{A+B})}{\varphi(e^A)},$$

as required. □

The elegant proofs of the Golden–Tompson and the Peierls–Bogoliubov inequalities for Hermitian $A, B \in \mathbb{M}_n$ are given in [23, Sect. 7].

8.3 Inequalities for Real Functions and the Trace

A *σ-algebra* on a set Ω is a nonempty collection $\mathfrak{A}(\Omega)$ of subsets of Ω such that it is closed under complement, countable unions, and contains $\emptyset$. Let $I_{\mathscr{A}}$ be the identity element of a von Neumann algebra $\mathscr{A}$. An $\mathscr{A}_+$-valued measure on the σ-algebra $\mathfrak{A}(\Omega)$ is a map $\mathbb{X} = \{X(Q)\} : \mathfrak{A}(\Omega) \to \mathscr{A}_+$ that is σ-additive in the sense of the σ-weak topology on $\mathscr{A}$. If, moreover, $X(\Omega) = I_{\mathscr{A}}$, then an $\mathscr{A}_+$-valued measure is called an *$\mathscr{A}_+$-valued decomposition of the identity*. For any $\phi \in \mathscr{A}_*$, the relation

$$\phi(\mathbb{X})(Q) = \phi(X(Q))$$

defines the scalar σ-additive measure on $\mathfrak{A}(\Omega)$ with a finite total variation. Let $g : \Omega \to \mathbb{R}$ be a bounded measurable function. The map

$$\phi \mapsto \int_\Omega g \, d\phi(\mathbb{X})$$

on $\mathscr{A}_*$ is linear and continuous, hence the integral

$$\int_\Omega g \, d\mathbb{X} \tag{8.3.1}$$

can be understood in the σ-weak sense, that is, as the Hermitian operator $\mathbf{I}(\mathbb{X}, g) \in \mathscr{A}$ such that

$$\phi(\mathbf{I}(\mathbb{X}, g)) = \int_\Omega g \, d\phi(\mathbb{X})$$

for all $\phi \in \mathscr{A}_*$. It is easy to verify that

- if $g \ge 0$, then $\mathbf{I}(\mathbb{X}, g) \ge 0$;
- $\mathbf{I}(\mathbb{X}, sg_1 + tg_2) = s\mathbf{I}(\mathbb{X}, g_1) + t\mathbf{I}(\mathbb{X}, g_2)$ for all $s, t \in \mathbb{R}$;
- $\|\mathbf{I}(\mathbb{X}, g)\| \le \sup\limits_{\omega \in \Omega} |g(\omega)|$;
- if $g \ge 0$ and ψ is a normal weight on $\mathscr{A}$, then

$$\psi(\mathbf{I}(\mathbb{X}, g)) = \int_{\Omega} g \, d\psi(\mathbb{X}).$$

If $A \in \mathscr{A}_{\text{sa}}$, and if $\Omega \subset \mathbb{R}$ contains the spectrum of A, then by $\mathbb{P}^A$ we denote the orthogonal decomposition of the identity on the Borel σ-algebra of the set Ω, corresponding to the operator A. Every operator $A \in \mathscr{A}_{\text{sa}}$ admits a representation in the form of relation (8.3.1), in particular, one of these represenattions is given by the spectral representation:

$$A = \int_{[-\|A\|, \|A\|]} \lambda \, d\mathbb{P}^A.$$

Let $\mathscr{A}$ and $\mathscr{B}$ be von Neumann algebras. Let $\pi : \mathscr{B} \to \mathscr{A}$ be a linear positive σ-weak continuous map, and let $\mathbb{X}$ be a $\mathscr{B}_+$-valued measure on $\mathfrak{A}(\Omega)$. Then the relation $\pi(\mathbb{X})(Q) = \pi((X(Q))$ defines the $\mathscr{A}_+$-valued measure on $\mathfrak{A}(\Omega)$. It is easy to verify that if $g : \Omega \to \mathbb{R}$ is a bounded measurable function, then $\pi(\mathbf{I}(\mathbb{X}, g)) = \mathbf{I}(\pi(\mathbb{X}), g)$.

In the rest of this section, let φ be a faithful normal tracial functional on a von Neumann algebra $\mathscr{A}$. The following theorem was proved in [24] by applying [25, Proposition 12].

Theorem 8.3.1 *Let f be a real convex function on an interval $[a, b] \subset \mathbb{R}$ and let $\mathbb{X}$ be a $\mathscr{A}_+$-valued decomposition of the identity on the σ-algebra $\mathfrak{A}(\Omega)$. Then*

$$\varphi(\mathbf{I}(\mathbb{X}, f \circ g)) \geq \varphi(f(\mathbf{I}(\mathbb{X}, g))). \tag{8.3.2}$$

Proof Step 1. We show that if for a self-adjoint element $A \in \mathscr{A}_{\varphi}$, we set

$$\|A\|_{\varphi} =: \inf\{\varphi(A_1 + A_2) : \ A = A_1 - A_2, \ A_1, A_2 \in \mathscr{A}_{\varphi+}\},$$

then equality $\varphi(|A|) = \|A\|_{\varphi}$ holds. Let $A \in \mathscr{A}_{\varphi}$ be self-adjoint and $A = A' - A''$ for some $A', A'' \in \mathscr{A}_{\varphi+}$. If $P = \int_0^{\infty} E_{\lambda}^A d\lambda$, then $PAP = A_+$, hence $PA'P \geq A_+$. Since φ is a trace,

$$\varphi(A') \geq \varphi(PA'P) \geq \varphi(A_+), \ \ \varphi(A'') \geq \varphi(A_-),$$

and $\varphi(|A|) \leq \|A\|_{\varphi}$. By the definition of $\|A\|_{\varphi}$, $\|A\|_{\varphi} \leq \varphi(|A|) = \varphi(A_+) + \varphi(A_-)$.

Step 2. For the case when $f(\lambda) = |\lambda|$ the assertion of the theorem reduces to the inequality

$$\varphi(A_1 + A_2) \geq \varphi(|A_1 - A_2|) \ \ (A_1, A_2 \in \mathscr{A}_+),$$

as shown in Step 1. Applying this inequality, it is easy to verify inequality (8.3.2) for any convex piecewise linear function f. By taking the limit, we can extend this inequality to any convex function. □

For the algebra $\mathbb{B}(\mathscr{H})$ and the canonical trace on it, an analogue of Theorem 8.3.1 was obtained in [26]. The following theorem was proved in [24].

Theorem 8.3.2 *Let $\mathscr{B}$ be a von Neumann algebra, let $\pi : \mathscr{B} \to \mathscr{A}$ be a linear positive σ-weak continuous map. Let f be a real convex function on an interval $[a, b]$, let the spectrum of $A \in \mathscr{A}_{\text{sa}}$ belong to $[a, b]$. Then, if*

(i) $\pi(I_{\mathscr{B}}) = I_{\mathscr{A}}$, *or*
(ii) $\pi(I_{\mathscr{B}}) \leq I_{\mathscr{A}}$, $0 \in [a, b]$ *and* $f(0) \leq 0$,

then $\varphi(\pi(f(A))) \geq \varphi(f(\pi(A)))$.

Proof We represent an operator A in the form

$$A = \int_{[a,b]} \lambda d\,\mathbb{P}^A = \mathbf{I}(\mathbb{P}^A, \lambda).$$

Then $\pi(A) = \mathbf{I}(\pi(\mathbb{P}^A), \lambda)$, $f(A) = \mathbf{I}(\mathbb{P}^A, f(\lambda))$, $\pi(f(A)) = \mathbf{I}(\pi(\mathbb{P}^A), f(\lambda))$.
Under conditions of item (i) we obtain

$$\varphi(\pi(f(A))) = \varphi(\mathbf{I}(\pi(\mathbb{P}^A), f(\lambda))) \geq \varphi(f(\mathbf{I}(\pi(\mathbb{P}^A), \lambda))) = \varphi(f(\pi(A))).$$

In order to prove the desired inequality under condition (ii), we define an $\mathscr{A}_+$-valued decomposition of the identity $\mathbb{X}$ by

$$X(Q) = \begin{cases} \pi(\mathbb{P}^A(Q)), & \text{if } 0 \notin Q, \\ \pi(\mathbb{P}^A(Q)) + I_{\mathscr{A}} - \pi(I_{\mathscr{B}}), & \text{if } 0 \in Q, \end{cases}$$

where Q is a Borel subset of $[a, b]$. Then, via conditions of item (ii), we get

$$\begin{aligned} \varphi(\pi(f(A))) &= \varphi(\mathbf{I}(\pi(\mathbb{P}^A), f(\lambda))) \geq \varphi(\mathbf{I}((\mathbb{X}, f(\lambda))) \\ &\geq \varphi(f(\mathbf{I}(\mathbb{X}, \lambda))) = \varphi(f(\mathbf{I}(\pi(\mathbb{P}^A), \lambda))) = \varphi(f(\pi(A))). \end{aligned}$$

□

Corollary 8.3.3 *Let f be a real convex function on an interval $[a, b]$, let the spectrum of $A \in \mathscr{A}_{\text{sa}}$ belong to $[a, b]$, and let operators $B_1, \ldots, B_n$ lie in $\mathscr{A}$. Then, if*

(i) $\sum_{i=1}^n B_i^* B_i = I_{\mathscr{A}}$, *or*
(ii) $\sum_{i=1}^n B_i^* B_i \leq I_{\mathscr{A}}$, $0 \in [a, b]$ *and* $f(0) \leq 0$,

then $\varphi(\sum_{i=1}^n B_i^* f(A) B_i) \geq \varphi(f(\sum_{i=1}^n B_i^* A B_i))$.

The following two theorems were proved in [24].

Theorem 8.3.4 *Let f be a real convex function on $\mathbb{R}_+$ with $f(0) \leq 0$. Then*

$$\varphi(f(A + B)) \geq \varphi(f(A)) + \varphi(f(B)) \ \textit{for all}\ A, B \in \mathscr{A}_+.$$

Proof Assume first that $f(0) = 0$. In this case, we can move analogously to the proof of [27, Lemma 3]. It is well known, that there exist operators $U, V \in \mathscr{A}$ such that

$$A^{1/2} = U(A+B)^{1/2}, \quad B^{1/2} = V(A+B)^{1/2},$$

and $U^*U + V^*V$ is the support of the operator $A + B$. By utilizing Corollary 8.3.3 we can derive the following:

$$\begin{aligned}\varphi(f(A)) + \varphi(f(B)) &= \varphi(f(U(A+B)U^*)) + \varphi(f(V(A+B)V^*)) \\ &\leq \varphi(Uf(A+B)U^*) + \varphi(Vf(A+B)V^*) \\ &= \varphi((U^*U + V^*V)f(A+B)) = \varphi(f(A+B)).\end{aligned}$$

Let $f(0) \leq 0$. Consider the function $f_1(\lambda) = f(\lambda) - f(0)$ on $\mathbb{R}^+$. Then

$$\begin{aligned}\varphi(f(A)) + \varphi(f(B)) &= \varphi(f_1(A)) + \varphi(f_1(B)) + 2f(0)\varphi(I_{\mathscr{A}}) \\ &\leq \varphi(f_1(A+B)) - f(0)\varphi(I_{\mathscr{A}}) = \varphi(f(A+B)).\end{aligned}$$

□

Via inequality (8.3.2) one can prove the following inequality. For other proofs, see [28] or [29].

Theorem 8.3.5 *Let f be a real convex function on $[a, b]$, and let the spectra of $A, B \in \mathscr{A}_{\text{sa}}$ lie in $[a, b]$. Then, for all $0 \leq \alpha \leq 1$,*

$$\varphi(f(\alpha A + (1-\alpha)B)) \leq \alpha\varphi(f(A)) + (1-\alpha)\varphi(f(B)).$$

Corollary 8.3.6 (see Theorem 1 of [30]).

(i) Let $f : \mathbb{R} \to \mathbb{R}_+$ be a convex function with the Δ_2-condition (that is, there exists a constant $c_f > 0$ such that $f(2\lambda) \leq c_f f(\lambda)$ for all $\lambda \in \mathbb{R}$), and let $A, B \in \mathscr{A}_{\text{sa}}$. Then

$$\varphi(f(A+B)) \leq \frac{1}{2}c_f(\varphi(f(A)) + \varphi(f(B)).$$

(ii) Let $f : \mathbb{R}_+ \to \mathbb{R}_+$ be a concave function such that there exists a constant $c_f \geq 0$ such that $f(2\lambda) \geq c_f f(\lambda)$ for all $\lambda \in \mathbb{R}_+$, and let $A, B \in \mathscr{A}_+$. Then

$$\varphi(f(A+B)) \geq \frac{1}{2}c_f(\varphi(f(A)) + \varphi(f(B)).$$

Proof For (i) we consider the relation

$$\begin{aligned}\varphi(f(A+B)) &= \varphi\Big(f\Big(\frac{1}{2}\cdot 2A + \frac{1}{2}\cdot 2B\Big)\Big) \leq \frac{1}{2}\varphi(f(2A)) + \frac{1}{2}\varphi(f(2B)) \\ &\leq \frac{1}{2}c_f(\varphi(f(A)) + \varphi(f(B))).\end{aligned}$$

For (ii) it can be seen that

$$\varphi(f(A+B)) \geq \frac{1}{2}\varphi(f(2A)) + \frac{1}{2}\varphi(f(2B)) \geq \frac{1}{2}c_f(\varphi(f(A)) + \varphi(f(B))).$$

□

The statements proved in Sects. 4.2 and 4.3 admit generalizations to the case of a semifinite trace, as well as to unbounded functions and unbounded operators affiliated with the von Neumann algebra $\mathscr{A}$. In order to prove these generalizations, it suffices to carry out appropriate passages to the limit. In [31], unbounded analogues of the inequalities obtained in Sect. 7.3 were proved by another method. We give the necessary definitions to describe this more modern and general method based on Segal's theory of noncommutative integration.

A linear operator $X : \mathfrak{D}(X) \to \mathscr{H}$, with the domain $\mathfrak{D}(X) \subseteq \mathscr{H}$ is said to be *affiliated with* $\mathscr{A}$ if $XY \subseteq YX$ for all $Y \in \mathscr{A}^c$ (equivalently, $XU \subseteq UX$ for any unitary U in $\mathscr{A}^c$).

Definition 8.3.7 Let $\mathscr{A}$ be a semifinite von Neumann algebra equipped with a faithful normal semifinite trace τ. A closed operator X, affiliated with $\mathscr{A}$ and possessing a domain $\mathfrak{D}(X)$ which is everywhere dense in $\mathscr{H}$ is said to be τ*-measurable* if, for any $\varepsilon > 0$, there exists a projection E in $\mathscr{A}$ such that $E\mathscr{H} \subset \mathfrak{D}(X)$ and $\tau(I - P) < \varepsilon$.

The set $S(\mathscr{A}, \tau)$ of all τ-measurable operators is a $*$-algebra under the adjoint operator, multiplication by a scalar, and operations of strong addition and strong multiplication resulting from the closure of ordinary operations. An unbounded operator X affiliated with $\mathscr{A}$ is τ-measurable (see [31]) if and only if

$$\tau\left(E^{|X|}(n, \infty)\right) \to 0 \ \text{as} \ n \to \infty,$$

where $E^{|X|}(A)$ is the spectral projection of the operator $|X|$, corresponding to the Borel set $A \subset \mathbb{R}$.

Let $\mathcal{L}_+$ and $\mathcal{L}_{\text{sa}}$ denote the positive and self-adjoint parts of a family $\mathcal{L} \subset S(\mathscr{A}, \tau)$, respectively. We denote by $\leq$ the partial order in $S(\mathscr{A}, \tau)_{\text{sa}}$ generated by its proper cone $S(\mathscr{A}, \tau)_+$. If $X \in S(\mathscr{A}, \tau)$ and $X = U|X|$ is the polar decomposition of X, then $U \in \mathscr{A}$ and $|X| = \sqrt{X^*X} \in S(\mathscr{A}, \tau)_+$.

If $X, Y \in S(\mathscr{A}, \tau)_+$, then $X \leq Y$ holds if and only if

$$\mathfrak{D}(Y^{1/2}) \subseteq \mathfrak{D}(X^{1/2}) \ \text{and} \ \|X^{1/2}\xi\| \leq \|Y^{1/2}\xi\|, \ \ \xi \in \mathfrak{D}(Y^{1/2}).$$

Definition 8.3.8 The generalized singular value function $\mu(X) : t \mapsto \mu(t, X)$ of the operator $X \in S(\mathscr{A}, \tau)$ is defined by

$$\mu(t, X) = \inf\{\|XP\| : \ P \in \mathscr{A} \text{ is a projection, } \tau(I - P) \leq t\}.$$

Obviously, $\mu(t, X)$ admits the following *minimax representation*:

$$\mu(t, X) = \inf_{P \in \mathscr{A}_{\text{pr}},\, \tau(I-P) \le t} \; \sup_{\xi \in P\mathscr{H},\, \|\xi\|=1} \|X\xi\|.$$

Thus, for $X \in S(\mathscr{A}, \tau)_+$, we get $\mu(t, X) = \mu(t, X^2)^{1/2}$ and

$$\mu(t, X) = \inf_{P \in \mathscr{A}_{\text{pr}},\, \tau(I-P) \le t} \; \sup_{\xi \in P\mathscr{H},\, \|\xi\|=1} \langle X\xi, \xi\rangle. \tag{8.3.3}$$

For $X \in S(\mathscr{A}, \tau)$, we define the *distribution function* of the operator X by

$$d_X(t) = \tau(E^{|X|}(t, \infty)), \quad t \ge 0.$$

Then $d_X(t) < \infty$ for t large enough and $d_X(t) \to 0$ as $t \to \infty$. The map $t \mapsto d_X(t)$ on $\mathbb{R}_+$ is nonincreasing and continuous from the right, since $E^{|X|}(t_n, \infty) \nearrow E^{|X|}(t, \infty)$ in the strong operator topology as $t_n \searrow t$ by normality of τ.

Lemma 8.3.9 $\mu(t, X) = \inf\{s \ge 0 : \; d_X(s) \le t\}$ *for all* $t > 0$*; the infimum is attained and* $d_X(\mu(t, X)) \le t,\; t > 0$.

Proof As the map $s \mapsto d_X(s)$ is continuous from the right, the second assertion is obvious. Denote the infimum in the lemma by b. The relation $d_X(b) \le t$ means $\tau(I - P) \le t$ with $P = E^{|X|}[0, b]$. But, $\|XP\| = \|\,|X|P\| \le b$ and $\mu(t, X) \le b$.

On the other hand, put $\varepsilon > 0$ and $P \in \mathscr{A}_{\text{pr}}$ such that $\tau(I - P) \le t$ and

$$\|XP\| < \mu(t, X) + \varepsilon = \beta.$$

If $\xi \in P\mathscr{H} \cap E^{|X|}(\beta, \infty)\mathscr{H}$, $\|\xi\| = 1$, then $\langle X^*X\xi, \xi\rangle \ge \beta^2$ and $\langle X^*X\xi, \xi\rangle < \beta^2$. Hence, there are no such vectors ξ, then $P \wedge E^{|X|}(\beta, \infty) = 0$ and

$$E^{|X|}(\beta, \infty) = E^{|X|}(\beta, \infty) - P \wedge E^{|X|}(\beta, \infty) \sim P \vee E^{|X|}(\beta, \infty) - P \le I - P$$

in the Murray–von Neumann sense. Thus,

$$d_X(\beta) \le \tau(I - P) \le t, \quad b \le \beta = \mu(t, X) + \varepsilon.$$

Since ε is arbitrary, the proof is completed. □

Remark 8.3.10 1. Let $\mathscr{B}$ be any von Neumann subalgebra of $\mathscr{A}$ containing the spectral projections of X. The above proof shows that

$$\mu(t, X) = \inf\{\|XP\| : \; P \;\text{ is a projection in } \mathscr{B} \text{ with } \tau(I - P) \le t\}.$$

2. If $\mathscr{A} = L_\infty(\Omega, \nu)$ with the trace $\tau(\cdot) = \int_\Omega \cdot\, d\nu$, given by integration with respect to ν, then $S(\mathscr{A}, \tau)$ consists of all ν-measurable functions on Ω bounded

everywhere but for a set of finite measure. For $\phi \in S(\mathscr{A}, \tau)$, the generalized singular value function

$$\mu(t, \phi) = \inf\{s > 0 : \ \nu(\{\omega \in \Omega : \ |\phi(\omega)| > s\}) \le t\}$$

is precisely the classical decreasing rearrangement of the function ϕ, which is usually denoted by $\phi^*(t)$ (see [32]).

These two remarks are useful tricks to reduce the analysis of a generalized singular value function for a single operator to the classical commutative situation.

For any closed and densely defined linear operator $X : \mathfrak{D}(X) \to \mathscr{H}$, the *null projection* $n(X) = n(|X|)$ is the projection onto its kernel, the *range projection* $P_{\mathrm{ran}(X)}$ is the projection onto the closure of its range, and the *support projection* $\mathrm{supp}(X)$ of X is defined by $\mathrm{supp}(X) = I - n(X)$.

For any $t > 0$, put $\mathcal{R}_t = \{Y \in S(\mathscr{A}, \tau) : \ \tau(\mathrm{supp}(|Y|) \le t\}$.

Lemma 8.3.11 *For any $X \in S(\mathscr{A}, \tau)$ and $t > 0$, $\mu(t, X)$ is exactly the "approximation number"*

$$\mathrm{dist}(X, \mathcal{R}_t) = \inf\{\|X - Y\| : \ Y \in \mathcal{R}_t\}.$$

Proof Let $X = U|X|$ be the polar decomposition of X and $Y = U\int_{(\beta,\infty)} \lambda dE_\lambda^{|X|}$ with $\beta = \mu(t, X)$. Then $\|X - Y\| \le \beta = \mu(t, X)$, and by Lemma 8.3.9 $\tau(\mathrm{supp}(|Y|) = d_X(\beta) \le t$. Hence, we get $\mathrm{dist}(X, \mathcal{R}_t) \le \mu(t, X)$.

On the other hand, put $Y \in \mathcal{R}_t$ and $P = I - \mathrm{supp}(|Y|)$. As $XP = (X - Y)P$, we obtain $\|XP\| \le \|X - Y\|$. But $\tau(I - P) \le t$ so that $\mu(t, X) \le \|X - Y\|$ and $\mu(t, X) \le \mathrm{dist}(X, \mathcal{R}_t)$. □

Lemma 8.3.12 *Let $X, \ Y \in S(\mathscr{A}, \tau)$.*

(i) The map $t \mapsto \mu(t, X)$ is nonincreasing and continuous from the right; moreover, $X \in \mathscr{A}$ if and only if $(\|X\| =) \lim_{t\to 0+} \mu(t, X) < \infty$;
(ii) $\mu(t, X) = \mu(t, |X|) = \mu(t, X^*)$ and $\mu(t, \lambda X) = |\lambda|\mu(t, X)$ for all $t > 0$ and $\lambda \in \mathbb{C}$;
(iii) $\mu(t, X) \le \mu(t, Y)$ for all $t > 0$ if $0 \le X \le Y$;
(iv) $\mu(t, f(|X|)) = f(\mu(t, X))$ for all continuous increasing functions $f : \mathbb{R}^+ \to \mathbb{R}^+$ with $f(0) \ge 0$ and $t > 0$;
(v) $\mu(s + t, X + Y) \le \mu(s, X) + \mu(t, Y)$ for all $s, t > 0$;
(vi) $\mu(t, AXB) \le \|A\|\|B\|\mu(t, X)$ for all $A, B \in \mathscr{A}$ and $t > 0$;
(vii) $\mu(s + t, XY) \le \mu(s, X)\mu(t, Y)$ for all $s, t > 0$.

Proof (i) The monotonicity seems clear. If $\mu(\cdot, X)$ is not continuous from the right at t, then

$$\mu(t, X) > \beta \ge \mu(t + \varepsilon, X) \ \text{ for all } \ \varepsilon > 0.$$

Then $d_X(\beta) \le d_X(\mu(t + \varepsilon, X)) \le t + \varepsilon$ by Lemma 8.3.9, and $d_X(\beta) \le t$. Therefore, $\mu(t, X) \le \beta$, a contradiction.

The relation $\mu(t, X) \le \|X\|$ follows from definition. Assume that $\|X\| > \beta \ge \mu(\varepsilon, X)$, $\varepsilon > 0$. Then we get $d_X(\beta) = 0$ as before and $\|X\| \le \beta$, a contradiction.

(ii) follows immediately from Lemma 8.3.9 or 8.3.11.

(iii) From $0 \le X \le Y$, we get

$$E^X(s,\infty) \wedge E^Y[0,s] = 0, \quad E^X(s,\infty) \overset{<}{\sim} E^Y(s,\infty)$$

in the Murray–von Neumann sense (as in the proof of Lemma 8.3.9; we write $P \overset{<}{\sim} Q$ if $P \le R \sim Q$ for some $R \in \mathscr{A}_{\text{pr}}$). Hence, $d_s(X) \le d_s(Y)$, $s \ge 0$, and the result follows.

(iv) Let $\mathscr{N}$ be the von Neumann subalgebra of $\mathscr{A}$ generated by the spectral projections of $|X|$. It obviously contains the spectral projections of $f(|X|)$. For any $P \in \mathscr{N}_{\text{pr}}$, we get $f(\||X|P\|) = \|f(|X|)P\|$ since f is continuous and increasing, and satisfies $f(0) \ge 0$. Using Remark 8.3.10, we get $\mu(t, f(|X|)) = f(\mu(t, X))$.

(v), (vi), and (vii) are deduced from Lemma 8.3.11. We only prove (vii) ((v) and (vi) are easily verified). Fix $\varepsilon > 0$. By Lemma 8.3.11 there exist $A, B \in S(\mathscr{A}, \tau)$ such that

$$\|X - A\| \le \mu(s, X) + \varepsilon, \quad \tau(\text{supp}(|A|)) \le s,$$

$$\|Y - B\| \le \mu(t, Y) + \varepsilon, \quad \tau(\text{supp}(|B|)) \le t.$$

For $C = (X - A)B + AY$ we get

$$\|XY - C\| = \|(X - A)(Y - B)\| \le (\mu(s, X) + \varepsilon)(\mu(t, Y) + \varepsilon).$$

We obtain $\tau(\text{supp}(|C|)) \le s + t$ thanks to the following three relations:

$$\text{supp}(|C|) \le \text{supp}(|(X - A)B|) \vee \text{supp}(|AY|),$$

$$\text{supp}(|(X - A)B|) \le \text{supp}(|B|),$$

$$\text{supp}(|AY|) \sim \text{supp}(|Y^*A^*|) \le \text{supp}(|A^*|) \sim \text{supp}(|A|).$$

Hence, Lemma 8.3.11 yields $\mu(s + t, XY) \le (\mu(s, X) + \varepsilon)(\mu(t, Y) + \varepsilon)$. □

Lemma 8.3.13 *Let $X \in S(\mathscr{A}, \tau)$ and $P \in \mathscr{A}_{\text{pr}}$. Then $\mu(t, XP) = 0$ for $t \ge \tau(P)$. In particular, if $\tau(I) = a$ is finite, then $\mu(t, X) = 0$ for $t \ge a$.*

Proof It follows from Lemma 8.3.11. □

For a positive self-adjoint operator $X = \int_0^\infty \lambda dE_\lambda^X$ affiliated with $\mathscr{A}$, we set

$$\tau(X) := \sup_n \tau\left(\int_0^n \lambda dE_\lambda^X\right) = \int_0^\infty \lambda d\tau(E_\lambda^X).$$

For $0 < p < \infty$, $L_p(\mathscr{A}, \tau)$ is defined as the set of all densely defined closed operators X affiliated with $\mathscr{A}$ such that $\|X\|_p = \tau(|X|^p)^{1/p} < \infty$.

Theorem 8.3.14 *If $X \in S(\mathscr{A}, \tau)_+$, then $\tau(X) = \int_0^\infty \mu(t, X)dt$.*

Proof If $X \in \mathscr{A}_+$, then using the spectral projections, one can write X as the norm limit of an increasing sequence (X_n) of operators having the form as in Exercise 8.8.4 below. Then $\mu(t, X_n) \nearrow \mu(t, X)$ as $n \to \infty$. In fact, it follows from Lemma 8.3.12, (i), (v) that $|\mu(t, A) - \mu(t, B)| \le \|A - B\|$ for all operators $A, B \in \mathscr{A}$. Hence, $\int_0^\infty \mu(t, X_n)dt \nearrow \int_0^\infty \mu(t, X)dt$ as $n \to \infty$ by the monotone convergence theorem. On the other hand, obviously that $\tau(X_n) \nearrow \tau(X)$ as $n \to \infty$.

Assume now that X is a general positive τ-measurable operator. For every $n \in \mathbb{N}$, put $P_n = E^X[0, n]$. By Lemma 8.3.12(iii)

$$\mu(t, XP_1) \le \mu(t, XP_2) \le \cdots \le \mu(t, X).$$

We show that $\lim_{n\to\infty} \mu(t, XP_n) = \mu(t, X)$. Assume that $s = \lim_{n\to\infty} \mu(t, XP_n) < \mu(t, X)$. Then the sequence $E^{TP_n}(\mu(t, XP_n), \infty) = E^T(\mu(t, XP_n), n])$ converges to $E^X[s, \infty)$ in the strong operator topology. Since $\tau(E^{TP_n}(\mu(t, XP_n), \infty)) \le t$, via the lower semicontinuity of τ, we obtain

$$\tau(E^X(s, \infty)) \le \tau(E^X[s, \infty)) \le \liminf_{n\to\infty} \tau(E^{TP_n}(\mu(t, XP_n), \infty)) \le t,$$

which contradicts $s < \mu(t, X)$. Hence,

$$\mu(t, XP_n) \nearrow \mu(t, X) \ \text{ and } \ \int_0^\infty \mu(t, XP_n)dt \nearrow \int_0^\infty \mu(t, X)dt$$

as $n \to \infty$ by the monotone convergence theorem. On the other hand, $\tau(XP_n) \nearrow \tau(X)$ as $n \to \infty$ (from the definition). Thus, the general case is reduced to the bounded case. □

If $X \in S(\mathscr{A}, \tau)_{\mathrm{sa}}$ and $f : \mathbb{R} \to \mathbb{C}$ is a Borel function, then the normal operator $f(X)$ is defined via functional calculus by

$$f(X) = \int_0^\infty f(\lambda)\, dE_\lambda^X,$$

see Sect. 1.5. Recall that the spectral measure $E^{f(X)}$ of $f(X)$ is given by

$$E^{f(X)}(B) = E^X(f^{-1}(B)), \quad B \in \mathrm{Bor}(\mathbb{C}), \tag{8.3.4}$$

where $\mathrm{Bor}(\mathbb{C})$ denotes the σ-algebra of all Borel subsets in $\mathbb{C}$. Since E^X takes values in the projections of $\mathscr{A}$, it is clear from (8.3.4) that $E^{f(X)}(B)$ lies in $\mathscr{A}$ for any $B \in \mathrm{Bor}(\mathbb{C})$, and so $f(X)$ is affiliated with $\mathscr{A}$.

Lemma 8.3.15 *If $X \in S(\mathscr{A}, \tau)_{\mathrm{sa}}$ and $f : \mathbb{R} \to \mathbb{C}$ is a Borel function bounded on bounded subsets of $\mathbb{R}$, then $f(X) \in S(\mathscr{A}, \tau)$.*

Proof We show that $\tau(E^{|f(X)|}(s,\infty)) < \infty$ for some $s > 0$. By (8.3.4) we get

$$E^{|f(X)|}(s,\infty) = E^X(\{\lambda \in \mathbb{R} : |f(\lambda)| > s\} \tag{8.3.5}$$

for all $s > 0$. Since $X \in S(\mathscr{A},\tau)_{\text{sa}}$, there exists $t > 0$ such that $\tau(E^{|X|}(t,\infty)) < \infty$ and, since f is bounded on the interval $[-t,t]$, there exists a constant $C > 0$ such that $|f(\lambda)| \le C$ for all $\lambda \in [-t,t]$. This implies that

$$\{\lambda \in \mathbb{R} : |f(\lambda)| > C\} \subseteq \{\lambda \in \mathbb{R} : |\lambda| > t\}$$

and so it follows from (8.3.5) that

$$\begin{aligned} E^{|f(X)|}(C,\infty) &= E^X(\{\lambda \in \mathbb{R} : |f(\lambda)| > C\}) \\ &\le E^X(\{\lambda \in \mathbb{R} : |\lambda| > t\}) = E^{|X|}(t,\infty). \end{aligned}$$

Therefore, $\tau(E^{|f(X)|}(C,\infty)) \le \tau(E^{|X|}(t,\infty)) < \infty$. □

Let $\mathbb{F}$ be the set of all continuous increasing functions $f : \mathbb{R}_+ \to \mathbb{R}_+$ with $f(0) = 0$. By Theorem 8.3.14 we get

Corollary 8.3.16 *For all $f \in \mathbb{F}$ and $X \in S(\mathscr{A},\tau)$,*

$$\tau(f(|X|) = \int_0^\infty f(\mu(t,X))dt.$$

In particular, $\|X\|_p = \left(\int_0^\infty \mu(t,X)^p dt\right)^{1/p}$ *for* $< p < \infty$.

Proof Apply Lemma 8.3.12(iv), and Theorem 8.3.14. □

Corollary 8.3.17 *For all $X, Y \in S(\mathscr{A},\tau)_+$ the following conditions are equivalent:*

(i) $\mu(t,X) \le \mu(t,Y)$, $t > 0$;
(ii) $d_X(s) \le d_Y(s)$, $s > 0$;
(iii) $\tau(f(X)) \le \tau(f(Y))$ *for any* $f \in \mathbb{F}$.

Proof (i) $\Rightarrow$ (iii) follows from Corollary 8.3.16.

(iii) $\Rightarrow$ (ii) By approximation of the indicator function $\chi_{(s,\infty)}$ from below by a sequence $(f_n) \subset \mathbb{F}$ we obtain $\tau(f_n(X)) \le \tau(f_n(Y))$ by the assuption. It follows from Lebesgue's dominated convergence theorem (applied to $d\|E_s^X\xi\|$, ξ is a vector) that $f_n(X) \nearrow E^X(s,\infty)$ and $f_n(Y) \nearrow E^Y(s,\infty)$ in the strong operator topology as $n \to \infty$. Then the normality of τ implies $d_X(s) \le d_Y(s)$.

(ii) $\Rightarrow$ (i) follows from Lemma 8.3.9. □

Define the two-sided ideal of τ-compact operators as

$$S_0(\mathscr{A},\tau) := \{X \in S(\mathscr{A},\tau) : \lim_{t\to\infty} \mu(t,X) = 0\}.$$

The two-sided ideal $\mathcal{F}(\tau)$ in $\mathscr{A}$ consisting of all elements of τ-finite range is defined by

$$\mathcal{F}(\tau) = \{X \in \mathscr{A} : \tau(P_{\mathrm{ran}(X)}) < \infty\} = \{X \in \mathscr{A} : \tau(\mathrm{supp}(X)) < \infty\}.$$

Equivalently, $\mathcal{F}(\tau) = \{X \in \mathscr{A} : \mu(t, X) = 0 \text{ for some } t > 0\}$.

Lemma 8.3.18 *If $X, Y \in S_0(\mathscr{A}, \tau)$ and $s, t > 0$, then*

(i) $\tau(I - E_{s+t}^{|X+Y|}) \le \tau(I - E_s^{|X|}) + \tau(I - E_t^{|Y|})$;
(ii) $\tau(I - E_{st}^{|XY|}) \le \tau(I - E_s^{|X|}) + \tau(I - E_t^{|Y|})$.

Proof (i). If $A \in S_0(\mathscr{A}, \tau)$, then $\tau(I - E_\lambda^{|A|}) < \infty$ for any $\lambda > 0$. Assume that $\tau(I - E_{s+t}^{|X+Y|}) > 0$. If $\tau(I - E_s^{|X|}) = \tau(I - E_t^{|Y|}) = 0$, the assertion follows from the triangle inequality $\|X + Y\| \le \|X\| + \|Y\|$, since in this case, $X, Y \in \mathscr{A}$. Without loss of generality, let $\tau(I - E_t^{|Y|}) > 0$. Fix $a > 0$ and put $b_s = \tau(I - E_s^{|X|})$ and $b_t = \tau(I - E_t^{|Y|})$. For

$$\alpha = \frac{b_s}{b_s + b_t} + \frac{a-1}{2} \quad \text{and} \quad \beta = \frac{b_t}{b_s + b_t} + \frac{a-1}{2},$$

it holds that

$$\begin{aligned}\tau(I - E_{s+t}^{|X+Y|}) &= \sup\{\varepsilon \ge 0 : \mu(\varepsilon, X + Y) > s + t\} \\ &\le a \sup\{\varepsilon \ge 0 : \mu(\alpha\varepsilon, X) + \mu(\beta\varepsilon, Y) > s + t\}.\end{aligned}$$

Since $b > b_s$ implies that $\mu(b, X) \le s$, and $b > b_t$ implies that $\mu(b, Y) \le t$, we obtain $\mu(\alpha(b_s + b_t), X) + \mu(\beta(b_s + b_t), Y) \le s + t$. Hence,

$$\tau(I - E_{s+t}^{|X+Y|}) \le a(\tau(I - E_s^{|X|}) + \tau(I - E_t^{|Y|})).$$

Letting $a \to 1$, we get the result.

The proof of assertion (ii) is similar to the one given above. □

We recall the definition of the measure topology t_τ on the algebra $S(\mathscr{A}, \tau)$. For every $\varepsilon, \delta > 0$, we define the set

$$V(\varepsilon, \delta) = \{X \in S(\mathscr{A}, \tau) | \text{ there exists } P \in \mathscr{A}_{\mathrm{pr}} \text{ such that } \|XP\|_\infty \le \varepsilon,\ \tau(I - P) \le \delta\}.$$

The topology generated by the sets $V(\varepsilon, \delta), \varepsilon, \delta > 0$, is called the *measure topology* t_τ on $S(\mathscr{A}, \tau)$ [31, 33]. It is well known that the algebra $S(\mathscr{A}, \tau)$ equipped with the measure topology is a complete metrizable topological $*$-algebra [33]. Note that a sequence $\{X_n\}_{n=1}^\infty \subset S(\mathscr{A}, \tau)$ converges to zero with respect to the measure topology t_τ (that is, $X_n \xrightarrow{\tau} 0$) if and only if $\tau\big(E^{|X_n|}(\varepsilon, \infty)\big) \to 0$ as $n \to \infty$ for all $\varepsilon > 0$. Moreover, $\mathscr{A}$ is t_τ-dense in $S(\mathscr{A}, \tau)$. In fact, if $X = U|X| \in S(\mathscr{A}, \tau)$ and $|X| = \int_0^\infty t dE_t^{|X|}$, then the sequence$(U \int_0^n t dE_t^{|X|})_{n=1}^\infty$ in $\mathscr{A}$ tends to X as $n \to \infty$ in the measure topology.

Lemma 8.3.19 *Let $X_n, X \in S(\mathscr{A}, \tau)$ for $n = 1, 2, \ldots$. Then $(X_n)_{n=1}^{\infty}$ converges to X in the measure topology if and only if $\lim_{n\to\infty} \mu(t, X - X_n) = 0$ for each $t > 0$.*

Proof We show that $Y \in V(\varepsilon, \delta)$ if and only if $\mu(\delta, Y) \leq \varepsilon$. If $Y \in V(\varepsilon, \delta)$, we clearly have $\mu(\delta, Y) \leq \varepsilon$ by Definition 8.3.8. Conversely, if $\mu(\delta, Y) \leq \varepsilon$, then $P = E^{|Y|}[0, \mu(\delta, Y)])$ satisfies $\tau(I - P) = d_Y(\mu(\delta, Y))$ (Lemma 8.3.9). Since

$$\|YP\| = \| \, |Y|P\| \leq \mu(\delta, Y) \leq \varepsilon,$$

we get $Y \in V(\varepsilon, \delta)$. □

Proposition 8.3.20 *For $X \in S(\mathscr{A}, \tau)$, the following conditions are equivalent:*

(i) $d_X(\varepsilon) < \infty$ *for all* $\varepsilon > 0$;
(ii) $X \in S_0(\mathscr{A}, \tau)$;
(iii) there exists a sequence (X_n) in $\mathcal{F}(\tau)$ converging to X in the measure topology.

Proof (iii) $\Rightarrow$ (ii) For any $\varepsilon > 0$, choose a natural n_0 with $\mu(1, X - X_{n_0}) \leq \varepsilon$ (by Lemma 8.3.19). Since $\mu(t, X_{n_0}) = 0$ for $t \geq \tau(\mathrm{supp}(|X_{n_0}|))$ (by Lemma 8.3.13), Lemma 8.3.12, (v) implies that

$$\mu(t, X) \leq \mu(t-1, X_{n_0}) + \mu(1, X - X_{n_0}) \leq \varepsilon$$

for $t \geq \tau(\mathrm{supp}(|X_{n_0}|)) + 1$.

(ii) $\Rightarrow$ (i) For any $\varepsilon > 0$, pick up $t_0 > 0$ such that $\mu(t_0, X) \leq \varepsilon$. Then we get

$$\infty > t_0 \geq d_X(\mu(t_0, X)) \geq d_X(\varepsilon) \qquad \text{(by Lemma 8.3.9)}$$

.

(i) $\Rightarrow$ (iii) If $X = U|X| = U \int_0^{\infty} \lambda \, dE_{\lambda}^{|X|}$, then the sequence $(U \int_{1/n}^{n} \lambda \, dE_{\lambda}^{|X|})_{n=1}^{\infty}$ does the job. □

Thus, $S_0(\mathscr{A}, \tau)$ is the closure of $\mathcal{F}(\tau)$ with respect to the measure topology. The following "Fatou's lemma" is useful.

Lemma 8.3.21 *If $(X_n)_{n=1}^{\infty}$ is a sequence of τ-measurable operators converging to $X \in S(\mathscr{A}, \tau)$ in the measure topology, then*

(i) $\mu(t, X) \leq \liminf_{n\to\infty} \mu(t, X_n)$ *for any* $t > 0$;
(ii) $\mu(t, X) = \lim_{n\to\infty} \mu(t, X_n)$ *if the map* $s \mapsto \mu(s, X)$ *is continuous at* $s = t$, *or if* $\mu(t, X_n) \leq \mu(t, X)$.

Proof (i) For any $\varepsilon > 0$, Lemma 8.3.12, (v) yields the inequality

$$\mu(t + \varepsilon, X) - \mu(\varepsilon, X - X_n) \leq \mu(t, X_n).$$

By taking the $\liminf_{n\to\infty}$ of both sides of this inequality and applying Lemma 8.3.19, we get

$$\mu(t+\varepsilon, X) \le \liminf_{n\to\infty} \mu(t, X_n).$$

Letting $\varepsilon \downarrow 0$, we conclude by Lemma 8.3.12, (i), that $\mu(t, X) \le \liminf_{n\to\infty} \mu(t, X_n)$.

(ii) For small $\varepsilon > 0$ $(0 < \varepsilon < t)$, we get as before

$$\mu(t, X_n) \le \mu(t-\varepsilon, X) + \mu(\varepsilon, X_n - X), \quad \limsup_{n\to\infty} \mu(t, X_n) \le \mu(t-\varepsilon, X).$$

If $s \mapsto \mu(s, X)$ is continuous at $s = t$, then by letting $\varepsilon \downarrow 0$ we get

$$\limsup_{n\to\infty} \mu(t, X_n) \le \mu(t, X) (\le \liminf_{n\to\infty} \mu(t, X_n)).$$

If $\mu(t, X_n) \le \mu(t, X)$ for all n, the result is clear. □

Theorem 8.3.22 *If $(X_n)_{n=1}^{\infty}$ is a sequence of positive τ-measurable operators converging to $X \in S(\mathscr{A}, \tau)$ in the measure topology, then*

(i) (Fatou's lemma) $\tau(X) \le \liminf_{n\to\infty} \tau(X_n)$.
(ii) (Monotone convergence theorem.) If $X_n \le X$ (or even $\mu(t, X_n) \le \mu(t, X)$, $t > 0$), then $\tau(X) = \lim_{n\to\infty} \tau(X_n)$.

Proof (i) Consider

$$\begin{aligned}
\tau(X) &= \int_0^\infty \mu(t, X)\, dt \quad \text{(by Theorem 8.3.14)} \\
&\le \int_0^\infty \liminf_{n\to\infty} \mu(t, X_n)\, dt \quad \text{(by Lemma 8.3.21)} \\
&\le \liminf_{n\to\infty} \int_0^\infty \mu(t, X_n)\, dt \quad \text{(by the usual Fatou's lemma)} \\
&= \liminf_{n\to\infty} \tau(X_n).
\end{aligned}$$

(ii) It follows from (i) and the relations

$$\tau(X_n) = \int_0^\infty \mu(t, X_n)\, dt \le \int_0^\infty \mu(t, X)\, dt = \tau(X).$$

□

The function $\mu(X)$ allows us to obtain several nontrivial and useful inequalities for the real function and the trace. We use τ to denote both the extension of τ from $\mathscr{A}_{\tau+}$ to a unique bounded linear functional on $\mathscr{A}_\tau$ $(= \mathscr{A} \cap L_1(\mathscr{A}, \tau))$ and the subsequent extension to the entire space $L_1(\mathscr{A}, \tau)$. If $X, Y \in S(\mathscr{A}, \tau)$ and $XY, YX \in L_1(\mathscr{A}, \tau)$, then $\tau(XY) = \tau(YX)$; see [34, Theorem 17]. It is known that $L_1(\mathscr{A}, \tau)$ is order-isomorphic to the predual $\mathscr{A}_*$.

If $\mathscr{A} = \mathbb{B}(\mathscr{H})$ and $\tau = \mathrm{tr}$ is the canonical trace, then $S(\mathscr{A}, \tau)$ coincides with $\mathbb{B}(\mathscr{H})$ and $L_p(\mathscr{A}, \tau)$ coincides with the Schatten ideal $\mathcal{C}_p$, $0 < p < \infty$. In this case,

$$\mu(t, X) = \sum_{n=1}^{\infty} s_n(X)\chi_{[n-1,n)}(t), \quad t > 0,$$

where $\{s_n(X)\}_{n=1}^{\infty}$ is the sequence of singular values of an operator X [35, Chap. 1]; here χ_A is the indicator function of a set $A \subset \mathbb{R}$.

Example 8.3.23 ([36]). Similar to (7.1), we can define the entropy of an operator $X \in L_1(\mathscr{A}, \tau)$, $X \geq 0$, by $H(X) = -\tau(X \log X)$. Then $H(X)$ is well-defined as finite values or $-\infty$, and is concave, that is,

$$H(tX + (1-t)Y) \geq tH(X) + (1-t)H(Y)$$

for all $X, Y \in L_1(\mathscr{A}, \tau)$, $X, Y \geq 0$, and $0 \leq t \leq 1$.

Let $\mathcal{X}$ be a metrizable topological vector space over the field $\mathbb{C}$, and c_0 be the set of all null sequences in $\mathbb{C}$. A map $\|\cdot\| : \mathcal{X} \to \mathbb{R}^+$ is called an *F-norm* if the following conditions hold:

(F1) $\|x\| = 0$ if and only if $x = 0$;
(F2) $\|ax\| = \|x\|$ for all $x \in \mathcal{X}$ and $a \in \mathbb{C}$, $|a| = 1$;
(F3) $\|x + y\| \leq \|x\| + \|y\|$ for all $x, y \in \mathcal{X}$;
(F4) if $(a_n) \in c_0$, then $(\|a_n x\|) \in c_0$ for all $x \in \mathcal{X}$.

For $f \in \mathbb{F}$, we define (see [37])

$$L_f(\mathscr{A}, \tau) = \{A \in S(\mathscr{A}, \tau) : \text{ there exists } l_A > 0 \text{ such that } f(l_A |A|) \in L_1(\mathscr{A}, \tau)\}.$$

To verify that $L_f(\mathscr{A}, \tau) \subset S_0(\mathscr{A}, \tau)$, put $A \in L_f(\mathscr{A}, \tau)$ and $l_A > 0$ with $a := \tau(f(l_A|A|)) < \infty$. Then

$$\begin{aligned} f(\mu(t, l_A |A|) &\leq t^{-1} \int_0^t f(\mu(s, l_A A))ds \leq t^{-1} \int_0^{\infty} f(\mu(s, l_A A))ds \\ &= t^{-1} a \to 0 \text{ as } t \to \infty \end{aligned}$$

and $f(\mu(t, l_A |A|)) = f(l_A \mu(t, A)) \to 0$ as $t \to \infty$. Thus, $A \in S_0(\mathscr{A}, \tau)$.

Theorem 8.3.24 *The map $\|\cdot\|_f : L_f(\mathscr{A}, \tau) \to \mathbb{R}_+$ defined by the equality*

$$\|A\|_f = \inf\{\varepsilon > 0 : \tau(f(\varepsilon^{-1}|A|)) \leq \varepsilon\}$$

is an F-norm on the $*$-linear space $L_f(\mathscr{A}, \tau)$.

Proof Let $A \in L_f(\mathscr{A}, \tau)$ and $l_A > 0$ satisfy the condition

$$\tau(f(l_A |A|)) = \int_0^{\infty} f(l_A \mu(t, A))dt < \infty.$$

It follows from item (ii) of Lemma 8.3.12 that $A^*, |A| \in L_f(\mathscr{A}, \tau)$. Assume that $B \in L_f(\mathscr{A}, \tau), 2l = \min\{l_A, l_B\}$, and $\mathbb{R}' = \{t > 0 : \mu(t/2, A) \geq \mu(t/2, B)\}$. Then

$$\begin{aligned}\tau(f(l\,|A+B|) &= \int_0^\infty f(l\,\mu(t, A+B))dt \leq \int_0^\infty f(l\,\mu(t/2, A) + l\,\mu(t/2, B))dt \\ &\leq \int_{\mathbb{R}'} f(2l\,\mu(t/2, A))dt + \int_{(0,\infty)\setminus\mathbb{R}'} f(2l\,\mu(t/2, B))dt \\ &\leq 2(\tau(f(l_A\,|A|)) + \tau(f(l_B\,|B|))) < \infty,\end{aligned}$$

so that $A + B \in L_f(\mathscr{A}, \tau)$. If $a \in \mathbb{C} \setminus \{0\}$, then $\tau(f(l_A\,|A|)) = \tau(f(|a|^{-1}l_A\,|aA|))$ $< \infty$ and $aA \in L_f(\mathscr{A}, \tau)$.

For $\|\cdot\|_f$, conditions (F1) and (F2) clearly hold. For (F3), let $A, B \in L_f(\mathscr{A}, \tau)$ and $\varepsilon > 0$ be arbitrary. By the definition of $\|\cdot\|_f$, there exist $a, b > 0$ such that $\tau(f(a^{-1}|A|)) < a$, $\tau(f(b^{-1}|A|)) < b$ and $\|A\|_f < a < \|A\|_f + \varepsilon$, $\|B\|_f < b < \|B\|_f + \varepsilon$. Then by item (vii) of Lemma 8.3.12 we obtain

$$\begin{aligned}\tau(f((a+b)^{-1}\,|A+B|) &= \int_0^\infty \tau(I - E_\lambda^{f((a+b)^{-1}\,|A+B|)})d\lambda \\ &= \int_0^\infty \tau(I - E_{(a+b)f^{-1}(\lambda)}^{|A+B|})d\lambda \\ &\leq \int_0^\infty (\tau(I - E_{af^{-1}(\lambda)}^{|A|})d\lambda + \tau(I - E_{bf^{-1}(\lambda)}^{|B|}))d\lambda \\ &= \int_0^\infty \tau(I - E_\lambda^{f(a^{-1}\,|A|)})d\lambda + \int_0^\infty \tau(I - E_\lambda^{f(b^{-1}\,|B|)})d\lambda \\ &= \tau(f(a^{-1}|A|)) + \tau(f(b^{-1}|B|)) < a + b\end{aligned}$$

and therefore (F3) is true.

To verify (F4), we choose $A \in L_f(\mathscr{A}, \tau)$, $(a_n) \in c_0$, $\varepsilon > 0$ arbitrarily, and, without loss of generality, we assume that $|a_n| < \varepsilon\, l_A$ $(n \in \mathbb{N})$. Then

$$\tau(f(\varepsilon^{-1}|a_n A|)) = \int_0^\infty f(\varepsilon^{-1}|a_n|\,\mu(t, A))dt < \infty.$$

Since f is nondecreasing, $f_n(t) := f(\varepsilon^{-1}|a_n|\,\mu(t, A)) \leq f(l_A\,\mu(t, A))$ for $t > 0$. By the continuity of f and the condition $f(0) = 0$ we obtain $(f_n(t)) \in c_0$ for all $t > 0$. Then by Lebesgue's Theorem $(\tau(f(\varepsilon^{-1}|a_n A|))) \in c_0$ and there exists $k \in \mathbb{N}$ such that $\tau(f(\varepsilon^{-1}|a_n A|)) < \varepsilon$ for all $n > k$. □

Theorem 8.3.25 *$L_f(\mathscr{A}, \tau)$ is a Frechet space for any $f \in \mathbb{F}$.*

Proof Let (A_n) be a $\|\cdot\|_f$-fundamental sequence in $L_f(\mathscr{A}, \tau)(\subset S_0(\mathscr{A}, \tau))$. Then it is t_τ-fundamental (see Exercise 8.8.8). Since $S_0(\mathscr{A}, \tau)$ is t_τ-closed, there exists $A \in S_0(\mathscr{A}, \tau)$ such that $A_n \xrightarrow{\tau} A$ as $n \to \infty$. It follows by theorem of O. E. Tikhonov [38] that

$$f(\varepsilon^{-1}|A_n - A_k|)) \xrightarrow{\tau} f(\varepsilon^{-1}|A_n - A|)) \text{ as } k \to \infty;$$

here $f(\varepsilon^{-1}|A_n - A_k|)) \in L_1(\mathscr{A}, \tau)$ ($\varepsilon > 0$ is arbitrary; $n, k > N(\varepsilon)$) and, applying Fatou's lemma (Lemma 8.3.21) to the sequence $(f(\varepsilon^{-1}|A_n - A_k|)))_k$, we get

$$\tau(f(\varepsilon^{-1}|A_n - A|))) \le \liminf_{k\to\infty} \tau(f(\varepsilon^{-1}|A_n - A_k|))) < \varepsilon.$$

Thus, $A \in L_f(\mathscr{A}, \tau)$ and $\|A_n - A\|_f \to 0$ as $n \to \infty$. □

For $f, g \in \mathbb{F}$, we write $f \prec g$ if there exists $C > 0$ such that $f(t) \le Cg(Ct)$ for all $t > 0$, and we write $f \asymp g$ (equivalence) if $f \prec g$ and $g \prec f$. It is easy to prove the following lemma.

Lemma 8.3.26 *(i) If $f \prec g$, then $L_g(\mathscr{A}, \tau) \subset L_f(\mathscr{A}, \tau)$. Hence, if $f \asymp g$, then $L_f(\mathscr{A}, \tau) = L_g(\mathscr{A}, \tau)$.*
(ii) $f \prec g$ if and only if $g^{-1} \prec f^{-1}$, where f^{-1} is the inverse of f. Hence, $f \asymp g$ if and only if $f^{-1} \asymp g^{-1}$.

Theorem 8.3.27 *(i) If*

$$f, g, h \in \mathbb{F} \text{ and } f^{-1} \cdot g^{-1} \prec h^{-1}, \tag{8.3.6}$$

then $L_f(\mathscr{A}, \tau)L_g(\mathscr{A}, \tau) \subset L_h(\mathscr{A}, \tau)$ and the following holds true:
(a) $\|AB\|_h \le C \max\{\|A\|_f \cdot \|B\|_g, \|A\|_f + \|B\|_g\}$ *($A \in L_f(\mathscr{A}, \tau)$, $B \in L_g(\mathscr{A}, \tau)$, and C is the constant in the definition of $\prec$ in* (8.3.6)*);*
(b) *for all $A \in L_f(\mathscr{A}, \tau)$ and for all $\varepsilon > 0$, there exists $\delta > 0$ such that $B \in L_g(\mathscr{A}, \tau)$ and $\|B\|_g < \delta \Rightarrow \|AB\|_h < \varepsilon$;*
(c) *the map $(A, B) \mapsto AB$ $(\langle L_f(\mathscr{A}, \tau), \|\cdot\|_f\rangle \times \langle L_g(\mathscr{A}, \tau), \|\cdot\|_g\rangle \to \langle L_h(\mathscr{A}, \tau), \|\cdot\|_h\rangle)$ is continuous.*
(ii) If $f, g, h \in \mathbb{F}$ and $f^{-1} \cdot g^{-1} \asymp h^{-1}$, then $L_f(\mathscr{A}, \tau)L_g(\mathscr{A}, \tau) = L_h(\mathscr{A}, \tau)$.

Proof (i)(a). Let $A \in L_f(\mathscr{A}, \tau)$, $B \in L_g(\mathscr{A}, \tau)$, $\varepsilon > 0$ be arbitrary, C be the constant in the definition of $\prec$ in (8.3.6), and $a, b > 0$ be such that

$$\|A\|_f < aC^{-1/2} < \|A\|_f + \varepsilon, \quad \|B\|_g < bC^{-1/2} < \|B\|_g + \varepsilon.$$

Then

$$\begin{aligned}\tau(I - E_{C\lambda}^{h((ab)^{-1}|AB|}) &= \tau(I - E_{abh^{-1}(\lambda)}^{|AB|}) \le \tau(I - E_{abC^{-1}f^{-1}(\lambda)g^{-1}(\lambda)}^{|AB|})\\ &\le \tau(I - E_{aC^{-1/2}f^{-1}(\lambda)}^{|A|}) + \tau(I - E_{bC^{-1/2}g^{-1}(\lambda)}^{|B|})\\ &= \tau(I - E_\lambda^{f(a^{-1}C^{1/2}|A|)}) + \tau(I - E_\lambda^{g(b^{-1}C^{1/2}|B|)}),\end{aligned}$$

$$C^{-1}\int_0^\infty \tau(I-E_\lambda^{h(ab)^{-1}|AB|})d\lambda \le \int_0^\infty \tau(I-E_\lambda^{f(a^{-1}C^{1/2}|A|)})d\lambda + \int_0^\infty \tau(I-E_\lambda^{g(b^{-1}C^{1/2}|B|)})d\lambda < (a+b)C^{-1/2}.$$

Therefore,

$$\tau(h(\max\{ab, C^{1/2}(a+b)\}^{-1}|AB|)) \le \tau(h((ab)^{-1}|AB|) < \max\{ab, C^{1/2}(a+b)\},$$

$AB \in L_h(\mathscr{A}, \tau)$, and (a) is obtained. To verify (b) we prove the inequality

$$h(st) \le C\max\{f(C^{1/2}s), g(C^{1/2}t)\} \quad \text{for all} \quad s, t > 0.$$

Put

$$\lambda_1 = f^{-1}(C^{-1}h(st)), \quad \lambda_2 = g^{-1}(C^{-1}h(st)).$$

If $\lambda_1 \le C^{1/2}s$, then $f(\lambda_1) = C^{1/2}h(st) \le f(C^{1/2}s)$. Since $\lambda_1\lambda_2 = Cst$, it follows from $\lambda_1 > C^{1/2}s$ that $\lambda_2 < C^{1/2}t$ and $C^{-1}h(st) = g(\lambda_2) < g(C^{1/2}t)$. Now assume that $A \in L_f(\mathscr{A}, \tau), \varepsilon > 0$ be arbitrary, $\beta := \varepsilon/4C$, and let $\delta \in (0, \beta)$ be chosen from the condition $\tau(f(C\delta\varepsilon^{-1}|A|)) < \beta$. It follows from item (v) of Lemma 8.3.12 and from the fact, proved above, that $\|B\|_g < \delta$ for $B \in L_g(\mathscr{A}, \tau)$, and that:

$$\int_0^\infty h(\varepsilon^{-1}\mu(t, AB))dt \le 2\int_0^\infty h(\varepsilon^{-1}\mu(t, A)\mu(t, B))dt \le 2C\Big(\int_0^\infty f(C\delta\varepsilon^{-1}\mu(t, A))dt + \int_0^\infty g(\delta^{-1}\mu(t, B))dt\Big) < 2C(\beta+\delta) < \varepsilon.$$

Thus, (b) is proved.

Since the proof of (b) and condition (8.3.6) are symmetric with respect f and g, we also have the following:

(b') for all $B \in L_g(\mathscr{A}, \tau)$ and for all $\varepsilon > 0$, there exists $\delta > 0$ such that $A \in L_f(\mathscr{A}, \tau)$ and $\|A\|_f < \delta \Rightarrow \|AB\|_h < \varepsilon$.

By applying the triangle inequality for $\|\cdot\|_h$ to the relation

$$A_nB_n - AB = (A_n - A)(B_n - B) + A(B_n - B) + (A_n - A)B$$

$(A, A_n \in L_f(\mathscr{A}, \tau), \; B, B_n \in L_g(\mathscr{A}, \tau), \; n \in \mathbb{N})$ and using (a), (b), and (b'), we obtain (c).

(ii). Put $h^{-1} = f^{-1} \cdot g^{-1}$, and let $A \in L_f(\mathscr{A}, \tau)$, $\|A\|_f < 1$. Then $B = h(|A|) \in L_1(\mathscr{A}, \tau)$, $B_1 = f^{-1}(B) \in L_f(\mathscr{A}, \tau)$, $B_2 = g^{-1}(B) \in L_g(\mathscr{A}, \tau)$, and $|A| = B_1 B_2 \in L_f(\mathscr{A}, \tau)L_g(\mathscr{A}, \tau)$. □

Corollary 8.3.28 *If $f \in \mathbb{F}$ and $(f^{-1})^2 \prec f^{-1}$, then $\langle L_f(\mathscr{A}, \tau), \|\cdot\|_f\rangle$ is a topological $*$-algebra.*

The concave function $f(t) = \log(1 + t)$, $t \geq 0$, lies in $\mathbb{F}$, $(f^{-1})^2 \prec f^{-1}$, and meets the Δ_2-condition.

Proposition 8.3.29 *If $\mathscr{A}$ is not of type I, and $\tau(P_{II}) = \infty$, then the converses of (i) of Lemma 8.3.26 and of Theorem 8.3.27 hold.*

Proof If the relation $f \prec g$ is false, then there exists a sequence (t_n) such that $0 < t_n < \infty$ and $f(t_n) > ng(nt_n)$ for all $n \in \mathbb{N}$. Let (a_n) be a sequence of positive numbers such that

$$\sum_{n=1}^{\infty} a_n g(nt_n) < \infty, \quad \sum_{n=1}^{\infty} a_n n g(nt_n) = \infty.$$

Put $A = \sum_{n=1}^{\infty} nt_n P_n$, where (P_n) is a pairwise orthogonal sequence in $\mathscr{A}_{\mathrm{pr}}$ and $\tau(P_n) = a_n$ for all $n \in \mathbb{N}$. Let $N_t = \{n \in \mathbb{N} : nt_n > t\}$ for any $t > 0$. Then

$$\tau(I_{\mathscr{A}} - E_t^{|A|}) = \sum_{n \in N_t} a_n \leq \sum_{n=1}^{\infty} a_n \frac{g(nt_n)}{g(t)} < \infty$$

and $A \in S(\mathscr{A}, \tau)$. It is clear that $A \in L_g(\mathscr{A}, \tau)$; but for every $l > 0$

$$\tau(f(l\,|A|)) = \sum_{n=1}^{\infty} f(lnt_n)\tau(P_n) \geq \sum_{n \geq l^{-1}} a_n n g(nt_n) = \infty,$$

therefore, $A \notin L_f(\mathscr{A}, \tau)$.

To prove the converse of Theorem 8.3.27, we set $\psi^{-1} := f^{-1} g^{-1}$. If $L_h(\mathscr{A}, \tau) = L_f(\mathscr{A}, \tau)L_g(\mathscr{A}, \tau)$, then $L_\psi(\mathscr{A}, \tau) = L_h(\mathscr{A}, \tau)L_g(\mathscr{A}, \tau) = L_h(\mathscr{A}, \tau)$ and $h^{-1} \asymp \psi^{-1} = f^{-1} g^{-1}$. If $L_f(\mathscr{A}, \tau)L_g(\mathscr{A}, \tau) \subset L_h(\mathscr{A}, \tau)$, then

$$L_f(\mathscr{A}, \tau)L_g(\mathscr{A}, \tau) = L_\psi(\mathscr{A}, \tau) \subset L_h(\mathscr{A}, \tau)$$

and, according to what was proved above, we conclude that $h \prec \psi$, that is, $\psi^{-1} \prec h^{-1}$ and $f^{-1} g^{-1} \prec h^{-1}$. □

Lemma 8.3.30 *Let X be a positive and Y be a self-adjoint τ-measurable operators, $t > 0$. If $\mu(t, Y_+) \leq \mu(t, X)$, $\mu(t, Y_-) \leq \mu(t, X)$, where Y_+ and Y_- are positive and negative parts of Y, respectively, then $\mu(2t, Y) \leq \mu(t, X)$.*

Proof If $\mathscr{N}$ is any commutative von Neuman subalgebra of $\mathscr{A}$ containing the spectral projections of Y, then $\mu(t, Y) = \inf\{\|YP\| : P \in \mathscr{N}_{\rm pr}, \tau(I - P) \le t\}$, see Remark 8.3.10. For arbitrary $\varepsilon > 0$, we consider the projections $P_1, P_2 \in \mathscr{N}$ such that

$$\mu(t, Y_+) + \frac{\varepsilon}{2} \ge \|Y_+ P_1\|, \quad \mu(t, Y_-) + \frac{\varepsilon}{2} \ge \|Y_- P_2\|$$

and $\tau(I - P_1) < t, \tau(I - P_2) < t$. Recall that the projection $P := P_1 \wedge P_2$ projects onto $P_1\mathscr{H} \cap P_2\mathscr{H}$, moreover, $P \in \mathscr{N}$ and $\tau(I - P) \le 2t$. Hence,

$$\begin{aligned}\|YP\| &= \|Y_+ P - Y_- P\| \le \max\{\|Y_+ P\|, \|Y_- P\|\}\\ &\le \max\left\{\mu(t, Y_+) + \frac{\varepsilon}{2},\ \mu(t, Y_-) + \frac{\varepsilon}{2}\right\} \le \mu(t, X) + \varepsilon,\end{aligned}$$

which implies that $\mu(2t, Y) \le \mu(t, X)$. □

We are ready to state our last result of this section; see [39].

Theorem 8.3.31 (see (2.4.31)). *Inequality* $\mu(t, X^*Y) \le \mu\left(t, \frac{XX^*+YY^*}{2}\right)$ *holds for all* $X, Y \in S(\mathscr{A}, \tau)$ *and* $t > 0$.

Proof Set $A = \begin{bmatrix} X & Y \\ 0 & 0 \end{bmatrix}$. Then $A^* = \begin{bmatrix} X^* & 0 \\ Y^* & 0 \end{bmatrix}$. By a straightforward calculation, we obtain

$$AA^* = \begin{bmatrix} X & Y \\ 0 & 0 \end{bmatrix} \cdot \begin{bmatrix} X^* & 0 \\ Y^* & 0 \end{bmatrix} = \begin{bmatrix} XX^* + YY^* & 0 \\ 0 & 0 \end{bmatrix},$$

$$A^*A = \begin{bmatrix} A^* & 0 \\ Y^* & 0 \end{bmatrix} \cdot \begin{bmatrix} X & Y \\ 0 & 0 \end{bmatrix} = \begin{bmatrix} X^*X & X^*Y \\ Y^*X & Y^*Y \end{bmatrix}.$$

If

$$U = \begin{bmatrix} 1 & 0 \\ 0 & -1 \end{bmatrix}, \ B = \begin{bmatrix} 0 & X^*Y \\ Y^*X & 0 \end{bmatrix},$$

then

$$\begin{aligned}\|Y \cdot P\| &= \|Y_+ \cdot P - Y_- \cdot P\| \le \max\{\|Y_+ \cdot P\|, \|Y_- \cdot P\|\}\\ &\le \max\left\{\mu(t, Y_+) + \frac{\varepsilon}{2}, \mu(t, Y_-) + \frac{\varepsilon}{2}\right\} \le \mu(t, X) + \varepsilon,\end{aligned}$$

$$
\begin{aligned}
B &= \begin{bmatrix} 0 & X^*Y \\ Y^*X & 0 \end{bmatrix} = \frac{1}{2}\begin{bmatrix} 0 & 2X^*Y \\ 2Y^*X & 0 \end{bmatrix} \\
&= \frac{1}{2}\begin{bmatrix} X^*X & X^*Y \\ Y^*X & Y^*Y \end{bmatrix} - \frac{1}{2}\begin{bmatrix} X^*X & -X^*Y \\ -Y^*X & Y^*Y \end{bmatrix} \\
&= \frac{1}{2}\begin{bmatrix} X^*X & X^*Y \\ Y^*X & Y^*Y \end{bmatrix} - \frac{1}{2}\begin{bmatrix} X^*X & X^*X \\ -Y^*X & -Y^*Y \end{bmatrix} \cdot \begin{bmatrix} 1 & 0 \\ 0 & -1 \end{bmatrix} \\
&= \frac{1}{2}\begin{bmatrix} X^*X & X^*Y \\ Y^*X & Y^*Y \end{bmatrix} - \frac{1}{2}\begin{bmatrix} 1 & 0 \\ 0 & -1 \end{bmatrix} \cdot \begin{bmatrix} X^*X & X^*Y \\ Y^*X & Y^*Y \end{bmatrix} \cdot \begin{bmatrix} 1 & 0 \\ 0 & -1 \end{bmatrix} \\
&= \frac{1}{2}(X^*X - UX^*XU^*).
\end{aligned}
$$

In other words, we conclude that

$$B = \frac{1}{2}A^*A - \frac{1}{2}UA^*AU^*.$$

By [34, Lemma 6], we obtain

$$B = B_+ - B_-, \quad \mu(t, B_+) \leq \mu\Big(t, \frac{1}{2}A^*A\Big) = \mu\Big(t, \frac{1}{2}AA^*\Big)$$

and

$$\mu(t, B_-) \leq \mu\Big(t, \frac{1}{2}UA^*AU^*\Big) \leq \|U\| \cdot \|U^*\| \mu\Big(t, \frac{1}{2}A^*A\Big) = \mu\Big(t, \frac{1}{2}A^*A\Big) = \mu\Big(t, \frac{1}{2}AA^*\Big).$$

By Lemma 8.3.30

$$\mu(2t, B) \leq \mu\Big(t, \frac{1}{2}AA^*\Big).$$

In other words, we can write down the relation

$$\mu\Big(2t, \begin{bmatrix} 0 & X^*Y \\ Y^*X & 0 \end{bmatrix}\Big) \leq \frac{1}{2}\mu\Big(t, \begin{bmatrix} XX^* + YY^* & 0 \\ 0 & 0 \end{bmatrix}\Big),$$

which implies that

$$\mu(t, X^*Y) \leq \frac{1}{2}\mu(t, XX^* + YY^*) = \mu\Big(t, \frac{XX^* + YY^*}{2}\Big).$$

□

8.4 Weak Majorization

Let τ be a faithful normal semifinite trace on a von Neumann algebra $\mathscr{A}$, and let L_1 and L_∞ be Lebesgue spaces on $(0, \tau(I))$. We say that $X \in (L_1 + L_\infty)(\mathscr{A}, \tau)$ if and only if $\mu(\cdot, X) \in (L_1 + L_\infty)(0, \infty)$. Here, we identify $\mathscr{A}$ with $L_\infty(\mathscr{A}, \tau)$. The space $(L_1 + L_\infty)(\mathscr{A}, \tau)$ can also be viewed as a sum of Banach spaces $L_1(\mathscr{A}, \tau)$ and $L_\infty(\mathscr{A}, \tau)$.

Definition 8.4.1 Let τ be a faithful normal semifinite trace on a von Neumann algebra $\mathscr{A}$. For $X, Y \in (L_1 + L_\infty)(\mathscr{A}, \tau)$, we write $X \prec_w Y$ and call the (Hardy–Littlewood–Pólya) *weak majorization*, whenever

$$\int_0^t \mu(s, X)\, ds \le \int_0^t \mu(s, Y)\, ds \quad \text{for all} \quad t > 0.$$

Observe that $X \prec_w Y$ if and only if $\mu(X) \prec_w \mu(Y)$. Sometimes, we also write $X \prec_w \phi$ instead of $\mu(X) \prec_w \mu(\phi)$ for a function ϕ.

Lemma 8.4.2 *Assume that $\mathscr{A}$ has no atoms. Then for any $X \in S(\mathscr{A}, \tau)$,*

$$\int_0^t \mu(s, X) ds = \sup\{\tau(E|X|E) : \; E \in \mathscr{A}_{\text{pr}} \;\text{ with }\; \tau(E) \le t\}.$$

Proof We can assume that $X \ge 0$. Let $X = \int_0^\infty \lambda\, dE_\lambda^X$ be the spectral representation. Take an abelian von Neumann subalgebra $\mathfrak{A} = L_\infty(\Omega, \nu)$ of $\mathscr{A}$ containing spectral projections of X with nonatomic measure ν (corresponding to τ). (By assumption such $\mathfrak{A}$ exists on $\mathscr{A}$.) Hence, for $X = \phi$ in $\mathfrak{A}$, we achieve the equality $\mu(t, X) = \phi^*(t)$ (Remark 8.3.10).

Since (Ω, ν) is nonatomic, we can apply the classical equality

$$\int_0^t \phi^*(s)\, ds = \sup\left\{\int_E |\phi|\, d\nu \; : \; E \subseteq \Omega \;\text{ with }\; \nu(E) \le t\right\}$$

(see [32]). Therefore,

$$\begin{aligned}\int_0^t \mu(s, X) ds &= \sup\{\tau(EXE) : \; E \in \mathfrak{A}_{\text{pr}} \;\text{ with }\; \tau(E) \le t\} \\ &\le \sup\{\tau(EXE) : \; E \in \mathscr{A}_{\text{pr}} \;\text{ with }\; \tau(E) \le t\}.\end{aligned}$$

Conversely, for $E \in \mathscr{A}_{\text{pr}}$ with $\tau(E) \le t$ consider

$$\begin{aligned}\tau(EXE) &= \int_0^\infty \mu(s, EXE)\,ds \qquad \text{(by Theorem 8.3.14)}\\ &= \int_0^t \mu(s, EXE)\,ds \qquad \text{(by Lemma 8.3.13)}\\ &\le \int_0^t \mu(s, X)\,ds \qquad \text{(by Lemma 8.3.12).}\end{aligned}$$

□

Remark 8.4.3 1. If $|X|$ does not have a point spectrum, the supremum can be taken over all projections (of trace at most t) in the von Neumann subalgebra generated by the spectral projections of $|X|$.

2. If $\mathscr{A}$ has atoms, we consider the tensor product $\mathscr{B} = \mathscr{A} \otimes L_\infty[0, 1]$ equipped with the trace φ given by the tensor product of τ with the trace

$$\phi \mapsto \int_0^1 \phi(t)\,dt, \quad \phi \in L_\infty[0, 1].$$

It is clear that $\mathscr{B}$ has no atoms and it is not hard to see that the embedding $X \mapsto X \otimes 1$, $X \in \mathscr{A}$ extends in a natural way to a generalized singular value function preserving the map of $S(\mathscr{A}, \tau)$ onto the $*$-subalgebra $S(\mathscr{A}, \tau) \otimes 1 \subseteq S(\mathscr{B}, \varphi)$. For details, see [40].

Example 8.4.4 Let $X, Y \in S(\mathscr{A}, \tau)_{\mathrm{sa}}$ and $X \le Y$. Then, $X_+ \prec_w Y_+$. Via Remark 8.4.3 it can be assumed that $\mathscr{A}$ has no atoms. Let $t > 0$ be given, and let $E \in \mathscr{A}_{\mathrm{pr}}$ with $\tau(E) \le t$. If $P = \mathrm{supp}(X_+)$, then

$$\begin{aligned}\tau(X_+E) &= \tau(EPXPE) \le \tau(EPYPE)\\ &\le \tau(EPY_+PE) \le \int_0^{\tau(E)} \mu(s, PY_+P)ds \le \int_0^t \mu(s, Y_+)ds.\end{aligned}$$

The result now follows by Lemma 8.4.2.

Let us discuss the Hölder inequality

$$\|XY\|_r \le \|X\|_p \|Y\|_q \quad (p, q, r > 0;\ p^{-1} + q^{-1} = r^{-1}), \tag{8.4.1}$$

see Theorem 8.4.5 and Problem 2.5.29. It seems that the only proof of (8.4.1) is based on the Weyl majorant inequality (see [41])

$$\Lambda_t(XY) \le \Lambda_t(X)\Lambda_t(Y), \quad t > 0,$$

where $\Lambda_t(X) := \exp \int_0^t \log \mu(s, X)\,ds,\ \ t > 0$. If, for example, X satisfies the "Lorentz space-Type condition"

$$X \in \mathscr{A} \ \text{ or } \ \mu(t, X) \le Ct^{-\alpha} \quad (C, \alpha > 0),\ t > 0, \tag{8.4.2}$$

then $\Lambda_t(X)$ is well-defined (that is, $\infty - \infty$ does not occur). Whenever $\Lambda_t(X)$ appears in what follows, we always assume that X meets (8.4.1).

Theorem 8.4.5 *Let $X, Y \in S(\mathscr{A}, \tau)$. The following conditions (i), (ii), and (iii) hold and are mutually equivalent:*

(i) $\|XY\|_r \le \|X\|_p \|Y\|_q, \quad (p, q, r > 0;\ p^{-1} + q^{-1} = r^{-1})$;
(ii) $\Lambda_t(XY) \le \Lambda_t(X)\Lambda_t(Y), \quad t > 0$;
(iii) $\int_0^t f(\mu(s, XY))ds \le \int_0^t f(\mu(s, X)\mu(s, Y))ds$, $t > 0$, *for any increasing function* $f : \mathbb{R}^+ \to \mathbb{R}$ *such that* $s \mapsto f(e^s)$ *is convex.*

This result for $X, Y \in \mathscr{A} \cap S_0(\mathscr{A}, \tau)$ was proved in [41]. We prove it for $X, Y \in S(\mathscr{A}, \tau)$, and show that each of (i)–(iii) can be deduced from the others.

Proof Assume that $\|X\|_p, \|Y\|_q < \infty$. Then the assertion follows from [41, Corollary 4.4, (iii)] and the following trick. Let $X = U|X|$ be the polar decomposition of X and $|X| = \int_0^\infty \lambda\, dE_\lambda^{|X|}$. Put $X_n = U \int_0^n \lambda\, dE_\lambda^{|X|}$ for any $n \in \mathbb{N}$. As $\|X\|_p < \infty$, we have $X_n \in \mathscr{A} \cap S_0(\mathscr{A}, \tau)$ because of

$$\mu(t, X_n) \le \mu(t, X) \to 0 \ \text{ as } \ t \to \infty.$$

Define Y_n similarly. By Corollary 4.4, [41] (that is, essentially (ii) and (iii) for τ-compact operators) we obtain

$$\begin{aligned}\left\{\int_0^\infty \mu(s, X_nY_n)^r\, ds\right\}^{1/r} &\le \left\{\int_0^\infty \mu(s, X_n)^p\, ds\right\}^{1/p} \left\{\int_0^\infty \mu(s, Y_n)^q\, ds\right\}^{1/q} \\ &\le \|X\|_p \|Y\|_q.\end{aligned}$$

Since $X_nY_n \xrightarrow{\tau} XY$ as $n \to \infty$,

$$\begin{aligned}\|XY\|_r &= \left\{\int_0^\infty \mu(s, XY)^r\, ds\right\}^{1/r} \\ &\le \left\{\int_0^\infty \liminf_{n\to\infty} \mu(s, X_nY_n)^r\, ds\right\}^{1/r} && \text{(by Lemma 8.3.21)} \\ &\le \left\{\liminf_{n\to\infty} \int_0^\infty \mu(s, X_nY_n)^r\, ds\right\}^{1/r} && \text{(by the usual Fatou's lemma).}\end{aligned}$$

Combining these two estimates, we get the desired result (i).

(i) $\Rightarrow$ (ii) We first show for $X, Y \in S(\mathscr{A}, \tau)$ that

$$\left\{\int_0^t \mu(s, XY)^r ds\right\}^{1/r} \le \left\{\int_0^t \mu(s, X)^{2r} ds\right\}^{1/2r} \left\{\int_0^t \mu(s, Y)^{2r} ds\right\}^{1/2r}, \quad t > 0, \quad (8.4.3)$$

based on (i). By Remark 8.4.3 we may and do assume that $\mathscr{A}$ has no atoms. Let $X = U|X|$ and $Y = V|Y|$ be the polar decompositions. Let $E \in \mathscr{A}_{\rm pr}$ commute with

$|XY|$ and satisfy $\tau(E) \le t$. Let $Q = \text{supp}(|EY^*|)$ and W be the partial isometry from the polar decomposition of EX^*. Thus,

$$W^*W = Q, \quad WW^* = \text{supp}(|XE|) \le E.$$

A straightforward check then yields

$$E|YX|^2E = (EX^*W^*E)(EW|X|^2W^*E)(EWXE).$$

Thus, (i) implies that

$$\begin{aligned}\|E|YX|^2E\|_r &\le \|EX^*W^*E\|_{4r}\|EW|X|^2W^*E\|_{2r}\|EWXE\|_{4r} \\ &\le \left\{\int_0^t \mu(s,X)^{4r}ds\right\}^{1/4r}\left\{\int_0^t \mu(s,Y)^{4r}ds\right\}^{1/2r}\left\{\int_0^t \mu(s,X)^{4r}ds\right\}^{1/4r}.\end{aligned}$$

Here, Lemma 8.3.13 and Lemma 8.3.12, (ii), (vi) were used. Note that $(E|XY|^2E)^r = E|XY|^{2r}E$. By taking $\mathfrak{A}$, which contains E of Lemma 8.4.2 (where X is replaced with $|XY|^{2r}$), we get (by Lemma 8.4.2)

$$\left\{\int_0^t \mu(s,XY)^{2r}ds\right\}^{1/r} \le \left\{\int_0^t \mu(s,X)^{4r}ds\right\}^{1/2r}\left\{\int_0^t \mu(s,Y)^{4r}ds\right\}^{1/2r},$$

which is precisely (8.4.3) since $r > 0$ is arbitrary. Dividing both sides of (8.4.3) by $t^{1/r}$, we obtain

$$\left\{\int_0^t \mu(s,XY)^r \frac{ds}{t}\right\}^{1/r} \le \left\{\int_0^t \mu(s,X)^{2r} \frac{ds}{t}\right\}^{1/2r}\left\{\int_0^t \mu(s,Y)^{2r} \frac{ds}{t}\right\}^{1/2r}.$$

Now assume that X, Y (hence XY) meet (8.4.2). By the well-known equality:

$$\exp\int_0^t \log|\phi(s)|\frac{ds}{t} = \lim_{r\to 0}\left\{\int_0^t |\phi(s)|^r \frac{ds}{t}\right\}^{1/r},$$

$$\text{if} \quad \int_0^t |\phi(s)|^r \frac{ds}{t} < \infty \quad \text{for some } r > 0$$

(see, for example, p. 74, [42]), we obtain $\Lambda_t(XY) \le \Lambda_t(X)\Lambda_t(Y)$ for any $t > 0$.

(ii) $\Rightarrow$ (iii) This follows from Corollary 4.2, [41], if X and Y are bounded. By the same trick as in the proof of (i), it is clear that (iii) remains valid for any X, Y.

(iii) $\Rightarrow$ (i) follows from the classical Hölder inequality [43, Chap. III] (with $f(s) = s^{1/r}$ and $t = \infty$). □

It is shown in [41] that the Minkowsky inequality

$$\|X+Y\|_p \le \|X\|_p + \|Y\|_p, \quad p \ge 1,$$

can be deduced from the von Neumann inequality

$$\Phi_t(X+Y) \le \Phi_t(X) + \Phi_t(Y), \quad t > 0,$$

where $\Phi_t(X) = \int_0^t \mu(s, X)ds$ (see Lemma 8.4.2); see also [44] for other versions of the Minkowsky inequality. If $X, Y \in S(\mathscr{A}, \tau)$, then there exist partial isometries $U, V \in \mathscr{A}$ such that

$$|X+Y| \le U|X|U^* + V|Y|V^*, \tag{8.4.4}$$

that is, the so-called *operator triangle inequality* holds [45].

Theorem 8.4.6 *Let $X, Y \in S(\mathscr{A}, \tau)$. The following conditions (i), (ii) and (iii) hold and are mutually equivalent:*

(i) $\|X+Y\|_p \le \|X\|_p + \|Y\|_p, \quad p \ge 1;$
(ii) $\Phi_t(X+Y) \le \Phi_t(X) + \Phi_t(Y), \quad t > 0;$
(iii) $\int_0^t f(\mu(s, X+Y))ds \le \int_0^t f(\mu(s, X) + \mu(s, Y))ds,\ t > 0,$ *for any convex continuous increasing function* $f : \mathbb{R}_+ \to \mathbb{R}$.

Proof (i) follows from the $L_p \times L_q$-duality; also it can be proved from the von Neumann inequality (see the proof below).

(ii) Assume that $\mathscr{A}$ has no atoms. Let $E \in \mathscr{A}_{\mathrm{pr}}$ with $\tau(E) \le t$. Put U, V in (8.4.4), and estimate

$$\begin{aligned}\tau(E|X+Y|E) &\le \tau(EU|X|U^*E) + \tau(EV|Y|V^*E)\\ &\le \int_0^t \mu(s, EU|X|U^*E)\,ds + \int_0^t \mu(s, EV|Y|V^*E)\,ds\\ &\le \int_0^t \mu(s, X)\,ds + \int_0^t \mu(s, Y)\,ds\end{aligned}$$

due to Lemma 8.3.13. Thus, Lemma 8.4.2 yields (ii).

(ii) $\Rightarrow$ (iii) By Lemma 4.3, (i), [41], for $X, Y \in \mathscr{A}$ (iii) follows from (ii). For general $X, Y \in S(\mathscr{A}, \tau)$, (iii) remains valid by the trick in the proof of (i) in Theorem 8.4.5.

(iii) $\Rightarrow$ (i) follows from the classical Minkowsky inequality [46, Chap. 9] (with $f(s) = s^p$ and $t = \infty$). □

We show that

$$\Phi_t(X) = \inf\{\|X_1\|_1 + t\|X_2\| \ :\ X = X_1 + X_2,\ X_1 \in L_1(\mathscr{A}, \tau),\ X_2 \in \mathscr{A}\} \tag{8.4.5}$$

for every $X \in S(\mathscr{A}, \tau)$ and $t > 0$. In particular, $X \in (L_1 + L_\infty)(\mathscr{A}, \tau)$ if and only if $\Phi_t(X) < \infty$ for some (hence all) $t > 0$. Formula (8.4.5) is well known in the abelian case: its right-hand side is called the *K-functional* in the theory of real interpolation and is denoted by $K(t, X) = K(t, X; L_1(\mathscr{A}, \tau), \mathscr{A})$ (see, for example, [32]).

Let $X = X_1 + X_2$ be an arbitrary decomposition. For $s > 0$ and $0 < a < 1$, we estimate

$$\begin{aligned}\mu(s, X) &\le \mu(as, X_1) + \mu((1-a)s, X_2) \quad \text{(by Lemma 8.3.12, (v))}\\ &\le \mu(as, X_1) + \|X_2\| \quad \text{(by Lemma 8.3.12, (i))},\end{aligned}$$

$$\begin{aligned}\Phi_t(X) &\le \int_0^t \mu(as, X_1)\,ds + \|X_2\| \le \int_0^\infty \mu(as, X_1)\,ds + \|X_2\|\\ &= a^{-1}\int_0^\infty \mu(s, X_1)\,ds + \|X_2\| = a^{-1}\|X_1\|_1 + t\|X_2\|.\end{aligned}$$

Letting $a \uparrow 1$, we obtain $\Phi_t(X) \le \inf\{\|X_1\|_1 + t\|X_2\|\}$.

To prove the reverse inequality, let $X = U|X|$ be the polar decomposition of X and $|X| = \int_0^\infty \lambda dE_\lambda^{|X|}$ be the spectral representation. Put $a = \mu(t, X)$,

$$X_1 = U\int_a^\infty (\lambda - a)dE_\lambda^{|X|}, \quad X_2 = X - X_1.$$

Since $|X_1| = U\int_a^\infty(\lambda - a)dE_\lambda^{|X|} = f(|X|)$ with

$$f(\lambda) = \begin{cases}0, & \text{if } 0 \le \lambda \le a,\\ \lambda - a, & \text{if } \lambda \ge a,\end{cases}$$

we get

$$\begin{aligned}\mu(s, X_1) &= f(\mu(s, X)) \quad \text{(by Lemma 8.3.12, (ii), (iv))}\\ &= \begin{cases}\mu(s, X) - a, & \text{if } 0 < s < t,\\ 0, & \text{if} s \ge t.\end{cases}\end{aligned}$$

Since $\|X_2\| \le a$, we obtain

$$\begin{aligned}\|X_1\|_1 + t\|X_2\| &\le \int_0^\infty \mu(s, X_1)\,ds + ta\\ &= \int_0^t (\mu(s, X) - a)\,ds + ta = \int_0^t \mu(s, X)\,ds.\end{aligned}$$

Thus, formula (8.4.5) is proved.

Theorem 8.4.7 *(Jensen-type inequality). Let $X \in S(\mathscr{A}, \tau)_+$ and U be a contraction in $\mathscr{A}$. For any convex function $f \in \mathbb{F}$, where $\mathbb{F}$ is the set of all continuous increasing functions $f : \mathbb{R}_+ \to \mathbb{R}_+$ with $f(0) = 0$,*

$$\mu(t, f(UXU^*)) = f(\mu(t, UXU^*) \le \mu(t, Uf(X)U^*) \ \textit{for all} \ t > 0.$$

If f is concave instead of convex, we obtain the reverse inequality.

Proof Consider only the convex case (the concave case can be proved similarly). Since f is convex, it is of the form

$$f(t)=\sup_{j\in J}(a_j t+b_j),\quad t\geq 0,$$

with $a_j\geq 0$ and $b_j\leq 0$. For any unit vector ξ and $j\in J$, we get

$$\begin{aligned}\langle Uf(X)U^*\xi,\xi\rangle &\geq \langle U(a_jX+b_jI)U^*\xi,\xi\rangle\\ &=a_j\langle UXU^*\xi,\xi\rangle+b_j\|U^*\xi\|^2\geq a_j\langle UXU^*\xi,\xi\rangle+b_j\end{aligned}$$

since $b_j\leq 0$ and $\|U^*\xi\|^2\leq 1$. Take the supremum over j and achieve

$$\langle Uf(X)U^*\xi,\xi\rangle\geq f(\langle UXU^*\xi,\xi\rangle).$$

Applying the expression on the right of (8.3.3), we obtain

$$\begin{aligned}\mu(t,Uf(X)U^*)&=\inf\sup\langle Uf(X)U^*\xi,\xi\rangle\geq\inf\sup f(\langle UXU^*\xi,\xi\rangle)\\ &=f(\inf\sup\langle UXU^*\xi,\xi\rangle)\qquad\text{(since } f \text{ is continuous and increasing)}\\ &=f(\mu(t,UTU^*)).\end{aligned}$$

□

Proposition 8.4.8 (see Corollary 8.3.3). *Let $f\in\mathbb{F}$, $X,Y\in S(\mathscr{A},\tau)$ and A and B be elements in $\mathscr{A}$ with $A^*A+B^*B\leq I$.*

(i) Let f be concave. If X,Y are positive, then

$$\mu(t,A^*f(X)A+B^*f(Y)B)\leq\mu(t,f(A^*XA+B^*YB))\ \textit{for all}\ t>0,$$

hence $\tau(A^*f(X)A)+\tau(B^*f(Y)B)\leq\tau(f(A^*XA+B^*YB))$.
Also, for general X,Y, we get $\tau(f(|X+Y|))\leq\tau(f(|X|))+\tau(f(|Y|))$.

(ii) Let f be convex. If X,Y are positive, then

$$\mu(t,A^*f(X)A+B^*f(Y)B)\geq\mu(t,f(A^*XA+B^*YB))\ \textit{for all}\ t>0,$$

hence $\tau(A^*f(X)A)+\tau(B^*f(Y)B)\geq\tau(f(A^*XA+B^*YB))$.

Also, for positive X,Y, we get $\tau(f(X+Y))\geq\tau(f(X))+\tau(f(Y))$.

Proof (i) Consider the von Neumann algebra $\mathscr{A}\otimes\mathbb{M}_2$ endowed with the trace

$$\tilde{\tau}=\begin{bmatrix}\tau & 0\\ 0 & \tau\end{bmatrix}.$$

Applying Theorem 8.4.7 to the contraction $\begin{bmatrix} A & 0 \\ B & 0 \end{bmatrix}$ and the positive ($\tilde{\tau}$-measurable) operator $\begin{bmatrix} X & 0 \\ 0 & Y \end{bmatrix}$, we obtain

$$\mu\left(t; \begin{bmatrix} A^*f(X)A + B^*f(Y)B & 0 \\ 0 & 0 \end{bmatrix}\right) \le \mu\left(t; \begin{bmatrix} f(A^*XA + B^*YB) & 0 \\ 0 & 0 \end{bmatrix}\right).$$

Here, the generalized singular value functions are taken with respect to $\tilde{\tau}$. However, this obviously means the first inequality of (i). To check the last inequality of (i), we first consider $X, Y \ge 0$. Then $X, Y \le X + Y$, hence there exist contractions $U, V \in \mathscr{A}$ such that

$$X^{1/2} = U(X+Y)^{1/2}, \quad Y^{1/2} = V(X+Y)^{1/2}, \quad U^*U + V^*V = \mathrm{supp}(X+Y).$$

Now we estimate

$$\begin{aligned} \tau(f(X)) + \tau(f(Y)) &= \tau(f(U(X+Y)U^*)) + \tau(f(V(X+Y)V^*)) \\ &\ge \tau(Uf(X+Y)U^*) + \tau(Vf(X+Y)V^*) \quad \text{(by Theorems 8.3.14 and 8.4.6)} \\ &= \tau(f(X+Y)^{1/2}U^*Uf(X+Y)^{1/2}) + \tau(f(X+Y)^{1/2}V^*Vf(X+Y)^{1/2}) \\ &= \tau(f(X+Y)^{1/2}\mathrm{supp}(X+Y)f(X+Y)^{1/2}) = \tau(f(X+Y)). \end{aligned}$$

For general $X, Y \in S(\mathscr{A}, \tau)$, we choose partial isometries $U, V \in \mathscr{A}$ such that $|X + Y| \le U|X|U^* + V|Y|V^*$; see (8.4.4). Since $U|X|U^*, V|Y|V^* \ge 0$, we get

$$\begin{aligned} \tau(f(|X+Y|)) &\le \tau(f(U|X|U^* + V|Y|V^*)) \quad \text{(by Lemma 8.3.12, (iii), and Corollary 8.3.16)} \\ &\le \tau(f(U|X|U^*)) + \tau(f(V|Y|V^*)) \quad \text{(by the previous case)} \\ &\le \tau(f(|X|)) + \tau(f(|Y|)). \end{aligned}$$

Here, the last inequality follows from relations

$$\mu(t, U|X|U^*) \le \mu(t, |X|), \quad \mu(t, V|Y|V^*) \le \mu(t, |Y|).$$

(ii) is proved by the same reasons as (i). (But note that the last part of (ii) based on (8.4.4) breaks down.) □

Lemma 8.4.9 *If* $X \in (L_1 + L_\infty)(\mathscr{A}, \tau)$, *then*

$$\int_0^t \mu(s, X)\, ds = \sup\{|\tau(XZ)| \mid Z \in \mathscr{A}, \tau(\mathrm{supp}(Z)) \le t, \ \|Z\| \le 1\}.$$

Proof For any $A \in S(\mathscr{A}, \tau)$ clearly $\tau(\mathrm{supp}(A)) = \inf\{s > 0 \mid \mu(s, A) = 0\}$. Hence, if $Z \in \mathscr{A}$ satisfies $\|Z\| \le 1$ and $\tau(\mathrm{supp}(Z)) \le t$, then $\mathrm{supp}(XZ) \le \mathrm{supp}(Z)$, and so

$$|\tau(XZ)| \le \tau(|XZ|) = \int_0^\infty \mu(s, XZ)\, ds = \int_0^t \mu(s, XZ)\, ds \le \int_0^t \mu(s, X)\, ds$$

by Lemma 8.3.12, (vi).

For the converse inequality, put $X \in (L_1 + L_\infty)(\mathscr{A}, \tau)$ and $E \in \mathscr{A}_{\text{pr}}$ with $\tau(E) \leq t$. Then $EXE, XE \in L_1(\mathscr{A}, \tau)$, and so $\tau(E|X|E) = \tau(|X|E)$. By Lemma 8.4.2 we get

$$\int_0^t \mu(s, X) ds = \sup\{\tau(|X|E)| \; E \in \mathscr{A}_{\text{pr}} \; \text{ with } \; \tau(E) \leq t\}.$$

Also, $|X|E = V^*XE$, where $X = V|X|$ is the polar decomposition of X, hence $\tau(|X|E) = \tau(V^*XE) = \tau(XEV^*)$. Setting $Z = EV^*$, we get $\mu(\cdot, Z) \leq \mu(\cdot, E)$ and so, $\mu(s, Z) = 0$ for any $s \geq \tau(E)$, that is, $\tau(\text{supp}(Z)) \leq \tau(E) \leq t$. □

Remark 8.4.10 Let $B \in \mathscr{A}$. The map $z \mapsto e^{zB}$ takes its values in $\mathscr{A}$ and is holomorphic on $\mathbb{C}$.

Lemma 8.4.11 *For any* $A, B \in \mathscr{A}_{\text{sa}}$ *and every* $\theta \in (0, 1)$,

$$e^{\theta B} A e^{(1-\theta)B} \prec_w Ae^B.$$

Proof Via Remark 8.4.10, the map $z \mapsto e^{zB} A e^{(1-z)B}$ takes values in $\mathscr{A}$ and is holomorphic on $\mathbb{C}$. Fix $Z \in \mathscr{A}$ such that $\|Z\| \leq 1$ and $\tau(\text{supp}(Z)) \leq t$. The $\mathbb{C}$-valued function $F(z) = \tau(e^{zB} A e^{(1-z)B})$ is also holomorphic on $\mathbb{C}$. For $X, Y \in (L_1 + L_\infty)(\mathscr{A}, \tau)$ the relation $\mu(XY) \prec_w \mu(X)\mu(Y)$ holds; see Theorem 8.4.5(iii), with $f(s) = s$, $s \geq 0$. Then for every $0 \leq \text{Re}z \leq 1$, we get

$$|F(z)| \leq t \|e^{zB} A e^{(1-z)B}\| \leq t \|e^{zB}\| \cdot \|A\| \cdot \|e^{(1-z)B}\| \leq t \|A\| e^{2\|B\|}.$$

Hence, F is a bounded function in the strip $0 \leq \text{Re}z \leq 1$. Since F is holomorphic on $\mathbb{C}$, it follows that F is continuous on the boundary of the strip $0 \leq \text{Re}z \leq 1$. By the Hadamard three-lines theorem, we obtain

$$|F(z)| \leq \sup_{y \in \mathbb{R}} \max\{|F(\text{i}y)|, |F(1 + \text{i}y)|\}.$$

To estimate $|F(\text{i}y)|$, we argue as follows:

$$|F(\text{i}y)| = |\tau(e^{\text{i}yB} A e^B e^{-\text{i}yB} Z)| \leq \tau(|Ae^B e^{-\text{i}yB} Z e^{\text{i}yB}|) = \tau(|Ae^B T|),$$

where $T = e^{-\text{i}yB} Z e^{\text{i}yB}$. Since $\mu(Ae^B T) \leq \|Ae^B\| \mu(T) = \|Ae^B\| \mu(Z)$, it is clear that $\mu(s, Ae^B T) = 0$ for $s \geq t$. Hence,

$$|\tau(Ae^B T)| = \int_0^\infty \mu(s, Ae^B T)\, ds \leq \int_0^t \mu(s, Ae^B T)\, ds \leq \int_0^t \mu(s, Ae^B)\, ds.$$

Similarly,

$$|F(1 + \text{i}y)| = |\tau(e^B A)| \leq \int_0^t \mu(s, e^B A)\, ds$$

and, appealing to the assumption $A = A^*$, we conclude that

$$|F(z)| = |\tau(e^{zB}Ae^{(1-z)B}Z)| \leq \int_0^t \mu(s, Ae^B)\,ds$$

for all $z \in \mathbb{C}$ with $0 \leq \mathrm{Re} z \leq 1$. Since this holds for all operators $Z \in \mathscr{A}$ such that $\|Z\| \leq 1$ and $\tau(\mathrm{supp}(Z)) \leq t$, we obtain from Lemma 8.4.9 that the estimate

$$\int_0^t \mu(s, e^{zB}Ae^{(1-z)B})\,ds \leq \int_0^t \mu(s, Ae^B)\,ds$$

holds for all $z \in \mathbb{C}$ with $0 \leq \mathrm{Re} z \leq 1$. Setting $z = \theta \in (0, 1)$, we conclude the proof. □

The assumption that $A \in \mathscr{A}$ is a self-adjoint operator was crucial for the above proof, where we concluded that $\mu(e^B A) = \mu(Ae^B)$. In fact, the assertion of the lemma does not hold for a nonself-adjoint operator A.

Example 8.4.12 There exist $A, B \in \mathbb{M}_2$ such that $B = B^*$ and the inequality $e^{B/2}Ae^{B/2} \prec_w Ae^B$ fails. Fix $x, y \in \mathbb{R}$ and consider

$$A = \begin{bmatrix} 0 & 1 \\ 0 & 0 \end{bmatrix}, \quad B = \begin{bmatrix} x & 0 \\ 0 & y \end{bmatrix}.$$

A direct computation yields $Ae^B = e^y A$ and $e^{B/2}Ae^{B/2} = e^{(x+y)/2}A$. Setting $x > y$, we obtain the assertion.

However, a quick analysis of the proof of Lemma 8.4.11 yields a following strengthening.

Lemma 8.4.13 *Let $A, B \in \mathscr{A}$ and $B = B^*$. Then*

$$e^{\theta B}Ae^{(1-\theta)B} \prec_w \max\{\mu(Ae^B), \mu(e^B A)\} \text{ for every } \theta \in (0, 1).$$

Lemma 8.4.14 *Let $A, B \in \mathscr{A}$.*

(i) If $A = A^*$ and $B \geq 0$, then

$$B^\theta AB^{1-\theta} \prec_w AB \text{ for every } \theta \in (0, 1).$$

(ii) If A is an arbitrary operator and $B \geq 0$, then

$$B^\theta AB^{1-\theta} \prec_w \max\{\mu(AB), \mu(BA)\} \text{ for every } \theta \in (0, 1).$$

Proof We only prove the first assertion. The proof of the second is the same via Lemma 8.4.13.

First, suppose that B is a positive invertible operator from $\mathscr{A}$. Then $\log(B)$ is a self-adjoint operator from $\mathscr{A}$. By applying Lemma 8.4.11 to the bounded operators A and $\log(B)$, we obtain the result.

In the general case, fix $n \in \mathbb{N}$, and consider the operator $B_n := B + \frac{1}{n} I_{\mathscr{A}}$, which is obviously invertible. It follows from the first part of the proof that

$$\int_0^t \mu(s, B_n^\theta A B_n^{1-\theta})\, ds \le \int_0^t \mu(s, AB_n)\, ds.$$

We can prove that

$$\mu(B_n^\theta A B_n^{1-\theta}) \to \mu(B^\theta A B^{1-\theta}), \quad \mu(AB_n) \to \mu(AB) \text{ as } n \to \infty$$

almost everywhere (see Exercise 8.8.6). Since the functions $\mu(B^\theta A B^{1-\theta})$ and $\mu(AB)$ are uniformly bounded, we easily infer from these convergences that for every $t \ge 0$, we obtain

$$\int_0^t \mu(s, B^\theta A B^{1-\theta})\, ds \le \int_0^t \mu(s, AB)\, ds.$$

□

Theorem 8.4.15 *Let $\mathscr{A}$ be a von Neumann algebra, and let $A, B \in S(\mathscr{A}, \tau)$ be such operators that $B \ge 0$ and $AB \in (L_1 + L_\infty)(\mathscr{A}, \tau)$.*

(i) If A is self-adjoint, then

$$B^\theta A B^{1-\theta} \prec_w AB \text{ for every } \theta \in (0, 1).$$

(ii) If A is an arbitrary operator, then

$$B^\theta A B^{1-\theta} \prec_w \max\{\mu(AB), \mu(BA)\} \text{ for every } \theta \in (0, 1).$$

Proof We prove the second assertion. Let $P_n = E^{|A|}[0, n)$, and let $Q_n = E^B[0, n)$. The operators AP_n and $Q_n B Qn = BQ_n$ are bounded and clearly, $AP_n \xrightarrow{\tau} A$ and $BQ_n \xrightarrow{\tau} B$ as $n \to \infty$. Hence, $(BQ_n)^\theta \xrightarrow{\tau} B^\theta$ (see for example, [38]) and, therefore,

$$(BQ_n)^\theta (AP_n)(BQ_n)^{1-\theta} \xrightarrow{\tau} B^\theta A B^{1-\theta} \text{ as } n \to \infty.$$

Also,

$$\mu((BQ_n)^\theta (AP_n)(BQ_n)^{1-\theta}) \to \mu(B^\theta A B^{1-\theta}) \text{ as } n \to \infty \tag{8.4.6}$$

almost everywhere (see Exercise 8.8.6). It follows now from Fatou's lemma that

$$\int_0^t \mu(s, B^\theta AB^{1-\theta})\,ds \le \liminf_{n\to\infty} \int_0^t \mu(s, (BQ_n)^\theta (AP_n)(BQ_n)^{1-\theta})ds.$$

By Lemma 8.4.14, we obtain

$$\int_0^t \mu(s, B^\theta AB^{1-\theta})\,ds \le \liminf_{n\to\infty} \int_0^t \max\{\mu(s, (BQ_n)(AP_n)),\ \mu(s, (AP_n)(BQ_n))\}ds.$$

Since $|AT| = |\,|A|T|$ for all operators $A, T \in S(\mathscr{A}, \tau)$, it follows that

$$\mu((AP_n)(BQ_n)) = \mu(|A|P_n(BQ_n)) = \mu(P_n(|A|B)Q_n) \le \mu(|A|B) = \mu(AB).$$

Also, we get

$$\mu((BQ_n)(AP_n)) = \mu(Q_n(BA)P_n) \le \mu(BA).$$

Theorem is proved. □

The following assertion is an extension of Lemma 8.4.11.

Proposition 8.4.16 *For any operators $A, B \in S(\mathscr{A}, \tau)_{\text{sa}}$, the relation*

$$e^{\theta B}e^{A}e^{(1-\theta)B} \prec_w e^{A}e^{B} \ \text{ holds for every } \ \theta \in (0, 1).$$

Proof By Lemma 8.3.15, we have $e^A, e^B \in S(\mathscr{A}, \tau)$. It suffices to prove the assertion only for the case of $e^A E^B \in (L_1 + L_\infty)(\mathscr{A}, \tau)$. Introducing projections $P_n := E^{|A|}[0, n)$, $Q_n := E^{|B|}[0, n)$, and operators $A_n := AP_n$, $B_n := BQ_n$ we infer from Lemma 8.4.11 that

$$e^{\theta B_n}e^{A_n}e^{(1-\theta)B_n} \prec_w e^{A_n}e^{B_n},\ n \ge 1.$$

The same argument as in the proof of Theorem 8.4.15 completes the proof. □

Remark 8.4.17 We give only few examples of weak majorization from a large number of well-known sets. Of those left out of the scope of this book, we consider few more examples.

1. If $A, B \in \mathscr{A}_+$ and $X \in (L_1 + L_\infty)(\mathscr{A}, \tau)$, then $2A^{1/2}XB^{1/2} \prec_w AX + XB$ [47, Proposition 5.1].

2. If $A \in S(\mathscr{A}, \tau)$ and $B \in S_0(\mathscr{A}, \tau)$, then $|\mu(A) - \mu(B)| \prec_w \mu(A - B)$ [48, Theorem 3.1].

3. For any operator monotone function $f : \mathbb{R}_+ \to \mathbb{R}_+$, it holds that [49, Theorem 1.1]

$$f(A) - f(B) \prec_w f(|A - B|),\quad A, B \in S(\mathscr{A}, \tau)_+.$$

4. Let $g \in \mathbb{F}$ with $g(\infty) = \infty$. If the inverse function of g is operator monotone, then [49, Theorem 1.9]

$$g(|A - B|) \prec_w g(A) - g(B),\quad A, B \in S(\mathscr{A}, \tau)_+.$$

Proof It follows from weak majorization 3 that for any $t > 0$ and $f = g^{-1}$ holds

$$\int_0^t \mu(s, X - Y)ds = \int_0^t \mu(s, f(g(X)) - f(g(Y)))ds \le \int_0^t \mu(s, f(|g(X) - g(Y)|))ds.$$

Since f is operator monotone, it follows that g is convex, hence (see, for example, the proof of [50, Theorem 2.1]) it follows that

$$\int_0^t \mu(s, g(X - Y))ds \le \int_0^t \mu(s, g \circ f(|g(X) - g(Y)|))ds = \int_0^t \mu(s, |g(X) - g(Y)|)ds$$

for all $t > 0$. □

5. If $\tau(I) < \infty$, $Z \in \mathscr{A}$ is an expansive operator (that is, $Z^*Z \ge I$), and

- if $f : \mathbb{R}_+ \to \mathbb{R}_+$ is a convex function with $f(0) = 0$, then

$$Z^* f(X)Z \prec_w f(Z^*XZ) \quad \text{for all} \quad X \in L_1(\mathscr{A}, \tau)_+;$$

- if $f : \mathbb{R}_+ \to \mathbb{R}_+$ is a concave function, then

$$f(Z^*XZ) \prec_w Z^* f(X)Z \quad \text{for all} \quad X \in L_1(\mathscr{A}, \tau)_+,$$

see [51].

8.5 Inequalities for Determinants and Characterization of the Trace

Let tr be the canonical trace on the matrix algebra $\mathbb{M}_n$ with the identity I. The *permanent of a matrix* $A = [a_{ij}] \in \mathbb{M}_n$ is defined to be $\mathrm{per}(A) = \sum_{\sigma \in \mathcal{S}_n} \prod_{i=1}^n a_{i\sigma(i)}$, where $\mathcal{S}_n$ denotes the symmetric group of all permutations of the numbers $\{1, 2, \dots, n\}$. We prove that if some analogues of the classical inequalities for the determinant and trace (or the permanent and the trace) of matrices hold for a positive functional φ on $\mathbb{M}_n$ with $\varphi(I) = n$ then $\varphi = \mathrm{tr}$. Also, we generalize Fischer's inequality for determinants and establish a new inequality for the trace of the matrix exponential.

Assume that $\varphi = \mathrm{tr}$ and $\lambda_t(A)$ $(t = 1, \dots, n)$ are the eigenvalues of $A \in \mathbb{M}_n$. Then the following relations hold:

- Schur's inequality [52, Chap. III, Sect. 1.4]

$$\sum_{t=1}^n |\lambda_t(A)|^2 \le \sum_{i,j=1}^n |a_{ij}|^2 (= \varphi(AA^*)) \quad \text{for all} \quad A \in \mathbb{M}_n,$$

equality is attained if and only if A is normal;

- the equality [52, Chap. I, Sect. 4.16, formula (1)]

$$\det(\exp(A)) = \exp(\varphi(A)) \quad \text{for all} \quad A \in \mathbb{M}_n; \tag{8.5.1}$$

• the inequality (see p. 33 or [53, Problem 3 on p. 163]) $\det(A)^{1/n} \leq \frac{1}{n}\varphi(A)$ for all positive semidefinite matrices $A \in \mathbb{M}_n$;

• the inequality [52, Chap. II, Sect. 4.4.12] $\operatorname{per}(A) \leq \frac{1}{n}\varphi(A^n)$ for all nonnegative Hermitian matrices $A \in \mathbb{M}_n$. By a *nonnegative matrix*, we mean a matrix whose all entities are nonengative real numbers.

We demonstrate that validity of each of these four relations implies that $\varphi = \operatorname{tr}$ (Theorems 8.5.6 and 8.5.8) for an arbitrary positive functional φ on $\mathbb{M}_n$ with $\varphi(I) = n$.

Let $\{P_k\}_{k=1}^m \subset \mathbb{M}_n$ with $P_k^2 = P_k$, $P_i P_k = 0$ for $i \neq k$, $i, k = 1, 2, \ldots, m$, and $\sum_{k=1}^m P_k = I$. Define the map $\mathcal{P} : \mathbb{M}_n \to \mathbb{M}_n$ by the formula

$$\mathcal{P}(A) = \sum_{k=1}^m P_k A P_k^* \quad \text{for all} \quad A \in \mathbb{M}_n.$$

If $P_k^2 = P_k = P_k^*$, $k = 1, 2, \ldots, m$, then $\mathcal{P}$ is a block projection operator whose properties are studied in [35, 54–56]. The formula $S = 2P - I$ $(P \in \mathbb{M}_n)$ establishes a bijection between the set of all idempotent matrices $(P^2 = P)$ and the set of all symmetries $(S^2 = I)$ from $\mathbb{M}_n$.

8.5.1 New Inequalities for Determinants and the Trace

Lemma 8.5.1 *Assume that* $\{P_k\}_{k=1}^m \subset \mathbb{M}_n$ *with* $P_k^2 = P_k$, $P_i P_k = 0$ *for* $i \neq k$, $i, k = 1, 2, \ldots, m$, *and* $\sum_{k=1}^m P_k = I$. *Then* $\operatorname{tr} A = \operatorname{tr}\left(\sum_{k=1}^m P_k A P_k\right)$ *for all* $A \in \mathbb{M}_n$. *In particular,* $\operatorname{tr}(\mathcal{P}(A)) = \operatorname{tr} A$, $A \in \mathbb{M}_n$, *for* $\{P_k\}_{k=1}^m \subset \mathbb{M}_n$ *with* $P_k^2 = P_k = P_k^*$, $k = 1, 2, \ldots, m$.

Proof If $A \in \mathbb{M}_n$, then

$$\operatorname{tr} A = \operatorname{tr}\left(\sum_{k=1}^m P_k A\right) = \sum_{k=1}^m \operatorname{tr}(P_k A) = \sum_{k=1}^m \operatorname{tr}(P_k A P_k) = \operatorname{tr}\left(\sum_{k=1}^m P_k A P_k\right).$$

□

Theorem 8.5.2 *The relation* $\det(\mathcal{P}(A)) \geq \det(A)$ *holds for all* $\{P_k\}_{k=1}^m \subset \mathbb{M}_n$ *with* $P_k^2 = P_k$, $P_i P_k = 0$ *for* $i \neq k$, $i, k = 1, 2, \ldots, m$, *and* $\sum_{k=1}^m P_k = I$ *for all positive semidefinite matrices* $A \in \mathbb{M}_n$.

Proof By the Determinant Product Theorem $\det(S) \in \{-1, +1\}$ for every symmetry $S \in \mathbb{M}_n$. If $X \in \mathbb{M}_n$ is positive semidefinite, then $\mathcal{P}(X)$ is positive semidefinite and $\det(X) \geq 0$. Hence, it suffices to verify the claim only for invertible matrices. The results of [57, 58] imply that the function

$$A \mapsto \log\det(A) \tag{8.5.2}$$

is concave on the set of positive definite matrices $A \in \mathbb{M}_n$ (see also see also [59, 15, Chap. 10, Sect. 2, Theorem 9']). By Lemma 2 from [55],

$$\mathcal{P}(A) = \frac{1}{2^{m-1}} \sum_{j=1}^{2^{m-1}} S_j A S_j^*, \tag{8.5.3}$$

for 2^{m-1} collections $\{t_{jk}\}_{k=1}^m$ with $t_{jk} \in \{-1, +1\}$, where $S_j = \sum_{k=1}^m t_{jk} P_k \in \mathbb{M}_n$ are symmetries for all $j = 1, 2, 3, \ldots, 2^{m-1}$. Therefore, $\det(S_j) = \det(S_j^*) \in \{-1, +1\}$ for all $j = 1, 2, 3, \ldots, 2^{m-1}$. The invertibility of $\mathcal{P}(A)$ for an positive definite (hence, invertible) $A \in \mathbb{M}_n$ follows from the representation of (8.5.3), where each summand $S_j A S_j^*$ is positive semidefinite and is invertible by the Invertible Product Theorem. Concavity of (8.5.2), the Determinant Product Theorem, and (8.5.3) imply that

$$\log\det(\mathcal{P}(A)) \ge \sum_{j=1}^{2^{m-1}} \frac{1}{2^{m-1}} \log\det(S_j A S_j^*) = \sum_{j=1}^{2^{m-1}} \frac{1}{2^{m-1}} \log\det(A) = \log\det(A).$$

Therefore,

$$\det(\mathcal{P}(A)) \ge \det(A) \tag{8.5.4}$$

via the strict monotonicity of the logarithmic function on the half-axis $(0, \infty)$. □

Relation (8.5.4) for a particular case when $\{P_k\}_{k=1}^m \subset \mathbb{M}_n$ are Hermitian idempotents is known as Fischer's inequality [60, Problem II.5.6]. Hence, by Lemma 8.5.1 and (8.5.1), we obtain

$$\det(\mathcal{P}(\exp(A))) \ge \det(\exp(A)) = \exp(\mathrm{tr}\, A) = \exp(\mathrm{tr}(\mathcal{P}(A)))$$

for all positive semidefinite matrices $A \in \mathbb{M}_n$.

Corollary 8.5.3 $\det(\mathcal{P}(A)) \ge \exp(\mathrm{tr}(\log A))$ *for each positive definite matrix* $A \in \mathbb{M}_n$.

Proof Indeed, $\det(\mathcal{P}(A)) = \det(\mathcal{P}(\exp(\log A))) \ge \det(\exp(\log A)) = \exp(\mathrm{tr}(\log A))$ for a positive definite matrix $A \in \mathbb{M}_n$. □

Lemma 8.5.4 *Assume that* $A \in \mathbb{M}_n$ *is positive semidefinite and* $B \in \mathbb{M}_n$ *is a contraction, and* $1 \le p < \infty$. *Then,*

$$\lambda_t((BAB^*)^p) \le \lambda_t(BA^pB^*) \;\; \text{for all} \;\; t = 1, 2, \ldots, n. \tag{8.5.5}$$

Proof Since the real function $s \mapsto s^q$ ($s \in \mathbb{R}_+$) is operator convex for $1 \le q \le 2$,

$$(BXB^*)^q \le BX^qB^*$$

for all positive semidefinite matrices $X \in \mathbb{M}_n$ and $B \in \mathbb{M}_n$ with $\|B\| \le 1$ by [61, Theorem 2.1]. By monotonicity of eigenvalues (that is, $\lambda_t(X) \le \lambda_t(Y)$ for all $t = 1, 2, \ldots, n$ for $0 \le X \le Y$) this matrix inequality leads us to the claim of the lemma for $1 \le q \le 2$. Let $t \in \{1, 2, \ldots, n\}$, and let $2 < p < \infty$ be fixed. Choose $j \in \mathbb{N}$ such that $2^{j-1} < p \le 2^j$ and put $q = \sqrt[j]{p}$. Then $j \ge 2$ and $1 < 2^{\frac{j-1}{j}} < q \le 2$. Hence,

$$\begin{aligned}\lambda_t(BA^pB^*) &= \lambda_t(B(A^{p/q})^qB^*) \ge \lambda_t((BA^{p/q}B^*)^q) = \lambda_t((BA^{p/q}B^*))^q\\ &= \lambda_t(B(A^{p/q^2})^qB^*)^q \ge \ldots \ge \lambda_t(BA^{p/q^j}B^*)^{q^j} = \lambda_t(BAB^*)^p\\ &= \lambda_t((BAB^*)^p)\end{aligned}$$

by the monotonicity of the power functions $s \mapsto s^b$ ($s \in \mathbb{R}_+$) and the equality $\lambda_t(X^b) = \lambda_t(X)^b$ for all positive semidefinite $X \in \mathbb{M}_n$ and the reals $b > 0$. □

Theorem 8.5.5 *Let* $\{P_k\}_{k=1}^m \subset \mathbb{M}_n$ *with* $P_k^2 = P_k = P_k^*$, $P_iP_k = 0$ *for* $i \ne k$, $i, k = 1, 2, \ldots, m$, *and let* $\sum_{k=1}^m P_k = I$. *Then* $\mathrm{tr}(\exp(\mathcal{P}(A))) \le \mathrm{tr}(\mathcal{P}(\exp(A)))$ *for all positive semidefinite matrices* $A \in \mathbb{M}_n$.

Proof Evidently,

$$\begin{aligned}\exp(\mathcal{P}(A)) &= I + \sum_{k=1}^m P_kAP_k + \sum_{k=1}^m \frac{(P_kAP_k)^2}{2!} + \cdots + \sum_{k=1}^m \frac{(P_kAP_k)^j}{j!} + \cdots\\ &= -(m-1)I + \Big(I + P_1AP_1 + \frac{(P_1AP_1)^2}{2!} + \cdots + \frac{(P_1AP_1)^j}{j!} + \cdots\Big) + \cdots\\ &+ \Big(I + P_mAP_m + \frac{(P_mAP_m)^2}{2!} + \cdots + \frac{(P_mAP_m)^j}{j!} + \cdots\Big),\end{aligned}$$

$$\begin{aligned}\mathcal{P}(\exp(A)) &= P_1\Big(I + A + \frac{A^2}{2!} + \cdots + \frac{A^j}{j!} + \cdots\Big)P_1 + \cdots\\ &+ P_m\Big(I + A + \frac{A^2}{2!} + \cdots + \frac{A^j}{j!} + \cdots\Big)P_m\\ &= -(m-1)I + \Big(I + P_1AP_1 + \frac{P_1A^2P_1}{2!} + \cdots + \frac{P_1A^JP_1}{j!} + \cdots\Big)\\ &\cdots + \Big(I + P_mAP_m + \frac{P_mA^2P_m}{2!} + \cdots + \frac{P_mA^jP_m}{j!} + \cdots\Big);\end{aligned}$$

the matrix series converges in norm (that is, elementwise). Since the matrix trace coincides with the spectral trace and is a continuous linear functional, Theorem 8.5.5 follows from Lemma 8.5.4. □

8.5.2 The Inequalities for Determinants that Characterize the Trace

Theorem 8.5.6 *The following conditions are equivalent for a positive functional φ on the algebra $\mathbb{M}_n$ with $\varphi(I) = n$:*

(i) $\varphi = \mathrm{tr}$;
(ii) $\det(\mathcal{P}(\exp(A))) \geq \exp(\varphi(A))$ *for all $\mathcal{P}$ and positive semidefinite $A \in \mathbb{M}_n$;*
(iii) $\det(A)^{1/n} \leq \frac{1}{n}\varphi(A)$ *for all positive semidefinite $A \in \mathbb{M}_n$;*
(iv) $per(A) \leq \frac{1}{n}\varphi(A^n)$ *for all nonnegative Hermitian matrices $A \in \mathbb{M}_n$;*
(v) $\det(I + \varepsilon A) = 1 + \varepsilon\varphi(A) + o(\varepsilon)$ *as* $\varepsilon \to 0+$ *for all positive semidefinite $A \in \mathbb{M}_n$.*

Moreover, if φ is faithful, then (i)-(v) are equivalent to the conditions:
(vi) $\det(\exp(A)) \leq \exp(\varphi(A))$ *for all positive semidefinite $A \in \mathbb{M}_n$;*
(vii) $\varphi(A^p)^{1/p} \leq \varphi(A^q)^{1/q}$ *for all positive semidefinite $A \in \mathbb{M}_n$ and $0 < q < p$.*

Proof The implication (i) $\Rightarrow$ (ii) follows from Theorem 8.5.5 and (8.5.1); see the implication (i) $\Rightarrow$ (v) in [62, Chap. 6, Sect. 9, Exercise 1].

Without loss of generality, assume that $\varphi(X) = \mathrm{tr}(S_\varphi X)$ for all $X \in \mathbb{M}_n$, where

$$S_\varphi = \mathrm{diag}(s_1, \ldots, s_n) \in \mathbb{M}_n$$

is positive semidefinite and $s_1 + \cdots + s_n = n$. We need to show that

$$s_1 = \cdots = s_n = 1. \tag{8.5.6}$$

(ii) $\Rightarrow$ (i). If (8.5.6) is not valid, then there exists $k \in \{1, \ldots, n\}$ such that $s_k > 1$. By the spectral theorem in finite dimensions, $\exp(A) = \exp(1) \cdot A + \exp(0) \cdot (I - A)$ for the projection

$$A = \mathrm{diag}(\underbrace{0, \ldots, 0}_{k-1 \text{ times}}, 1, 0, \ldots, 0) \in \mathbb{M}_n, \ A^2 = A = A^* \tag{8.5.7}$$

while, by (ii), $\exp(1) \geq \exp(s_k)$ for the map $\mathcal{P}$ associated with all projections of the form (8.5.7) with $k = 1, 2, \ldots, n$.

Consequently, $s_k \leq 1$; a contradiction.

(iii) $\Rightarrow$ (i). If (8.5.6) is not valid, then there exists $k \in \{1, \ldots, n\}$ such that $s_k > 1$. Given a real $\varepsilon > 0$ introduce the matrix $A_\varepsilon = (1+\varepsilon)I - \varepsilon A$, where A is from (8.5.7). Inserting $A_\varepsilon (\in \mathbb{M}_n^+)$ in (iii), we obtain

$$\begin{aligned}(1+\varepsilon)^{\frac{n-1}{n}} &\leq \frac{1}{n}((1+\varepsilon)s_1 + \cdots + (1+\varepsilon)s_{k-1} + s_k + (1+\varepsilon)s_{k+1} + \cdots + (1+\varepsilon)s_n) \\ &= \frac{1}{n}((1+\varepsilon)n - \varepsilon s_k) = 1 + \varepsilon - \frac{s_k}{n}\varepsilon.\end{aligned}$$

Recall the Taylor formula with Peano remainder:

$$(1+\varepsilon)^{\frac{n-1}{n}} = 1 + \frac{n-1}{n}\varepsilon + o(\varepsilon) \quad \text{as} \quad \varepsilon \to 0+.$$

Now, (iii) takes the form

$$1 + \frac{n-1}{n}\varepsilon + o(\varepsilon) \leq 1 + \varepsilon - \frac{s_k}{n}\varepsilon \quad \text{as} \quad \varepsilon \to 0+.$$

Consequently, $s_k \leq 1$, which leads to a contradiction.

(iv) $\Rightarrow$ (i). If (8.5.6) is not valid, then there exists $k \in \{1, \ldots, n\}$ such that $s_k > 1$. Given $1 > \varepsilon > 0$, introduce the matrix $A_\varepsilon = I - \varepsilon A$, where A is from (8.5.7). Put A_ε in (iv) and obtain

$$1 - \varepsilon \leq \frac{1}{n}(s_1 + \cdots + s_{k-1} + (1-\varepsilon)^n s_k + s_{k+1} + \cdots + s_n).$$

Write the Taylor formula with Peano remainder:

$$(1-\varepsilon)^n = 1 - n\varepsilon + o(\varepsilon) \quad \text{as} \quad \varepsilon \to 0+.$$

Now, (iv) takes the form $1 - \varepsilon \leq 1 - s_k\varepsilon + o(\varepsilon)$ as $\varepsilon \to 0+$. Consequently, $s_k \leq 1$; a contradiction.

(v) $\Rightarrow$ (i). If (8.5.6) is not valid, then there exists $k \in \{1, \ldots, n\}$ such that $s_k > 1$. By (v), we obtain

$$1 + \varepsilon = 1 + s_k\varepsilon + o(\varepsilon) \quad \text{as} \quad \varepsilon \to 0+.$$

for the projection A from (8.5.7). Consequently, $s_k = 1$; a contradiction.

(vi) $\Rightarrow$ (i). If (8.5.6) is not valid, then there exists $k \in \{1, \ldots, n\}$ such that $0 < s_k < 1$. By (vi), $\exp(1) \leq \exp(s_k)$ for the projection A from (8.5.7). Consequently, $s_k \geq 1$; a contradiction.

(i) $\Rightarrow$ (vii). Without loss of generality, assume that $A = \mathrm{diag}(a_1, \ldots, a_n)$ with $a_j \geq 0$ for all $j = 1, \ldots, n$. Then $A^r = \mathrm{diag}(a_1^r, \ldots, a_n^r)$ for all $r > 0$. By Jensen's inequality (see (1.4.1) and [63, Theorem 19]),

$$\varphi(A^p)^{1/p} = (a_1^p + \cdots + a_n^p)^{1/p} \leq (a_1^q + \cdots + a_n^q)^{1/q} = \varphi(A^q)^{1/q}$$

for all $0 < q < p$.

(vii) $\Rightarrow$ (i). If (8.5.6) is not valid, then there exists $k \in \{1, \ldots, n\}$ such that $0 < s_k < 1$. By (vii), $s_k^q \leq s_k^p$ for the projection A from (8.5.7). Consequently, $s_k \geq 1$; a contradiction. Recall that if $1 < p < \infty$ and φ is a positive functional on $\mathbb{M}_n$ with $\varphi(A^p) \leq \varphi(A^p)$ for $0 \leq A \leq B$, then $\varphi = \lambda\,\mathrm{tr}$ with some $\lambda \in \mathbb{R}_+$ [64, Theorem]. □

Corollary 8.5.7 *For a positive functional φ on $\mathbb{M}_n$ with $\varphi(I) = n$, the following conditions are equivalent:*

(i) $\varphi = \mathrm{tr}$;

(ii) $\det(\exp(A)) \geq \exp(\varphi(A))$ *for all positive semidefinite* $A \in \mathbb{M}_n$.

Theorem 8.5.8 *For a positive functional* φ *on* $\mathbb{M}_n$ *with* $\varphi(I) = n$, *the following conditions are equivalent:*

(i) $\varphi = \mathrm{tr}$;
(ii) $\sum_{t=1}^{n} \lambda_t(A)^2 \leq \varphi(A^2)$ *for all positive semidefinite* $A \in \mathbb{M}_n$;
(iii) $\left|\lambda_t(A) - \dfrac{\varphi(A^*A)}{n}\right| \leq \left(\dfrac{n-1}{n}\left(\varphi(A^*A) - \dfrac{|\varphi(A)|^2}{n}\right)\right)^{1/2}$ *for all* $A \in \mathbb{M}_n$ *and* $t = 1, \ldots, n$;
(iv) $\sum_{i=1}^{n} a_{ii}^2 \leq \varphi(A^2)$ *for all positive semidefinite* $A = [a_{ij}] \in \mathbb{M}_n$;
(v) $\varphi(A^2) \leq \mathrm{tr}(A)^2$ *for all positive semidefinite* $A \in \mathbb{M}_n$;
(vi) $\sqrt{\mathrm{tr}\, A} \leq \varphi(\sqrt{A})$ *for all positive semidefinite* $A \in \mathbb{M}_n$;
(vii) $\varphi(\sqrt{A}) \leq \sum_{i=1}^{n} \sqrt{a_{ii}}$ *for all positive semidefinite* $A = [a_{ij}] \in \mathbb{M}_n$.

Proof The implication (i) $\Rightarrow$ (ii) is the aforementioned Schur's inequality; see the implication (i) $\Rightarrow$ (iii) in [60, Problem I.6.16, p. 172] and the implications (i) $\Rightarrow$ (iv)-(vii) in [53, Problem 16, p. 24].

Let us show the converse implications. Without loss of generality, assume that $\varphi(X) = \mathrm{tr}(S_\varphi X)$ for all $X \in \mathbb{M}_n$, where $S_\varphi = \mathrm{diag}(s_1, \ldots, s_n) \in \mathbb{M}_n$ is positive semidefinite and $s_1 + \cdots + s_n = n$. We need to verify relations (8.5.6). If (8.5.6) is not valid, then there exist $m, j \in \{1, \ldots, n\}$ such that $s_m < 1$ and $s_j > 1$.

(ii) $\Rightarrow$ (i): By (ii), $1 = \sum_{t=1}^{n} \lambda_t(A)^2 > \varphi(A^2) = s_j$ for the projection A (with $j = k$) from (8.5.7); a contradiction.

(v) $\Rightarrow$ (i) and (vii) $\Rightarrow$ (i): For the matrix A indicated above, inequality (v) (or (vii)) gives $s_j \leq 1$; a contradiction.

(iii) $\Rightarrow$ (i): Inequality (iii) for $t = 1$ implies that $s_m \geq 1$ for the projection A (with $m = k$) from (8.5.7); a contradiction.

(iv) $\Rightarrow$ (i) and (vi) $\Rightarrow$ (i): Inequality (iv) (or (vi)) gives $s_m \geq 1$ for the projection A (with $m = k$) from (8.5.7); a contradiction. □

8.6 Noncommutative Probability Spaces

The noncommutative probability theory is inspired by certain philosophical aspects of quantum theory, where random variables are not modeled as measurable functions but as operators on a Hilbert space. Its history goes back to the works of von Neumann in the early thirties; see [65, 66].

Let us illustrate how classical probability spaces, as defined through measure theory, are related to commutative von Neumann algebras. Consider a probability space (Ω, Σ, μ) and the associated Hilbert space $\mathscr{H} = L_2(\Omega, \Sigma, \mu)$ equipped with the inner product $\langle X, Y\rangle = \int_\Omega \overline{X} Y d\mu$. For each $X \in L_\infty(\Omega, \Sigma, \mu)$, the multiplication operator $M_X : \mathscr{H} \to \mathscr{H}$ is defined as follows:

$$M_X Y(\omega) = X(\omega)Y(\omega) \qquad (Y \in \mathscr{H}).$$

It is straightforward to verify that the map $X \longmapsto M_X$ is an isometric $*$-homomorphism from $L_\infty(\Omega, \Sigma, \mu)$ into $\mathbb{B}(\mathscr{H})$, and its range is equal to its double commutant. Consequently, $L_\infty(\Omega, \Sigma, \mu)$ can be seen as a commutative von Neumann algebra on the Hilbert space $L_2(\Omega, \Sigma, \mu)$ equipped with the faithful normal finite trace $\mathbb{E}$, defined by

$$\mathbb{E}(f) := \int_\Omega f d\mu.$$

It is worth noting that the projections in $L_\infty(\Omega, \Sigma, \mu)$ correspond precisely to the characteristic functions of measurable subsets of Ω.

The Gelfand Theorem [67] in the setting of von Neumann algebras ensures that the converse holds. More precisely, if $\mathscr{A}$ is a commutative von Neumann algebra equipped with a faithful normal finite trace τ, then there exists a probability space (Ω, Σ, μ) such that $\mathscr{A}$ is of the form $L_\infty(\Omega, \Sigma, \mu)$. In fact, the map $X \longmapsto f_X$ is an isometric $*$-isomorphism from $\mathscr{A}$ onto $L_\infty(\Omega, \Sigma, \mu)$ such that

$$\mathbb{E}(f_X) := \int_\Omega f_X d\mu = \tau(X).$$

Now, we present the notion of a noncommutative probability space as a noncommutative measure theory.

Definition 8.6.1 If $\mathscr{A}$ is a von Neumann algebra on a Hilbert space $\mathscr{H}$ equipped with a faithful normal finite trace τ, then the pair $(\mathscr{A}, \tau)$ is said to be a *noncommutative probability space* (or *quantum probability space*).

In this section, we assume that τ is unital. Similar to the usual measure spaces, we can introduce the noncommutative L_p-spaces associated with $(\mathscr{A}, \tau)$. Let $1 \leq p < \infty$. Recall that the Banach space $L_p(\mathscr{A}) = L_p(\mathscr{A}, \tau)$ is the completion of $\mathscr{A}$ in $S(\mathscr{A}, \tau)$ with respect to the p-norm $\|X\|_p := \tau(|X|^p)^{1/p}$. The elements of $L_1(\mathscr{A})$ are called *noncommutative random variables*.

One may seek the results in commutative probability theory that remain valid in the noncommutative setting. Due to the significance of inequalities for estimating the probability of an event or the expectation of, say, the sum of some classical random variables, it is interesting to know whether it is possible to establish some of their versions in the setting of noncommutative probability spaces.

The strategy used to investigate a specific property in noncommutative probability spaces can be sated in three steps:

(i) We create an algebra of functions within a probability space so that the gained algebra stores information about the space.
(ii) We consider the commutative algebra obtained in the first step and determine the specific property.
(iii) We replace the obtained algebra with a noncommutative algebra having this specific property except for the commutativity.

While some tools used in the study of classical probability spaces are applied to the noncommutative case, nevertheless, many difficulties may appear in obtaining non-commutative results. Among them, we refer to the invalidity of the triangle inequality for the modulus of operators, the product of two self-adjoint operators may not be self-adjoint, and there may not exist a pointwise maximum of two noncommutative random variables. These and other reasons require us to adopt new techniques.

Let $X \in \mathscr{A}$ be a self-adjoint operator. By the Borel functional calculus, for every Borel function $f : \mathrm{sp}(X) \to \mathbb{C}$, the operator $f(X)$ is defined by $f(X) = \int f(\lambda) dE_\lambda^X$. In particular,

$$\chi_B(X) = \int_B dE_\lambda^X = E^X(B),$$

in which χ_B stands for the characteristic function of B. We frequently use $\chi_B(X)$ instead of $E^X(B)$. As in the classical setting, we adopt the following notation:

$$\mathrm{Prob}(x \geq t) := \tau(\chi_{[t,\infty)}(X)).$$

Clearly,

$$\tau\left(\chi_{[t,\infty)}(|X|)\right) = \tau\left(\chi_{[t,\infty)}(X)\right) + \tau\left(\chi_{[t,\infty)}(-X)\right). \tag{8.6.1}$$

When $X \geq 0$ and $t > 0$, the relation $\chi_{[t,\infty)}(X)t \leq X$ holds true. Therefore, we arrive at the inequality

$$\mathrm{Prob}(X \geq t) = \tau(\chi_{[t,\infty)}(X)) \leq t^{-1}\tau(X), \tag{8.6.2}$$

which is called the *Markov inequality*. For a self-adjoint element $X \in \mathscr{A}$, it follows from the Markov inequality that

$$\mathrm{Prob}(X \geq t) = \mathrm{Prob}(e^X \geq e^t) = \tau(\chi_{[e^t,\infty)}(e^X)) \leq e^{-t}\tau(e^X)$$

from which we get the following *exponential Chebyshev inequality*:

$$\tau(\chi_{[t,\infty)}(X)) \leq e^{-t}\tau(e^X). \tag{8.6.3}$$

We refer the reader to [68] for more information on noncommutative probability spaces.

Many of the basic concepts in classical probability have a natural noncommutative analogue. For example, associated with $\{\xi \in \mathscr{H} : \langle |X|\xi, \xi\rangle > t\}$, one can correspond the spectral projection represented by $E^{|X|}(t, \infty)$ (see for example, [67]). So as we already mentioned, $\tau\left(E^{|X|}[0, t]\right)$ can be defined as the noncommutative distribution function.

Another example is the noncommutative variance. The *variance* of an element $X \in \mathscr{A}$ is defined by

$$\operatorname{var}(X) := \tau\left((X - \tau(X))^2\right).$$

For any self-adjoint element $X \in \mathscr{A}$ and $\lambda > 0$, it follows from the Markov inequality that

$$\tau\left(E^{|X-\tau(X)|}[\lambda, \infty)\right) \leq \lambda^{-2}\operatorname{var}(X). \tag{8.6.4}$$

The noncommutative counterpart of the notion of median is given as follows.

Definition 8.6.2 For a self-adjoint element $X \in \mathscr{A}$, we say that a real number m is the *median* of X if the following inequalities hold

$$\tau\left(E^X(-\infty, m]\right) \geq \frac{1}{2} \quad \text{and} \quad \tau\left(E^X[m, \infty)\right) \geq \frac{1}{2}.$$

The median of X is denoted by $\operatorname{med}(X)$. The median of any self-adjoint element X always exists. In fact, the real number $m := \sup\{\alpha : \tau\left(E^X[\alpha, \infty)\right) \geq \frac{1}{2}\}$ is a median of X; see [69]. Some properties of the median are presented in the next proposition.

Proposition 8.6.3 *Let $X \in \mathscr{A}_{\text{sa}}$, $p \geq 1$, and α be a positive real number. Then*

(i) *if $\tau\left(E^{|X|}[\alpha, \infty)\right) < \frac{1}{2}$, then $|\operatorname{med}(X)| \leq \alpha$.*
(ii) *$|\operatorname{med}(X)| \leq 2^{\frac{1}{p}}\|X\|_p$.*
(iii) *$|\operatorname{med}(X) - \tau(X)| \leq \sqrt{2\operatorname{var}(X)}$.*

Proof (i) It follows from $E^{|X|}[\alpha, \infty) = E^X[\alpha, \infty) + E^{-X}[\alpha, \infty)$ and the assumption that $\tau(E^X[\alpha, \infty)) < \frac{1}{2}$ and $\tau(E^{-X}[\alpha, \infty)) < \frac{1}{2}$. If $\operatorname{med}(X) > \alpha$, then we arrive at $\tau(E^X[m, \infty)) \leq \tau(E^X[\alpha, \infty)) < \frac{1}{2}$, which is impossible; hence $\operatorname{med}(X) \leq \alpha$. Similarly, $\operatorname{med}(X) \geq -\alpha$.
(ii) The desired assertion is concluded from (i) and Chebyshev inequality (8.8.1)

$$\tau\left(E^{|X|}(2^{\frac{1}{p}}\|x\|_p, \infty)\right) \leq \frac{\|X\|_p^p}{(2^{\frac{1}{p}}\|X\|_p)^p} = \frac{1}{2}.$$

(iii) It can be concluded from (ii) with $X - \tau(X)$ instead of X, and the fact that $\| X - \tau(X) \|_2^2 = \operatorname{var}(X)$. □

Assume that $\mathscr{B}$ is a von Neumann subalgebra of $\mathscr{A}$. Then there exists a map $\mathcal{E}_{\mathscr{B}} : \mathscr{A} \longrightarrow \mathscr{B}$ satisfying the following properties:

(i) $\mathcal{E}_{\mathscr{B}}$ is a normal positive contractive projection from $\mathscr{A}$ onto $\mathscr{B}$;
(ii) $\mathcal{E}_{\mathscr{B}}(AXB) = A\mathcal{E}_{\mathscr{B}}(X)B$ for every $X \in \mathscr{A}$ and $A, B \in \mathscr{B}$;
(iii) $\tau \circ \mathcal{E}_{\mathscr{B}} = \tau$.

Moreover, $\mathcal{E}_{\mathscr{B}}$ is the unique map satisfying (ii) and (iii). It is known that $\mathcal{E}_{\mathscr{B}}$ can be extended to a contractive positive projection, denoted by the same $\mathcal{E}_{\mathscr{B}}$, from $L_p(\mathscr{A})$ onto $L_p(\mathscr{B})$.

A filtration of $\mathscr{A}$ is an increasing sequence $(\mathscr{A}_j, \mathcal{E}_j)_{0\le j\le n}$ of von Neumann subalgebras of $\mathscr{A}$ together with the conditional expectations $\mathcal{E}_j$ of $\mathscr{A}$ with respect to $\mathscr{A}_j$ such that $\bigcup_j \mathscr{A}_j$ is σ-weak-dense in $\mathscr{A}$. It follows from $\mathscr{A}_j \subseteq \mathscr{A}_{j+1}$ that

$$\mathcal{E}_i \circ \mathcal{E}_j = \mathcal{E}_j \circ \mathcal{E}_i = \mathcal{E}_{\min\{i,j\}} \tag{8.6.5}$$

for all $i, j \ge 0$.

A sequence $(X_j)_{j\ge 0}$ in $L_1(\mathscr{A})$ is called a *martingale* with respect to the filtration $(\mathscr{A}_j)_{0\le j\le n}$ if $X_j \in L_1(\mathscr{A}_j)$ and $\mathcal{E}_j(X_{j+1}) = X_j$ for every $j \ge 0$. It follows from (8.6.5) that $\mathcal{E}_j(X_i) = X_j$ for all $i \ge j$. For such a martingale X in $L_p(\mathscr{A})$ $(1 \le p \le \infty)$, we set $\|X\|_p := \sup_{n\ge 1} \|X_n\|_p$. If $\|X\|_p < \infty$, then X is called a *bounded L_p-martingale*. The difference sequence $dX = (dX_n)_{n\ge 1}$ of a martingale $X = (X_n)_{n\ge 1}$ is defined by $dX_n = X_n - X_{n-1}$ $(n \ge 1)$ with the usual convention that $X_0 = 0$. This martingale is called the *martingale difference* of (X_j). For other types of martingales and their inequalities; see [70–72].

The notion of independence is one of the main notions in classical probability. In the noncommutative probability setting, the concept of independence also plays an essential role. There are several noncommutative notions of independence; see [73]. Recall that two commutative random variables X and Y are independent if

$$\mathbb{E}\,(f(X)\,g(Y)) = \mathbb{E}(f(X))\,\mathbb{E}(g(Y))$$

for all bounded continuous complex functions f and g. In addition, conditional independence with respect to a sub-σ-algebra can be defined. Here, we state two types of independence for noncommutative random variables.

Let $\mathscr{B}$ and $\mathscr{B}_j$ $(1 \le j \le n)$ be von Neumann subalgebras of $\mathscr{A}$ with $\mathscr{B} \subseteq \bigcap_{j=1}^n \mathscr{B}_j$. The family $\{\mathscr{B}_j\,;\, 1 \le j \le n\}$ is called *successively independent* over $\mathscr{B}$ if for every $j > 1$,

$$\mathcal{E}_{\mathscr{B}}(A\,B) = \mathcal{E}_{\mathscr{B}}(A)\mathcal{E}_{\mathscr{B}}(B)$$

for all $A \in \mathscr{B}_j$ and B in the von Neumann algebra generated by $\bigcup_{1\le k<j} \mathscr{B}_k$. This is an extension of the classical independence of commutative random variables. Indeed, two elements X and Y of a commutative von Neumann algebra $L_\infty(\mu)$ in which μ is a probability measure are independent if the algebras they generate are independent over the complex field $\mathbb{C}$.

A sequence $(X_n)_{n=1}^\infty$ is said to be *successively independent* [74] if the subalgebras $W^*(X_j)$ and $W^*\left(X_1, \dots, X_{j-1}\right)$ are successively independent (over $\mathbb{C}$) for all $1 < j$, where $W^*(F)$ denotes the von Neumann algebra generated by the spectral projections of the real and imaginary parts of any member of $F \subset \mathscr{A}$.

Two von Neumann subalgebras $\mathscr{B}_1$ and $\mathscr{B}_2$ of $\mathscr{A}$ are said to be *Boolean independent* if

$$\tau(AB) = \tau(A)\tau(B)$$

for any $A \in \mathscr{B}_1$ and $B \in \mathscr{B}_2$. Two random variables X and Y are Boolean independent if $W^*(X)$ and $W^*(Y)$ are Boolean independent. The next result reads as follows; see [75].

Theorem 8.6.4 *Let $X \in \mathscr{A}_{\text{sa}}$. Then*

$$\operatorname{var}(X) = \inf\{\tau\left((X-Y)^2\right) : Y \in \mathscr{A}_{\text{sa}} \text{ and } X, Y \text{ are Boolean independent}\}. \tag{8.6.6}$$

Proof Let $Y \in \mathscr{A}_{\text{sa}}$ and $\tau(XY) = \tau(X)\tau(Y)$. First note that

$$\tau(Y)^2 = \tau(YI)^2 \leq \tau\left(Y^2\right)\tau(I) = \tau\left(Y^2\right).$$

Then consider

$$\begin{aligned}
\operatorname{var}(X) =& \tau\left(X^2\right) - \tau(X)^2 \\
\leq& \tau\left(X^2\right) - \tau(X)^2 + (\tau(X) - \tau(Y))^2 \\
=& \tau\left(X^2\right) - 2\tau(X)\tau(Y) + \tau(Y)^2 \\
\leq& \tau\left(X^2\right) - 2\tau(XY) + \tau(Y^2) \\
=& \tau\left((X-Y)^2\right).
\end{aligned}$$

It follows from the definition of variance that (8.6.6) is valid. □

Among several important inequalities, we mention few that can be interesting for general public; see also [76–78] for other types of such inequalities.

Burkholder–Gundy inequalities: The exploration of noncommutative versions of classical martingale inequalities began with the noncommutative Burkholder–Gundy inequalities. The classical Burkholder–Gundy inequalities state that for every bounded L_p-martingale $f = (f_n)$ on a probability space, the L_p norm of the square function $\left(\sum_n |df_n|^2\right)^{1/2}$ is equivalent to the L_p norm of $f = (f_n)$ when $1 < p < \infty$. The simplified version of noncommutative Burkholder–Gundy inequality says that if $p \geq 2$ and $(X_j)_{1\leq j\leq n}$ is a bounded L_p-martingale, then for some positive numbers c_p, c_p' depending only on p, it holds that

$$c_p \sup_j \|X_j\|_p \leq \max\left\{\left\|\left(\sum_j |D_j|^2\right)^{\frac{1}{2}}\right\|_p, \left\|\left(\sum_j |D_j^*|^2\right)^{\frac{1}{2}}\right\|_p\right\} \leq c_p' \sup_j \|X_j\|_p, \tag{8.6.7}$$

where $D_j := X_j - X_{j-1}$ $(1 \leq j \leq n)$ with the convention that $X_0 = 0$. Note that in the commutative case, $|D_j| = |D_j^*|$, and so the two norms in the middle phrase in (8.6.7) coincide; see also [79].

Rosenthal inequality: The Rosenthal inequalities give us two-sided estimates for the p-th moments of the sum of a sequence of classical independent random variables with zero mean. A simplified version of the noncommutative Rosenthal inequality is as follows: Let $\mathscr{B}_j$ $(1 \leq j \leq n)$ be successively independent von Neumann subalgebras over $\mathscr{B}$. If $p \geq 2$ and $X_j \in L_p(\mathscr{B}_j)$ with $\mathcal{E}_{\mathscr{B}}(X_j) = 0$ for all j, then

$$\frac{c}{p^2}\left\|\sum_j X_j\right\|_p \leq \max\left\{\left(\sum_j \|X_j\|_p^p\right)^{\frac{1}{p}}, \left\|\left(\sum_j \mathcal{E}_{\mathscr{B}}|X_j|^2\right)^{\frac{1}{2}}\right\|_p, \left\|\left(\sum_j \mathcal{E}_{\mathscr{B}}|X_j^*|^2\right)^{\frac{1}{2}}\right\|_p\right\}$$
$$\leq 2\left\|\sum_j X_j\right\|_p$$

For the general form of this inequality; see [74, 80].

Kolmogorov inequality: The Kolmogorov inequality provides an upper bound for the probability that the sum of a finite number of classically independent random variables is greater than a given value. Its commutative version asserts that if X_j's are n independent random variables with zero mean and finite variance, then for every $\lambda > 0$,

$$P\left(\max_{1\leq k\leq n}\left|\sum_{j=1}^{k} X_j\right| \geq \lambda\right) \leq \frac{1}{\lambda^2}\sum_{j=1}^{n}\mathbb{E}X_j^2.$$

Batty [81] established a noncommutative Kolmogorov strong law of large numbers by proving a quantum form of Kolmogorov inequality. Another noncommutative form of Kolmogorov inequality was stated in [77] as follows:
Assume that $X_1, \ldots, X_n \in \mathscr{A}$ are successively independent so that $\tau(X_k) = 0$ for all k. Then, for each $\lambda > 0$ there exists a projection E such that

$$1 - \frac{2(\lambda + \max_{1\leq k\leq n}\|X_k\|)^2}{\sum_{k=1}^{n}\mathrm{var}(X_k)} \leq \tau(E) \leq \frac{\tau\left(\left(\sum_{k=1}^{n} X_k\right)^2\right)}{\lambda^2}.$$

In what follows, we provide some inequalities in the framework of noncommutative probability spaces.

8.6.1 Bennett Inequality

The next result, proved in [82], is a generalization of the *noncommutative Bennett inequality* due to Junge and Zeng [83] .

Theorem 8.6.5 (Noncommutative Bennett inequality) *Let $\mathscr{B} \subseteq \mathscr{A}_j \subseteq \mathscr{A}$ be successively independent over $\mathscr{B}$ and $X_j \in \mathscr{A}_j$ be self-adjoint and $1 \leq r \leq 2$ such that (i) $\mathcal{E}_{\mathscr{B}}(X_j) \leq 0$, (ii) $\mathcal{E}_{\mathscr{B}}(|X_j|^r) \leq b_j^r$, and (iii) $\|X_j\| \leq M$ for some $M > 0$ and all $1 \leq j \leq n$. Then for each $t \geq 0$,*

$$\text{Prob}\left(\sum_{j=1}^{n} X_j \geq t\right) = \tau\left(\chi_{[t,\infty)}\left(\sum_{j=1}^{n} X_j\right)\right) \leq \exp\left(\frac{-b}{M^r}\Phi\left(\frac{tM^{r-1}}{b}\right)\right), \tag{8.6.8}$$

where $\Phi(\alpha) = (1+\alpha)\log(1+\alpha) - \alpha$ *and* $b = \sum_{j=1}^{n} b_j^r$.

Proof Let $\lambda \geq 0$. Then

$$\begin{aligned}
\mathcal{E}_{\mathscr{B}}\left(e^{\lambda X_n}\right) &= \mathcal{E}_{\mathscr{B}}\left(\sum_{k=0}^{\infty}\frac{(\lambda X_n)^k}{k!}\right) = \sum_{k=0}^{\infty}\frac{\lambda^k}{k!}\mathcal{E}_{\mathscr{B}}(X_n^k) \\
&= 1 + \mathcal{E}_{\mathscr{B}}(X_n) + \sum_{k=2}^{\infty}\frac{\lambda^k}{k!}\mathcal{E}_{\mathscr{B}}(X_n^k) \\
&\leq 1 + \sum_{k=2}^{\infty}\frac{\lambda^k}{k!}\mathcal{E}_{\mathscr{B}}(X_n^k) \qquad (\text{since } \mathcal{E}_{\mathscr{B}}(X_n) \leq 0) \\
&\leq 1 + \sum_{k=2}^{\infty}\frac{\lambda^k}{k!}\mathcal{E}_{\mathscr{B}}(|X_n|^k) \qquad (\text{since } X_n^k \leq |X_n|^k) \\
&= 1 + \sum_{k=2}^{\infty}\frac{\lambda^k}{k!}\mathcal{E}_{\mathscr{B}}(|X_n|^r|X_n|^{k-r}) \\
&\leq 1 + \sum_{k=2}^{\infty}\frac{\lambda^k}{k!}\|X_n\|^{k-r}\mathcal{E}_{\mathscr{B}}(|X_n|^r) \\
&\qquad (\text{by } |X_n|^r|X_n|^{k-r} \leq \|X_n\|^{k-r}|X_n|^r) \\
&\leq 1 + \sum_{k=2}^{\infty}\frac{\lambda^k}{k!}\|X_n\|^{k-r}b_n^r \\
&= 1 + \frac{b_n^r}{\|X_n\|^r}(e^{\lambda\|X_n\|} - 1 - \lambda\|X_n\|) \\
&\leq \exp\left(\frac{b_n^r}{\|X_n\|^r}(e^{\lambda\|X_n\|} - 1 - \lambda\|X_n\|)\right).
\end{aligned}$$

Since the function $f(s) := \exp\left(\frac{e^{\lambda s}-1-\lambda s}{s^r}\right)$ is increasing for $s > 0$, we get

$$\mathcal{E}_{\mathscr{B}}\left(e^{\lambda X_n}\right) \leq \exp\left(\frac{b_n^r}{M^r}(e^{\lambda M} - 1 - \lambda M)\right). \tag{8.6.9}$$

From (8.6.4) we infer that

$$\tau\left(\chi_{[t,\infty)}\left(\sum_{j=1}^{n} X_j\right)\right) = \tau\left(\chi_{[\lambda t,\infty)}\left(\sum_{j=1}^{n} \lambda X_j\right)\right)$$

$$\leq \exp(-\lambda t)\tau\left(e^{\sum_{j=1}^{n}\lambda X_j}\right). \tag{8.6.10}$$

It follows from the Golden–Thompson inequality (8.1.1) that

$$\begin{aligned}
\tau\left(e^{\sum_{j=1}^{n}\lambda X_j}\right) \leq & \tau\left(e^{\sum_{j=1}^{n-1}\lambda X_j} e^{\lambda X_n}\right) \\
= & \tau\left(\mathcal{E}_{\mathscr{B}}\left(e^{\sum_{j=1}^{n-1}\lambda X_j} e^{\lambda X_n}\right)\right) \\
= & \tau\left(\mathcal{E}_{\mathscr{B}}\left(e^{\sum_{j=1}^{n-1}\lambda X_j}\right)\mathcal{E}_{\mathscr{B}}\left(e^{\lambda X_n}\right)\right) \\
\leq & \exp\left(\frac{b_n^r}{M^r}(e^{\lambda M} - 1 - \lambda M)\right)\tau\left(\mathcal{E}_{\mathscr{B}}\left(e^{\sum_{j=1}^{n-1}\lambda X_j}\right)\right) \\
& \text{(by (8.6.9) and traciality of } \tau) \\
\leq & \exp\left(\frac{\sum_{j=1}^{n} b_j^r}{M^r}(e^{\lambda M} - 1 - \lambda M)\right). \\
& \text{(by iterating } n-2 \text{ times)}
\end{aligned}$$

From the latter inequality together with (8.6.10) we conclude that

$$\tau\left(\chi_{[t,\infty)}\left(\sum_{j=1}^{n} X_j\right)\right) \leq \exp\left(-\lambda t + \frac{\sum_{j=1}^{n} b_j^r}{M^r}(e^{\lambda M} - 1 - \lambda M)\right).$$

Applying the basic calculus method, we reach the minimizing value

$$\lambda = \frac{1}{M}\log\left(1 + \frac{tM^{r-1}}{\sum_{j=1}^{n} b_j^r}\right),$$

that yields (8.6.8). □

Since both inequalities $\Phi(t) \geq \frac{t^2}{2+2t/3}$ and $\Phi(t) \geq \frac{t}{2}\operatorname{arcsinh}(\frac{t}{2})$ are valid for all $t \geq 0$, we can deduce the *Bernstein inequality* and the *Prohorov inequality* from Bennett's inequality as follows.

Corollary 8.6.6 *Under the hypothesis of Theorem 8.6.5,*

$$\tau\left(\chi_{[t,\infty)}\left(\sum_{j=1}^{n} X_j\right)\right) \leq \exp\left(-\frac{t^2 M^{r-2}}{2b + (2/3)tM^{r-1}}\right)$$

and

$$\tau\left(\chi_{[t,\infty)}\left(\sum_{j=1}^{n} X_j\right)\right) \leq \exp\left(-\frac{t}{2M}\operatorname{arcsinh}\left(\frac{tM}{2b}\right)\right).$$

We immediately deduce the following commutative Bernstein's Inequality.

Corollary 8.6.7 *Let $X_1, \ldots, X_n$ be independent Bernoulli commutative random variables taking values* 1 *and* -1 *with probability* $1/2$. *Then*

$$\text{Prob}\left(\left|\frac{1}{n}\sum_{i=1}^{n} X_i\right| \geq \varepsilon\right) \leq 2e^{\frac{-n\varepsilon^2}{2+(2\varepsilon/3)}}.$$

8.6.2 *Hoeffding Inequality*

The Hoeffding inequality is used to constrain the sum of independent bounded commutative random variables. The classical Hoeffding inequality states that if $a \leq X_1, \ldots, X_n \leq b$ are independent commutative random variables with $\mathbb{E}X_i = \mu$ $(1 \leq i \leq n)$, then

$$P\left(|\overline{X}_n - \mu| \geq t\right) \leq 2e^{-2nt^2/(b-a)^2},$$

where $\overline{X}_n = (\sum_{i=1}^{n} X_i)/n$; see [84]. Mimicking [82], the *noncommutative Hoeffding* and some consequences are given.

Theorem 8.6.8 (Noncommutative Hoeffding inequality) *Let $\mathscr{B} \subseteq \mathscr{A}_j \subseteq \mathscr{A}$ be successively independent over $\mathscr{B}$, and let $X_j \in \mathscr{A}_j$ be self-adjoint such that $a_j \leq X_j \leq b_j$ for some real numbers $a_j < b_j$ and $\mathcal{E}_{\mathscr{B}}(X_j) = \mu$ for some $\mu \geq 0$ and all $1 \leq j \leq n$. Then*

$$\text{Prob}(|S_n - n\mu| \geq t) \leq 2\exp\left\{\frac{-2t^2}{\sum_{j=1}^{n}(b_j - a_j)^2}\right\} \tag{8.6.11}$$

for any $t > 0$, where $S_n = \sum_{j=1}^{n} X_j$.

Proof We first prove that if $X \in \mathscr{A}_j$ is self-adjoint such that $a \leq X \leq b$ and $\mathcal{E}_{\mathscr{B}}(X) = 0$, then

$$\mathcal{E}_{\mathscr{B}}(e^{sX}) \leq \exp\left\{\frac{s^2(b-a)^2}{8}\right\} \tag{8.6.12}$$

for all $s > 0$.
Assume that $s > 0$. Since $t \mapsto e^{ts}$ is convex, the inequality

$$e^{s\alpha} \leq e^{sb}\frac{\alpha - a}{b-a} + e^{sa}\frac{b-\alpha}{b-a}$$

holds for any $a \leq \alpha \leq b$. It follows from the functional calculus for self-adjoint operators that

$$e^{sX} \leq e^{sb}\frac{X-a}{b-a} + e^{sa}\frac{b-X}{b-a}.$$

Since $\mathcal{E}_{\mathscr{B}}$ is a positive map and $\mathcal{E}_{\mathscr{B}}(X) = 0$, we achieve

$$\mathcal{E}_{\mathscr{B}}(e^{sX}) \leq \frac{-a}{b-a}e^{sb} + \frac{b}{b-a}e^{sa} = e^{h(\alpha)},$$

where $\alpha = s(b-a)$, $h(\alpha) = -\gamma\alpha + \log(1-\gamma+\gamma e^{\alpha})$ and $\gamma = -a/(b-a)$. In addition, $h(0) = h'(0) = 0$ and $h''(\alpha) \leq \frac{1}{4}$ for all $\alpha > 0$. By Taylor's Theorem there exists a real number $\xi \in (0, \alpha)$ such that

$$h(\alpha) = h(0) + \alpha h'(0) + \frac{\alpha^2}{2}h''(\xi) \leq \frac{\alpha^2}{8} = \frac{s^2(b-a)^2}{8}.$$

Hence,

$$\mathcal{E}_{\mathscr{B}}(e^{sX}) \leq \exp\left\{\frac{s^2(b-a)^2}{8}\right\}.$$

Next, for arbitrary value of $\mathcal{E}_{\mathscr{B}}(X)$, by taking $Y := X - \mu$, we obtain $a - \mu \leq Y \leq b - \mu$ and $\mathcal{E}_{\mathscr{B}}(Y) = 0$. Via (8.6.12) we arrive at

$$\mathcal{E}_{\mathscr{B}}\left(e^{s(X-\mu)}\right) \leq \exp\left\{\frac{s^2(b-a)^2}{8}\right\}. \tag{8.6.13}$$

Next, by the same reasoning as in the proof of Theorem 8.6.5, we get from (8.6.13) that

$$\begin{aligned}\tau\left(e^{\lambda\sum_{j=1}^{n}(X_j-\mu)}\right) \leq& \tau\left(\mathcal{E}_{\mathscr{B}}\left(e^{\lambda\sum_{j=1}^{n-1}(X_j-\mu)}\right)\mathcal{E}_{\mathscr{B}}\left(e^{\lambda(X_n-\mu)}\right)\right)\\ \leq& e^{\frac{\lambda^2(b_n-a_n)^2}{8}}\tau\left(\mathcal{E}_{\mathscr{B}}\left(e^{\lambda\sum_{j=1}^{n-1}(X_j-\mu)}\right)\right) \leq \cdots\\ \leq& \exp\left\{\frac{\lambda^2\sum_{j=1}^{n}(b_j-a_j)^2}{8}\right\}.\end{aligned}$$

Now, (8.6.1) and the exponential Chebyshev inequality (8.6.4) yield that

$$\begin{aligned}\mathrm{Prob}(|S_n - n\mu| \geq t) =& 2\mathrm{Prob}(S_n - n\mu \geq t)\\ \leq& 2e^{-\lambda t}\exp\left\{\frac{\lambda^2\sum_{j=1}^{n}(b_j-a_j)^2}{8}\right\}.\end{aligned}$$

This is minimized when $\lambda = \frac{4t}{\sum_{j=1}^{n}(b_j-a_j)^2}$. Thus,

$$\text{Prob}(|S_n - n\mu| \geq t) \leq 2\exp\left\{\frac{-2t^2}{\sum_{j=1}^{n}(b_j - a_j)^2}\right\},$$

which is the desired inequality. □

8.6.3 Azuma Inequality

The *Azuma inequality* [85] provides a concentration result for martingales with bounded differences. It states that if (X_j) is a martingale and $|X_j - X_{j-1}| < c_j$ almost surely, then

$$\text{Prob}(X_n - X_0 \geq \lambda) \leq \exp\left(\frac{-\lambda^2}{2\sum_{j=1}^{n} c_j^2}\right)$$

for all positive integers n and all $\lambda > 0$. Our purpose is to explore an Azuma-type inequality under the Lipschitz condition for martingales in the context of noncommutative probability spaces and derive a noncommutative Hoeffding inequality. We follow the approach outlined in [86].

Theorem 8.6.9 *(Noncommutative Azuma inequality) Let $(X_j)_{0\leq j\leq n}$ be a self-adjoint martingale with respect to a filtration $(\mathscr{A}_j, \mathcal{E}_j)_{0\leq j\leq n}$, and let $dX_j = X_j - X_{j-1}$ be its associated martingale difference. Suppose that $-c_j \leq dX_j \leq c_j$ for some constants $c_j > 0$ $(1 \leq j \leq n)$. Then*

$$\text{Prob}\left(\left|\sum_{j=1}^{n} dX_j\right| \geq \lambda\right) \leq 2\exp\left\{\frac{-\lambda^2}{2\sum_{j=1}^{n} c_j^2}\right\}$$

for all $\lambda > 0$.

Proof Fix a number $t > 0$. The convexity of the function $f(s) = e^{ts}$ ensures that

$$e^{ts} \leq \frac{1}{2c}(e^{tc} - e^{-tc})s + \frac{1}{2}(e^{tc} + e^{-tc})$$

for all $-c \leq s \leq c$.
Since $-c_j \leq dX_j \leq c_j$, from the functional calculus we conclude that

$$e^{tdX_j} \leq \frac{1}{2c_j}(e^{tc_j} - e^{-tc_j})dX_j + \frac{1}{2}(e^{tc_j} + e^{-tc_j}).$$

Therefore,

$$\begin{aligned}
\mathcal{E}_{j-1}\left(e^{tdX_j}\right) \leq& \mathcal{E}_{j-1}\left(\frac{1}{2c_j}(e^{tc_j}-e^{-tc_j})dX_j+\frac{1}{2}(e^{tc_j}+e^{-tc_j})\right)\\
=&\frac{1}{2}(e^{tc_j}+e^{-tc_j}) \qquad\qquad (\text{by } \mathcal{E}_{j-1}(dX_j)=0,\ j\geq 2)\\
=&\sum_{n=0}^{\infty}\frac{(tc_j)^{2n}}{(2n)!}\leq\sum_{n=0}^{\infty}\frac{(tc_j)^{2n}}{2^n n!}=e^{\frac{t^2c_j^2}{2}}.
\end{aligned}$$

Let $\lambda \geq 0$. Inequality (8.6.4) yields that

$$\begin{aligned}
\text{Prob}\left(\sum_{j=1}^{n} dX_j \geq \lambda\right) \leq& e^{-t\lambda}\tau\left(e^{t\sum_{j=1}^{n} dX_j}\right)\\
\leq& e^{-t\lambda}\tau\left(e^{t\sum_{j=1}^{n-1} dX_j}e^{tdX_n}\right)\\
=& e^{-t\lambda}\tau\left(\mathcal{E}_{n-1}\left(e^{t\sum_{j=1}^{n-1} dX_j}e^{tdX_n}\right)\right)\\
=& e^{-t\lambda}\tau\left(e^{t\sum_{j=1}^{n-1} dX_j}\mathcal{E}_{n-1}\left(e^{tdX_n}\right)\right)\\
\leq& e^{-t\lambda}e^{t^2c_n^2/2}\tau\left(e^{t\sum_{j=1}^{n-1} dX_j}\right).
\end{aligned}$$

Iterating $n-2$ times, we get

$$\text{Prob}\left(\sum_{j=1}^{n} dX_j \geq \lambda\right) \leq \exp\left(-t\lambda+\frac{t^2}{2}\sum_{j=1}^{n}c_j^2\right).$$

The minimum value of $\exp\left(-t\lambda+\frac{t^2}{2}\sum_{j=1}^{n}c_j^2\right)$ occurs at $t=\frac{\lambda}{\sum_{j=1}^{n}c_j^2}$. Hence,

$$\text{Prob}\left(\sum_{j=1}^{n} dX_j \geq \lambda\right) \leq \exp\left(\frac{-\lambda^2}{2\sum_{j=1}^{n}c_j}\right). \tag{8.6.14}$$

Thus, the symmetry and inequality (8.6.14) imply that

$$\text{Prob}\left(\left|\sum_{j=1}^{n} dX_j\right| \geq \lambda\right) = 2\,\text{Prob}\left(\sum_{j=1}^{n} dX_j \geq \lambda\right) \leq 2\exp\left(\frac{-\lambda^2}{2\sum_{j=1}^{n}c_j}\right).$$

□

The next results present some *noncommutative McDiarmid type inequalities*.

Corollary 8.6.10 (Noncommutative McDiarmid inequality) *Let* $(\mathscr{A}_j, \mathcal{E}_j)_{0 \leq j \leq n}$ *be a filtration of* $\mathscr{A}$, *let* $X_j \in \mathscr{A}_j$ $(1 \leq j \leq n)$ *be self-adjoint, and let there exist maps* $g_j : \mathscr{A}_{1\,\mathrm{sa}} \times \cdots \times \mathscr{A}_{j\,\mathrm{sa}} \to \mathscr{A}_{\mathrm{sa}}$ *such that the sequence*

$$g_0(X_1, \ldots, X_n), g_1(X_1, \cdots, X_n), \cdots, g_n(X_1, \cdots, X_n)$$

constitute a martingale satisfying

$$-c_j \leq g_j(X_1, \cdots, X_n) - g_{j-1}(X_1, \cdots, X_n) \leq c_j$$

for any $1 \leq j \leq n$. *Then*

$$\mathrm{Prob}\,(|g_n(X_1, \ldots, X_n) - g_0(X_1, \ldots, X_n))| \geq t) \leq 2 \exp\left\{\frac{-t^2}{2\sum_{j=1}^n c_j^2}\right\}.$$

Proof The result is immediately concluded from Theorem 8.6.9 since the martingale consisting of $Y_j = g_j(X_1, \cdots, X_n)$, $0 \leq j \leq n$ satisfies the conditions of the theorem. □

If we take $c_j = 1$ and $g_j(X_1, \cdots, X_n) = \sum_{i=1}^j X_i$ in the previous Corollary, then the following *Chernoff type inequality* for commutative random variables is obtained.

Corollary 8.6.11 *Suppose that* $X_1, \cdots, X_n$ *are independent classical random variables with* $\mathbb{E}(X_j) = 0$ *and* $|X_j| \leq 1$ *for all* j. *Then*

$$\mathrm{Prob}\left(\left|\sum_{j=1}^n X_j\right| \geq t\right) \leq 2e^{-t^2/2n}$$

for all $t \geq 0$.

8.7 Concluding Remarks

The *universal representation* of a C^*-algebra $\mathscr{A}$ is the pair

$$\{\pi, \mathscr{H}\} = \sum_{\varphi \in \mathcal{S}(\mathscr{A})}^{\oplus} \{\pi_\varphi, \mathscr{H}_\varphi\},$$

where $\mathcal{S}(\mathscr{A})$ is the set of all states on $\mathscr{A}$, $(\pi_\varphi, \mathscr{H}_\varphi)$ is the Gelfand–Naimark–Segal representation of a C^*-algebra $\mathscr{A}$, associated with φ. Here we say that the von Neumann algebra $\mathscr{M} = \pi(\mathscr{A})^{cc}$, generated by $\pi(\mathscr{A})$, is the *universal enveloping von Neumann algebra* of a C^*-algebra $\mathscr{A}$ [1, Chap. III, Definition 2.3].

Consider a C^*-algebra $\mathscr{A}$. Let φ be a positive linear functional on $\mathscr{A}$ and π be the universal representation of $\mathscr{A}$. Then, an arbitrary state on $\mathscr{A}$ by the construction of π

turns into a vector state on $\pi(\mathscr{A})$, hence it extends to a normal state on the universal enveloping algebra $\mathscr{M} = \pi(\mathscr{A})^{cc}$. Then φ yields such a positive normal functional $\widehat{\varphi}$ on the universal enveloping von Neumann algebra that $\widehat{\varphi}(\pi(A)) = \varphi(A) \ (A \in \mathscr{A}_+)$.

A *representation with a trace* of a C^*-algebra $\mathscr{A}$ is a pair (π, ν) with the following properties:

- π is a nondegenerate representation of a C^*-algebra $\mathscr{A}$ on some Hilbert space;
- ν is a faithful normal trace on the von Neumann algebra $\pi(\mathscr{A})^{cc}$;
- $\pi(\mathscr{A}) \cap \mathfrak{N}_\nu$ generates the von Neumann algebra $\pi(\mathscr{A})^{cc}$, where

$$\mathfrak{N}_\nu = \{A \in \pi(\mathscr{A})^{cc} : \nu(A^*A) < \infty\}.$$

Let $\mathscr{A}$ be a C^*-algebra and let (π, ν) be a representation of $\mathscr{A}$ with a trace. Then ν is semifinite and $\varphi = (\nu \circ \pi)|_{\mathscr{A}_+}$ is a lower semicontinuous trace on $\mathscr{A}$. Conversely, let φ be a lower semicontinuous semifinite trace on $\mathscr{A}$. Then there exists a representation π of $\mathscr{A}$ with a trace ν ([87, 6.6]), which is called an *associated representation* to φ, and ν is called a *natural trace*. In this case, the relation

$$\nu(\pi(A)) = \varphi(A) \quad \text{for all} \quad A \in \mathscr{A}_+$$

holds (see [87, Proposition 6.6.5(i)]). Relations $\pi(X^*) = \pi(X)^*$, $\pi(xX + yY) = x\pi(X) + y\pi(Y)$ and $\pi(XY) = \pi(X)\pi(Y)$ hold for all $X, Y \in \mathscr{A}$ and $x, y \in \mathbb{C}$.

This information allows us to extend the well-known inequalities for normal semifinite traces on von Neumann algebras to lower semicontinuous traces on C^*-algebras (see, for example, Theorem 3.11 of [88]). Moreover, if $X \in \mathscr{A}_\varphi$, then $\pi(X) \in \mathscr{A}_\nu$ and $\nu(\pi(X)) = \varphi(X)$ [89, item (iii) of Lemma 2.1].

8.8 Exercises and Problems

Exercise 8.8.1 Let $P, Q \in \mathbb{M}_n$. Prove that $P \sim Q$ if and only if $\operatorname{tr} P = \operatorname{tr} Q$.

Exercise 8.8.2 Let P and Q be two projections in $\mathbb{B}(\mathscr{H})$. Prove that
(i) $P \vee Q - Q \sim P - P \wedge Q$.
(ii) If $P \wedge Q = 0$ then $P \prec Q^\perp$.

Exercise 8.8.3 Suppose that $\{P_n\}$ and $\{Q_n\}$ are sequences in $\mathbb{B}(\mathscr{H})$ that increasingly converge to the projections P and Q in the strong operator topology. Show that $P_n \vee Q_n \overset{sot}{\to} P \vee Q$. If the sequences are decreasing, what can we say about the convergence of $P_n \wedge Q_n$?

Exercise 8.8.4 Show that a trace φ is semifinite if the set $\mathscr{A}_\varphi$ is σ-weakly dense in $\mathscr{A}$.

Exercise 8.8.5 Show that if

$$X = \sum_{k=1}^{n} a_k P_k,$$

$a_k > 0$ and P_k are mutually orthogonal projections in $\mathscr{A}$, then $\tau(X) = \int_0^\infty \mu(t, X)dt$.

Exercise 8.8.6 If $X_n, X \in S(\mathscr{A}, \tau)$ and $X_n \xrightarrow{\tau} X$ as $n \to \infty$, then explore whether $\mu(X_n) \to \mu(X)$ as $n \to \infty$ almost everywhere.

Exercise 8.8.7 If $f \in \mathbb{F}$ is convex (respectively, concave), then show that the map $\|\cdot\|_f^*: L_f(\mathscr{A}, \tau) \to \mathbb{R}_+$ defined by the equality

$$\|A\|_f^* = \inf\{\varepsilon > 0 : \tau(f(\varepsilon^{-1}|A|)) \le 1\} \text{ (respectively, } \|A\|_f^* = \tau(f(|A|)))$$

is a norm (respectively, F-norm) on $L_f(\mathscr{A}, \tau)$, and the $\|\cdot\|_f^*$-convergence coincides with the $\|\cdot\|_f$-convergence.

Exercise 8.8.8 For $A_n, A \in L_f(\mathscr{A}, \tau)$, we say that $A_n \to A$ in *f-mean* as $n \to \infty$ if $\tau(f(|A_n - A|)) \to 0$ as $n \to \infty$. If $\|A_n - A\|_f \to 0$ as $n \to \infty$, then $A_n \to A$ in f-mean as $n \to \infty$. If $A_n \to A$ in f-mean as $n \to \infty$, then $A_n \xrightarrow{\tau} A$ as $n \to \infty$.

If $f \in \mathbb{F}$ with the Δ_2-condition, then prove that the convergence in f-mean coincides with the $\|\cdot\|_f$-convergence.

Exercise 8.8.9 The nonconcave function

$$f(t) = \begin{cases} 9t, & \text{if } 0 \le t \le 0.1, \\ \frac{1}{9}t + \frac{8}{9}, & \text{if } 0.1 < t < 1, \\ t, & \text{if } t \ge 1 \end{cases}$$

lies in $\mathbb{F}$ and $f(s+t) \le f(s) + f(t)$ for all $s, t \in \mathbb{R}_+$. Show that in $\mathbb{M}_2$ for the projections

$$A = \begin{bmatrix} 1 & 0 \\ 0 & 0 \end{bmatrix} \quad \text{and} \quad B = \frac{1}{2}\begin{bmatrix} 1 & 1 \\ 1 & 1 \end{bmatrix},$$

the inequality $\mathrm{tr}(f(A+B)) \le \mathrm{tr}(f(A)) + \mathrm{tr}(f(B))$ is false.

Exercise 8.8.10 Prove the noncommutative *Chebyshev inequality*

$$\tau\left(E^{|X|}[t, \infty)\right) \le t^{-p}\tau(|X|^p) \tag{8.8.1}$$

for $X \in \mathscr{A}$, $1 \le p < \infty$ and $t > 0$.

Exercise 8.8.11 Show that

$$\mathrm{var}(X) = \inf_{\lambda \in \mathbb{R}} \tau\left((X - \lambda I)^2\right).$$

Exercise 8.8.12 Show that for $1 \le p \le \infty$, if $X_\infty \in L_p(\mathscr{A})$, then $(\mathcal{E}_n(X_\infty))$ is a bounded L_p-martingale and $\mathcal{E}_n(X_\infty)$ converges to X_∞ in $L_p(\mathscr{A})$ (in w^*-topology for $p = \infty$).

Problem 8.8.13 Let P and Q be projections in a von Neumann algebra $\mathscr{A}$ such that $\|P - Q\| < 1$. Show that there exists a unitary $U \in \mathscr{A}$ such that $UPU^* = Q$ and $\|I - U\| \le \sqrt{2}\|P - Q\|$.

Problem 8.8.14 Let P, P', Q, and Q' be projections in a von Neumann algebra $\mathscr{A}$. Show that there is a unitary operator $U \in \mathscr{A}$ such that $UPU^* = P'$ and $UQU^* = Q'$ if and only if there is $\lambda > 1$ and a unitary V such that such that $V(\lambda P + Q)V^* = (\lambda P' + Q')$.

Problem 8.8.15 If P and Q are projections in $\mathbb{B}(\mathscr{H})$, establish that

$$\|P - Q\| = \max\{\|P(I - Q)\|, \|Q(I - P)\|\} = \max\{\|P - PQP\|^{1/2}, \|Q - QPQ\|^{1/2}\}.$$

Problem 8.8.16 Let P, Q be projections in a von Neumann algebra $\mathscr{A}$ and $U \in \mathscr{A}$ be such that $U^*U = P$, $UU^* = Q$. Show that

(i) $\|P - U\| = \|Q - U\| \ge \|P - Q\|$;

(ii) $P_1 \le P \Rightarrow Q_1 = UP_1U^* \sim P_1$, $Q_1 \le Q$, $\|P_1 - Q_1\| \le \|P - Q\|$.

8.9 Notes, Hints, and References

This chapter provides condensed information about noncommutative integration theory, spaces of τ-measurable operators, and noncommutative probability theory. It may also be useful for specialists in quantum statistical mechanics and quantum information. For those seeking more detailed explanations on the topics discussed here, we recommend books authored by Takesaki [1, 90], Sherstnev [91], and Dodds, de Pagter, and Sukochev [92] for noncommutative integration theory and Holevo [93] for quantum information. Here are some hints and references for the exercises and problems.

One side of the equivalence in Exercise 8.8.1 is easily established by using the tracial property of $\mathrm{tr}(\cdot)$. A suitable reference for Exercise 8.8.2 is [1, Chap. V, Proposition 1.6]. Indeed, if $P \wedge Q = 0$, then $P^\perp \vee Q^\perp = I$. Thus, $P = P^\perp \vee Q^\perp - P^\perp \sim Q^\perp - Q^\perp \wedge P^\perp \le Q^\perp$. For Exercise 8.8.3, see [94, Exercise 8.7.34]. For Exercise 8.8.4 consult [90, Definition VII.1.1] and [67, Proposition III.2.2.20].

Exercise 8.8.5 is in the "commutative setting" and Example 8.1.2 (ii) applies. For Exercise 8.8.6, refer to Lemma 8.3.19 and remember that the set of discontinuity points for the function $\mu(X)$ has measure zero. Exercise 8.8.6 can be solved using Theorem 8.3.5; note that for a convex function $f \in \mathbb{F}$, $\|\cdot\|_f^*$ represents the norm of the corresponding noncommutative Orlicz space.

Exercise 8.8.8 can be solved using definitions and simple estimates. To solve Exercise 8.8.9, Lemma 8.3.12(iv) can be applied. In fact, Exercise 8.8.10 is in the "commutative setting" and Example 8.1.2 (ii) applies. Exercise 8.8.11 can be solved by employing definitions and simple estimates. We refer the reader to [95] for Problems 8.8.12 and 8.8.13. A solution to Exercise 8.8.2 is presented in [68, Proposition 2.10].

For Problem 8.8.15, $\|P(I-Q)\|^2 = \|(I-Q)P\|^2 = \|P(I-Q)P\| = \|P - PQP\|$ and $(P-Q)^2 = A + B$ hold, where $A = P(I-Q)P$, $B = (I-P)Q(I-P)$. Since $AB = 0$, we have

$$\begin{aligned}\|P-Q\|^2 &= \|(P-Q)^2\| = \|A+B\| = \lim_{n\to\infty} \|(A+B)^n\|^{1/n} \\ &= \lim_{n\to\infty} \|A^n + B^n\|^{1/n} \leq \lim_{n\to\infty} [2\max\{\|A^n\|, \|B^n\|\}]^{1/n} \\ &= \max\{\|A\|, \|B\|\} \leq \|A+B\|.\end{aligned}$$

In Problem 8.8.16(i), the operator U is a partial isometry, hence $U = UU^*U$ and $\|P-U\| = \|U^*U - UU^*U\| \leq \|U^* - UU^*\| = \|U^* - Q\| = \|Q - U\|$. Analogously, $\|Q-U\| \leq \|P-U\|$. If, for example (see Problem 8.8.15) $\|P-Q\| = \|P(I-Q)\|$, then $\|P-Q\| = \|U^*U(I-Q)\| \leq \|U(I-Q)\| = \|(U-Q)(I-Q)\| \leq \|Q-U\|$.

For Part (ii), put $U_1 = UP_1$. Then $U_1^*U_1 = P_1$ and $U_1U_1^* = Q_1$. By (i) one has $\|P_1 - Q_1\| \leq \|P_1 - U_1\| = \|(P-U)P_1\| \leq \|P-U\|$.

References

1. M. Takesaki, *Theory of Operator Algebras*, vol. I. Operator Algebras and Non-commutative Geometry (Springer, Berlin, 2002), p. 5
2. H. Upmeier, Automorphism groups of Jordan C^* -algebras. Math. Z. **176**, 21–34 (1981)
3. S.A. Ayupov, Classification and representation of ordered Jordan algebras (Russian) Tashkent: "Fan" (1986)
4. L.T. Gardner, An inequality characterizes the trace. Canad. J. Math. **31**, 1322–1328 (1979)
5. O.E. Tikhonov, Subadditivity inequalities in von Neumann algebras and characterization of tracial functionals. Positivity **9**, 259–264 (2005)
6. D. Petz, J. Zemánek, Characterizations of the trace. Linear Algebra Appl. **111**, 43–52 (1988)
7. S.M. Manjegani, Inequalities in operator algebras, Ph.D. thesis, Regina University, Canada, Regina (2004), 95 p
8. M. Ruskai, Inequalities for traces on von Neumann algebras. Commun. Math. Phys. **26**, 280–289 (1972)
9. D.R. Farenick, S.M. Manjegani, Young's inequality in operator algebras. J. Ramanujan Math. Soc. **20**, 107–124 (2005)
10. H. Araki, On an inequality of Lieb and Thirring. Lett. Math. Phys. **19**, 167–170 (1990)
11. H. Kosaki, On an inequality of Araki-Lieb-Thirring (von Neumann algebra case). Proc. Am. Math. Soc. **114**, 477–481 (1992)
12. B. Simon, *Trace Ideals and Their Applications, Second Edition, Mathematical Surveys and Monographs*, vol. 120 (American Mathematical Society, Providence, RI, 2005)
13. H. Alhasan, K. Fawwaz, Characterization of tracial functionals on von Neumann algebras. Lobachevskii J. Math. **42**, 2273–2279 (2021)
14. A.M. Bikchentaev, O.E. Tikhonov, Characterization of the trace by Young's inequality. JIPAM. J. Inequal. Pure Appl. Math. **6**(49), 3 (2005)

15. A.M. Bikchentaev, Characterization of certain traces on von Neumann algebras, in *Infinite Dimensional Analysis, Quantum Probability and Applications*, ed. by L. Accardi et al. ICQPRT 2021 (Springer Proceedings in Mathematics & Statistics 390): QP41 Conference, Al Ain, UAE, March 28–April 1, 2021 (Springer, Berlin, 2022), pp. 279–289
16. K. Cho, T. Sano, Young's inequality and trace. Linear Algebra Appl. **431**, 1218–1222 (2009)
17. D.T. Hoa, H. Osaka, H.M. Toan, On generalized Powers-Størmer's inequality. Linear Algebra Appl. **438**, 242–249 (2013)
18. D.T. Hoa, O.E. Tikhonov, Weighted monotonicity inequalities for traces on operator algebras. Math. Notes **88**, 177–182 (2010)
19. A.I. Stolyarov, O.E. Tikhonov, A.N. Sherstnev, Characterization of normal traces on von Neumann algebras by inequalities for the modulus. Math. Notes **72**, 411–416 (2002)
20. M. Reed, B. Simon, *Methods of Modern Mathematical Physics* (Academic Press, New York-London, I. Functional Analysis, 1972)
21. S. Golden, Lower bounds for the Helmholtz function. Phys. Rev. **137**(2), 1127–1128 (1965)
22. C.J. Thompson, Inequality with applications in statistical mechanics. J. Math. Phys. **6**, 1812–1813 (1965)
23. H. Araki, Relative Hamiltonian for states of von Neumann algebras. Proc. Steklov Inst. Math. **135**, 18–25 (1975)
24. O.E. Tikhonov, Convex functions and inequalities for a trace (Russian), in *Constructive Theory of Functions and Functional Analysis*, vol. 6 (Kazan State University, Kazan, 1987), pp. 77–82
25. A.N. Sherstnev, A noncommutative analogue of the space L_1, (Russian), in *Mathematical Analysis* (Kazan Federal University, Kazan', 1978), pp. 112–123
26. F.A. Berezin, Convex functions on operators (Russian). Mat. Sb., New Ser. **88**, 268–276 (1972)
27. H. Kosaki, On the continuity of the map $\varphi \mapsto |\varphi|$ from the predual of a W^* -algebra. J. Funct. Anal. **59**, 123–131 (1984)
28. A. Lieberman, Entropy of states of a gage space. Acta Sci. Math. **40**, 99–105 (1978)
29. D. Petz, Spectral scale of selfadjoint operators and trace inequalities. J. Math. Anal. Appl. **109**, 74–82 (1985)
30. P.K. Tam, Isometries of L_p -spaces associated with semifinite von Neumann algebras. Trans. Am. Math. Soc. **254**, 339–354 (1979)
31. T. Fack, H. Kosaki, Generalized s -numbers of τ -measurable operators. Pacific J. Math. **123**, 269–300 (1986)
32. S.G. Kreĭn, Y.I. Petunin, E.M. Semënov, *Interpolation of Linear Operators, Translations of Mathematical Monographs*, vol. 54 (American Mathematical Society, Providence, R. I., 1982)
33. E. Nelson, Notes on non-commutative integration. J. Funct. Anal. **15**, 103–116 (1974)
34. L.G. Brown, H. Kosaki, Jensen's inequality in semi-finite von Neuman algebras. J. Oper. Theory **23**, 3–19 (1990)
35. I.C. Gohberg, M.G. Kreĭn, *Introduction to the Theory of Linear Nonselfadjoint Operators, Translations of Mathematical Monographs*, vol. 18 (American Mathematical Society, Providence, R. I., 1969)
36. H. Umegaki, Conditional expectation in an operator algebra. IV, Entropy and Information, Kōdai Math. Sem. Rep. **14**, 59–85 (1962)
37. A.M. Bikchentaev, On noncommutative function spaces. Selected Papers in K-theory. Am. Math. Soc. Transl. **154**(2), 179–187 (1992)
38. O.E. Tikhonov, Continuity of operator functions in topologies connected with a trace on a von Neumann algebra. (Russian) Izv. Vyssh. Uchebn. Zaved. Mat. 77–79 (1987) translated in Soviet Math. (Iz. VUZ), **31**, 110–114 (1987)
39. D. Dauitbek, N.E. Tokmagambetov, K.S. Tulenov, Commutator inequalities associated with polar decompositions of τ-measurable operators. Russian Math. (Iz. VUZ) **58**, 48–52 (2014)
40. W.F. Stinespring, Integration theorems for gages and duality for unimodular groups. Trans. Am. Math. Soc. **90**, 15–56 (1959)
41. T. Fack, Sur la notion de valeur caractéristique. J. Oper. Theory **7**, 307–333 (1982)
42. W. Rudin, *Real and Complex Analysis*, 2nd edn. (McGraw-Hill Series in Higher Mathematics. McGraw-Hill Book Co., New York-Düsseldorf-Johannesburg, 1974)

43. P.S. Bullen, *Handbook of Means and Their Inequalities, Mathematics and its Applications, 560* (Kluwer Academic Publishers Group, Dordrecht, 2003)
44. M.J. Cloud, B.C. Drachman, L.P. Lebedev, *Inequalities with Applications to Engineering* (Springer, Berlin, 1998)
45. C.A. Akemann, J. Anderson, G.K. Pedersen, Triangle inequalities in operator algebras. Linear Multilinear Algebra **11**, 167–178 (1982)
46. Z. Cvetkovski, *Inequalities* (Theorems, Techniques and Selected Problems), Springer, Heidelberg, 2012)
47. P.G. Dodds, T.K. Dodds, F.A. Sukochev, D. Zanin, Arithmetic-geometric mean and related submajorisation and norm inequalities for τ-measurable operators: Part II. Integr. Equ. Oper. Theory **92**(32), 60 (2020)
48. F. Hiai, Y. Nakamura, Majorizations for generalized s-numbers in semifinite von Neumann algebras. Math. Z. **195**, 17–27 (1987)
49. P.G. Dodds, K. Dodds, K. Theresa, On a submajorization inequality of T. Ando, in *Operator Theory in Function Spaces and Banach Lattices*, Operator Theory: Advances and Applications, vol. 75 (Birkhäuser, Basel, 1995), pp. 113–131
50. K.M. Chong, Some extensions of a theorem of Hardy, Littlewood and Pólya and their applications. Canad J. Math. **26**, 1321–1340 (1974)
51. N.T. Bekjan, D. Dauitbek, Submajorization inequalities of τ-measurable operators for concave and convex functions. Positivity **19**, 341–345 (2015)
52. M. Marcus, H. Mink, *A Survey of Matrix Theory and Matrix Inequalities* (Allyn and Bacon, Boston, 1964)
53. F. Zhang, *Matrix Theory*, 2nd edn. (Universitext, Springer, New York, Basic Results and Techniques, 2011)
54. V.I. Chilin, A.V. Krygin, Ph.A. Sukochev, Extreme points of convex fully symmetric sets of measurable operators. Integr. Equ. Oper. Theory **15**, 186–226 (1992)
55. A.M. Bikchentaev, Block projection operators in normed solid spaces of measurable operators. Russian Math. (Iz. VUZ) **56**, 75–79 (2012)
56. A.M. Bikchentaev, F. Sukochev, Inequalities for the block projection operators. J. Funct. Anal. **280**, 108851, 18 (2021)
57. R. Bellman, Notes on matrix theory. II. Am. Math. Monthly **60**, 173–175 (1953)
58. L. Mirsky, An inequality for positive definite matrices. Am. Math. Monthly **62**, 428–430 (1955)
59. P.D. Lax, *Linear Algebra and its Applications, Enlarged second edition, Pure and Applied Mathematics (Hoboken), Wiley-Interscience [John Wiley & Sons]* (Hoboken, NJ, 2007)
60. R. Bhatia, *Matrix Analysis* (Springer, New York, 1997)
61. F. Hansen, K.G. Pedersen, Jensen's inequality for operators and Löwner's theorem. Math. Ann. **258**, 229–241 (1982)
62. R. Bellman, *Introduction to Matrix Analysis*, 2nd edn. (SIAM, Philadelphia, 1997)
63. G.H. Hardy, J.E. Littlewood, G. Pólya, *Inequalities* (Cambridge University Press, Cambridge, 1988)
64. A.M. Bikchentaev, O.E. Tikhonov, Characterization of the trace by monotonicity inequalities. Linear Algebra Appl. **422**, 274–278 (2007)
65. M. Rédei, S.J. Summers, Quantum probability theory. Stud. Hist. Philos. Sci. B Stud. Hist. Philos. Modern Phys. **38**, 390–417 (2007)
66. P. Jorgensen, F. Tian, Non-commutative analysis, in With a Foreword by Wayne Polyzou (World Scientific Publishing Co. Pte. Ltd., Hackensack, NJ, 2017)
67. B. Blackadar, *Operator Algebras: Theory of C^* -algebras and von Neumann algebras, Encyclopaedia of Mathematical Sciences*, vol. 122 (Springer, Berlin Heidelberg, 2006)
68. Q. Xu, Operator spaces and noncommutative L^p, in *Lectures in the Summer School on Banach spaces and Operator Spaces* (Nankai University China, 2007)
69. Gh. Sadeghi, M.S. Moslehian, A. Talebi, Maximal inequalities in noncommutative probability spaces. Stochastics **94**, 212–225 (2022)
70. M.S. Moslehian, G. Sadeghi, M. Pliev, Inequalities for acceptable noncommutative random variables. Infin. Dimens. Anal. Quantum Probab. Relat. Top. **23**, 2050012, 8 (2020)

71. G. Sadeghi, M.S. Moslehian, Mixing sequences, and mixingales in quantum probability spaces. Rev. R. Acad. Cienc. Exactas Fis. Nat. Ser. A Mat . RACSAM **116**, 89, 21 (2022)
72. A. Talebi, G. Sadeghi, M. . Moslehian, Freedman inequality in noncommutative probability spaces. Complex Anal. Oper. Theory **16**, 22, 18 (2022)
73. Y. Jia, F. Sukochev, G. Xie, D. Zanin, Φ-moment inequalities for independent and freely independent random variables. J. Funct. Anal. **270**, 4558–4596 (2016)
74. M. Junge, Q. Xu, Noncommutative Burkholder/Rosenthal inequalities, II, Applications. Israel J. Math. **167**, 227–282 (2008)
75. M.S. Moslehian, A. Talebi, A variance bound for a general function of independent noncommutative random variables. Quaest. Math. **242**, 307–318 (2019)
76. A. Talebi, M.S. Moslehian, M. Pliev, G. Sadeghi, Band projections and decomposition of normal traces on von Neumann algebras. Positivity **26**, 71, 24 (2022)
77. A. Talebi, M.S. Moslehian, Gh. Sadeghi, Etemadi and Kolmogorov inequalities in noncommutative probability spaces. Michigan Math. J. **68**, 57–69 (2019)
78. A. Talebi, M.S. Moslehian, G. Sadeghi, Noncommutative Blackwell-Ross martingale inequality. Infin. Dimens. Anal. Quantum Probab. Relat. Top. **21**, 1850005, 9 (2018)
79. A. Talebi, M.S. Moslehian, Gh. Sadeghi, A representation of noncommutative BMO spaces. Statist. Probab. Lett. **153**, 65–70 (2019)
80. M. Junge, Q. Xu, Noncommutative Burkholder/Rosenthal inequalities. Ann. Probab. **31**, 948–995 (2003)
81. C.J.K. Batty, The strong law of large numbers for states and traces of a W^* -algebra. Z. Wahrsch. Verw. Gebiete **48**, 177–191 (1979)
82. M.S. Moslehian, Gh. Sadeghi, Inequalities for sums of random variables in noncommutative probability spaces. Rocky Mountain J. Math. **46**, 309–323 (2016)
83. M. Junge, Q. Zeng, Noncommutative Bennett and Rosenthal inequalities. Ann. Probab. **41**, 4287–4316 (2013)
84. W. Hoeffding, Probability inequalities for sums of bounded random variables. J. Am. Statist. Assoc. **58**, 13–30 (1963)
85. K. Azuma, Weighted sums of certain dependent random variables. Tôhoku Math. J. **19**, 357–367 (1967)
86. Gh. Sadeghi, M.S. Moslehian, Noncommutative martingale concentration inequalities. Illinois J. Math. **58**, 561–575 (2014)
87. J. Dixmier, *C^*-algebras and their Representations* (Russian), Translated from the French by A. I. Štern, ed. by A.A. Kirillova. Izdat. "Nauka" (Moscow, 1974). 399 pp
88. A.M. Bikchentaev, Trace inequalities for Rickart C^* -algebras. Positivity **25**, 1943–1957 (2021)
89. A.M. Bikchentaev, Differences and commutators of projections on a Hilbert space. Internat. J. Theor. Phys. **61**, 2, 10 (2022)
90. M. Takesaki, *Theory of Operator Algebras*, vol. II. Operator Algebras and Non-Commutative Geometry, vol. 6 (Springer, New York, 2003)
91. A.N. Sherstnev, *Methods of Bilinear Forms in Non-commutative Measure and Integral Theory (Russian)* (Fizmatlit, Moscow, 2008)
92. P.G. Dodds, B. de Pagter, F.A. Sukochev, *Noncommutative Integration and Operator Theory*. Progress in Mathematics, vol. 349 (Birkhaäuser, Cham, 2023)
93. A.S. Holevo, *Quantum Systems, Channels, Information: A Mathematical Introduction*, Texts and Monographs in Theoretical Physics (De Gruyter, Berlin, 2019)
94. R.V. Kadison, J.R. Ringrose, *Fundamentals of the Theory of Operator Algebras, vol. II: Advanced Theory*, Graduate Studies in Mathematics, vol. 16 (American Mathematical Society, Providence, RI, 1997)
95. I. Raeburn, A.M. Sinclair, The C^* -algebra generated by two projections. Math. Scand. **65**, 278–290 (1989)

Index

A. M. Bikchentaev et al., *Trace Inequalities*, Forum for Interdisciplinary Mathematics,
https://doi.org/10.1007/978-981-97-6520-1

MIX
Papier aus verantwortungsvollen Quellen
Paper from responsible sources
FSC® C105338

If you have any concerns about our products,
you can contact us on
ProductSafety@springernature.com

In case Publisher is established outside the EU,
the EU authorized representative is:
Springer Nature Customer Service Center GmbH
Europaplatz 3, 69115 Heidelberg, Germany

Printed by Libri Plureos GmbH
in Hamburg, Germany